Applied Biophysics

Applied Biophysics
A Molecular Approach for Physical Scientists

Tom A. Waigh
University of Manchester, Manchester, UK

John Wiley & Sons, Ltd

Copyright © 2007   John Wiley & Sons Ltd, The Atrium, Southern Gate, Chichester,
                      West Sussex PO19 8SQ, England

                      Telephone (+44) 1243 779777

Email (for orders and customer service enquiries): cs-books@wiley.co.uk
Visit our Home Page on www.wileyeurope.com or www.wiley.com

All Rights Reserved. No part of this publication may be reproduced, stored in a retrieval system or transmitted in any form or by any means, electronic, mechanical, photocopying, recording, scanning or otherwise, except under the terms of the Copyright, Designs and Patents Act 1988 or under the terms of a licence issued by the Copyright Licensing Agency Ltd, 90 Tottenham Court Road, London W1T 4LP, UK, without the permission in writing of the Publisher. Requests to the Publisher should be addressed to the Permissions Department, John Wiley & Sons Ltd, The Atrium, Southern Gate, Chichester, West Sussex PO19 8SQ, England, or emailed to permreq@wiley.co.uk, or faxed to (+44) 1243 770620.

Designations used by companies to distinguish their products are often claimed as trademarks. All brand names and product names used in this book are trade names, service marks, trademarks or registered trademarks of their respective owners. The Publisher is not associated with any product or vendor mentioned in this book.

This publication is designed to provide accurate and authoritative information in regard to the subject matter covered. It is sold on the understanding that the Publisher is not engaged in rendering professional services. If professional advice or other expert assistance is required, the services of a competent professional should be sought.

The Publisher and the Author make no representations or warranties with respect to the accuracy or completeness of the contents of this work and specifically disclaim all warranties, including without limitation any implied warranties of fitness for a particular purpose. The advice and strategies contained herein may not be suitable for every situation. In view of ongoing research, equipment modifications, changes in governmental regulations, and the constant flow of information relating to the use of experimental reagents, equipment, and devices, the reader is urged to review and evaluate the information provided in the package insert or instructions for each chemical, piece of equipment, reagent, or device for, among other things, any changes in the instructions or indication of usage and for added warnings and precautions. The fact that an organization or Website is referred to in this work as a citation and/or a potential source of further information does not mean that the author or the publisher endorses the information the organization or Website may provide or recommendations it may make. Further, readers should be aware that Internet Websites listed in this work may have changed or disappeared between when this work was written and when it is read. No warranty may be created or extended by any promotional statements for this work. Neither the Publisher nor the Author shall be liable for any damages arising herefrom.

Other Wiley Editorial Offices

John Wiley & Sons Inc., 111 River Street, Hoboken, NJ 07030, USA

Jossey-Bass, 989 Market Street, San Francisco, CA 94103-1741, USA

Wiley-VCH Verlag GmbH, Boschstr. 12, D-69469 Weinheim, Germany

John Wiley & Sons Australia Ltd, 42 McDougall Street, Milton, Queensland 4064, Australia

John Wiley & Sons (Asia) Pte Ltd, 2 Clementi Loop #02-01, Jin Xing Distripark, Singapore 129809

John Wiley & Sons Ltd, 6045 Freemont Blvd, Mississauga, Ontario L5R 4J3, Canada

Wiley also publishes its books in a variety of electronic formats. Some content that appears in print may not be available in electronic books.

Anniversary Logo Design: Richard J. Pacifico

Library of Congress Cataloging-in-Publication Data

Waigh, Tom A.
  Applied biophysics : a molecular approach for physical scientists / Tom A. Waigh.
    p. cm.
  Includes index.
  ISBN 978-0-470-01717-3 (alk. paper)
  1. Biophysics. I. Title.
  QH505.W35 2007
  571.4–dc22   2007011017

British Library Cataloguing in Publication Data

A catalogue record for this book is available from the British Library

ISBN 9780470017173 Cloth, 9780470017180 Paper

Typeset in 10.5/13 Sabon by Thomson Digital, India
Printed and bound in Great Britain by TJ International Ltd, Padstow, Cornwall
This book is printed on acid-free paper responsibly manufactured from sustainable forestry
in which at least two trees are planted for each one used for paper production.

Contents

Preface xi
Acknowledgements xiii

1 **The Building Blocks** 1
   1.1 Proteins 1
   1.2 Lipids 11
   1.3 Nucleic Acids 12
   1.4 Carbohydrates 15
   1.5 Water 18
   1.6 Proteoglycans and Glycoproteins 20
   1.7 Cells (Complex Constructs of Biomolecules) 21
   1.8 Viruses (Complex Constructs of Biomolecules) 22
   1.9 Bacteria (Complex Constructs of Biomolecules) 23
   1.10 Other Molecules 23
   Further Reading 23
   Tutorial Questions 24

2 **Mesoscopic Forces** 25
   2.1 Cohesive Forces 25
   2.2 Hydrogen Bonding 28
   2.3 Electrostatics 30
       2.3.1 Unscreened Electrostatic Interactions 30
       2.3.2 Screened Electrostatic Interactions 32
       2.3.3 The Force Between Charged Spheres in Solution 36
   2.4 Steric and Fluctuation Forces 38
   2.5 Depletion Forces 42
   2.6 Hydrodynamic Interactions 44
   2.7 Direct Experimental Measurements of Intermolecular and Surface Forces 44

	Further Reading	47
	Tutorial Questions	48

3 Phase Transitions — 49
- 3.1 The Basics — 49
- 3.2 Helix–Coil Transition — 53
- 3.3 Globule–Coil Transition — 59
- 3.4 Crystallisation — 64
- 3.5 Liquid–Liquid Demixing (Phase Separation) — 68
- Further Reading — 74
- Tutorial Questions — 74

4 Liquid Crystallinity — 77
- 4.1 The Basics — 77
- 4.2 Liquid–Nematic–Smectic Transitions — 92
- 4.3 Defects — 95
- 4.4 More Exotic Possibilities for Liquid Crystalline Phases — 100
- Further Reading — 103
- Tutorial Questions — 104

5 Motility — 107
- 5.1 Diffusion — 108
- 5.2 Low Reynold's Number Dynamics — 116
- 5.3 Motility — 119
- 5.4 First Passage Problem — 121
- 5.5 Rate Theories of Chemical Reactions — 125
- Further Reading — 127
- Tutorial Questions — 127

6 Aggregating Self-Assembly — 129
- 6.1 Surfactants — 133
- 6.2 Viruses — 137
- 6.3 Self-Assembly of Proteins — 139
- 6.4 Polymerisation of Cytoskeletal Filaments (Motility) — 142
- Further Reading — 148
- Tutorial Questions — 149

7 Surface Phenomena — 151
- 7.1 Surface Tension — 151
- 7.2 Adhesion — 154
- 7.3 Wetting — 156

CONTENTS vii

7.4	Capillarity	160
7.5	Experimental Techniques	164
7.6	Friction	165
7.7	Other Surface Phenomena	168
	Further Reading	168
	Tutorial Question	169

8 Biomacromolecules — 171
- 8.1 Flexibility of Macromolecules — 171
- 8.2 Good/Bad Solvents and the Size of Polymers — 177
- 8.3 Elasticity — 183
- 8.4 Damped Motion of Soft Molecules — 187
- 8.5 Dynamics of Polymer Chains — 191
- 8.6 Topology of Polymer Chains – Super Coiling — 199
- Further Reading — 201
- Tutorial Questions — 202

9 Charged Ions and Polymers — 205
- 9.1 Electrostatics — 207
- 9.2 Debye–Huckel Theory — 213
- 9.3 Ionic Radius — 214
- 9.4 The Behaviour of Polyelectrolytes — 218
- 9.5 Donnan Equilibria — 221
- 9.6 Titration Curves — 223
- 9.7 Poisson–Boltzmann Theory for Cylindrical Charge Distributions — 227
- 9.8 Charge Condensation — 228
- 9.9 Other Polyelectrolyte Phenomena — 232
- Further Reading — 234
- Tutorial Questions — 235

10 Membranes — 237
- 10.1 Undulations — 238
- 10.2 Bending Resistance — 240
- 10.3 Elasticity — 243
- 10.4 Intermembrane Forces — 248
- Further Reading — 250
- Tutorial Questions — 251

11 Continuum Mechanics — 253
- 11.1 Structural Mechanics — 254

11.2 Composites	258
11.3 Foams	261
11.4 Fracture	263
11.5 Morphology	265
Further Reading	265
Tutorial Questions	266

12 Biorheology — 267

12.1 Storage and Loss Moduli	270
12.2 Rheological Functions	274
12.3 Examples from Biology	276
12.3.1 Neutral Polymer Solutions	276
12.3.2 Polyelectrolytes	280
12.3.3 Gels	283
12.3.4 Colloids	287
12.3.5 Liquid Crystalline Polymers	288
12.3.6 Glassy Materials	290
12.3.7 Microfluidics in Channels	291
Further Reading	291
Tutorial Questions	291

13 Experimental Techniques — 293

13.1 Static Scattering Techniques	294
13.2 Dynamic Scattering Techniques	297
13.3 Osmotic Pressure	303
13.4 Force Measurement	306
13.5 Electrophoresis	314
13.6 Sedimentation	321
13.7 Rheology	325
13.8 Tribology	333
13.9 Solid Properties	334
Further Reading	335
Tutorial Questions	336

14 Motors — 339

14.1 Self-assembling Motility – Polymerisation of Actin and Tubulin	341
14.2 Parallelised Linear Stepper Motors – Striated Muscle	346
14.3 Rotatory Motors	350
14.4 Ratchet Models	350
14.5 Other Systems	352

Further Reading	353
Tutorial Question	353

15 Structural Biomaterials — 355
15.1 Cartilage – Tough Shock Absorber in Human Joints — 355
15.2 Spider Silk — 368
15.3 Elastin and Resilin — 369
15.4 Bone — 371
15.5 Adhesive Proteins — 372
15.6 Nacre and Mineral Composites — 373
Further Reading — 375
Tutorial Questions — 375

16 Phase Behaviour of DNA — 377
16.1 Chromatin – Naturally Packaged DNA Chains — 377
16.2 DNA Compaction – An Example of Polyelectrolyte Complexation — 380
16.3 Facilitated Diffusion — 383
Further Reading — 387

Appendix — 389
Answers to Tutorial Questions — 391

Index — 407

Preface

The field of molecular biophysics is introduced in the following pages. The presentation focuses on the simple underlying concepts and demonstrates them using a series of up to date applications. It is hoped that the approach will appeal to physical scientists who are confronted with biological questions for the first time as they become involved in the current biotechnological revolution.

The field of biochemistry is vast and it is not the aim of this textbook to encompass the whole area. The book functions on a reductionist, nuts and bolts approach to the subject matter. It aims to explain the constructions and machinery of biological molecules very much as a civil engineer would examine the construction of a building or a mechanical engineer examine the dynamics of a turbine. Little or no recourse is taken to the chemical side of the subject, instead modern physical ideas are introduced to explain aspects of the phenomena that are confronted. These ideas provide an alternative, complementary set of tools to solve biophysical problems. It is thus hoped that the book will equip the reader with these new tools to approach the subject of biological physics.

A few rudimentary aspects of medical molecular biophysics are considered. In terms of the statistics of the cause of death, heart disease, cancer and Alzheimer's are some of the biggest issues that confront modern society. An introduction is made to the action of striated muscle (heart disease), DNA delivery for gene therapy (cancers and genetic diseases), and self-assembling protein aggregates (amyloid diseases such as Alzheimer's). These diseases are some of the major areas of medical research, and combined with food (agrochemical) and pharmaceutics, provide the major industrial motivation encouraging the development of molecular biophysics.

Please try to read some of the highlighted books, they will prove invaluable to bridge the gap between undergraduate studies and active areas of research science.

TOM WAIGH
Manchester, UK
February 2007

Acknowledgements

I would like to thank my family (Roger, Sally, Cathy, Paul, Bronwyn and Oliver) and friends for their help and support. The majority of the book was written in the physics department of the Universities of Manchester and Leeds. The PhD and undergraduate students (the umpa lumpas etc.) who weathered the initial course and the rough drafts of the lecture notes on which this book was based should be commended. I am indebted to the staff at the University of Edinburgh, the University of Cambridge and the Collège de France who helped educate me concerning the behaviour of soft condensed matter and molecular biophysics.

Acknowledgements

1
The Building Blocks

It is impossible to pack a complete biochemistry course into a single introductory chapter. Some of the basic properties of the structure of simple biological macromolecules, lipids and micro organisms are covered. The aim is to give a basic grounding in the rich variety of molecules that life presents, and some respect for the extreme complexity of the chemistry of biological molecules that operates in a wide range of cellular processes.

1.1 PROTEINS

Polymers consist of a large number of sub-units (monomers) connected together with covalent bonds. A protein is a special type of polymer. In a protein there are up to twenty different amino acids (Figure 1.1) that can function as monomers, and all the monomers are connected together with identical peptide linkages (C–N bonds, Figure 1.2). The twenty amino acids can be placed in different families dependent on the chemistry of their different side groups. Five of the amino acids form a group with lipophilic (fat-liking) side-chains: glycine, alanine, valine, leucine, and isoleucine. Proline is a unique circular amino acid that is given its own separate classification. There are three amino acids with aromatic side-chains: phenylalanine, tryptophan, and tyrosine. Sulfur is in the side-chains of two amino acids: cysteine and methionine. Two amino acids have hydroxyl (neutral) groups that make them water loving: serine and threonine. Three amino acids have very polar positive side-chains: lysine, arginine and histidine. Two amino acids form a family with acidic

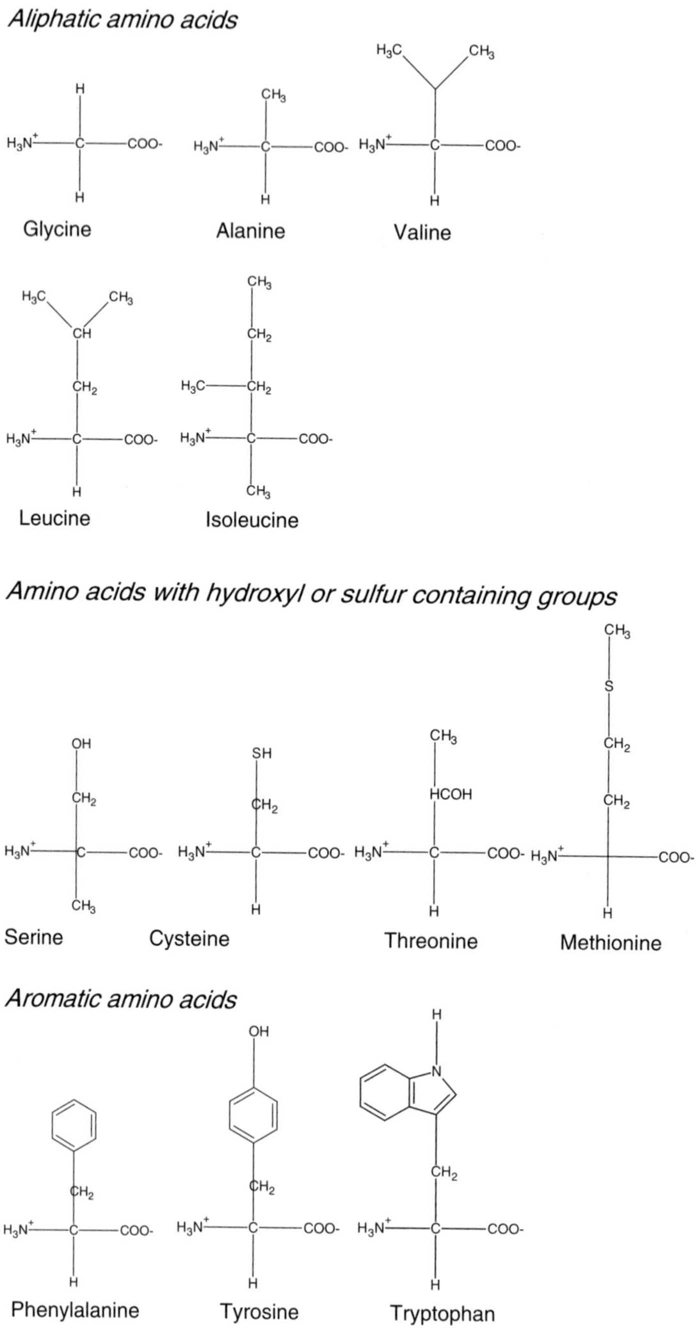

THE BUILDING BLOCKS

Aliphatic amino acids

Amino acids with hydroxyl or sulfur containing groups

Aromatic amino acids

Figure 1.1 The chemical structure of the twenty amino acids found in nature

Cyclic amino acid

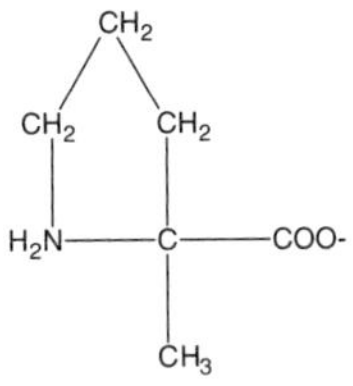

Proline

Basic amino acids

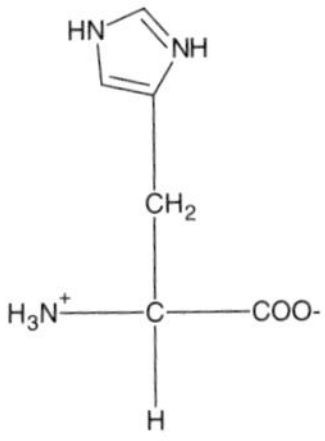

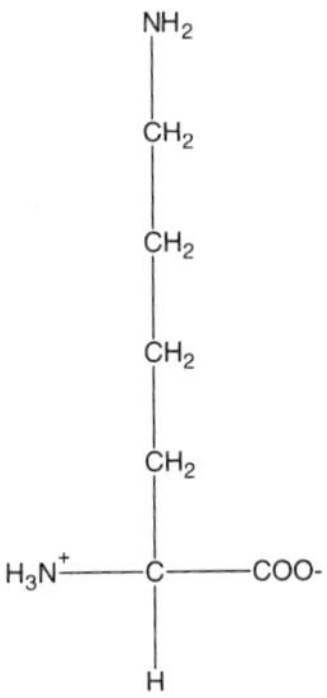

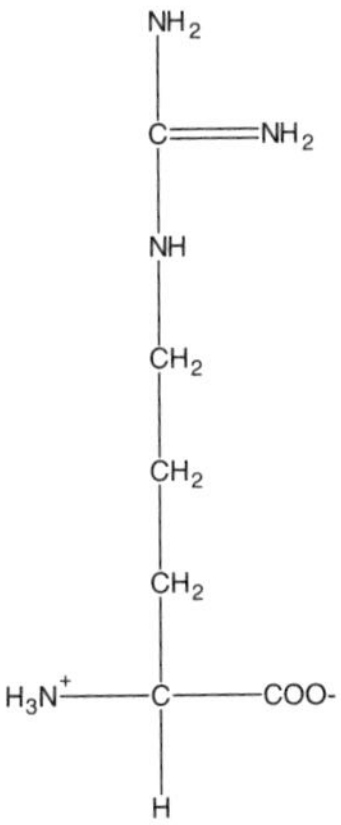

Histidine Lysine Arginine

Acidic amino acids and amides

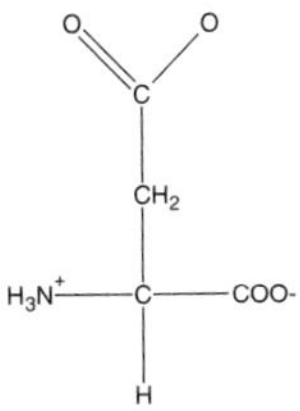

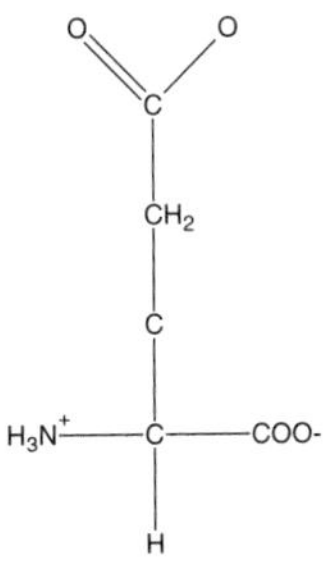

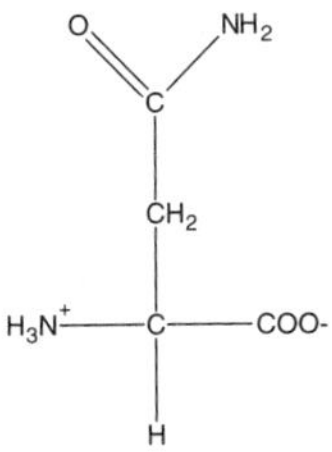

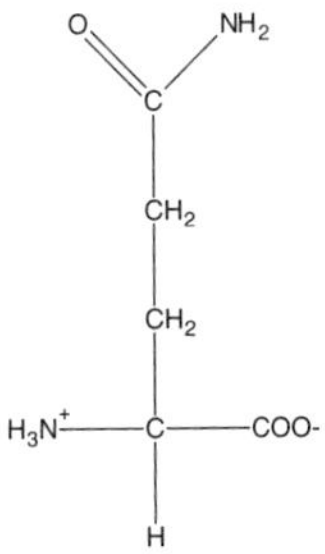

Aspartic acid Glutamic acid Asparagine Glutamine

Figure 1.1 (*Continued*)

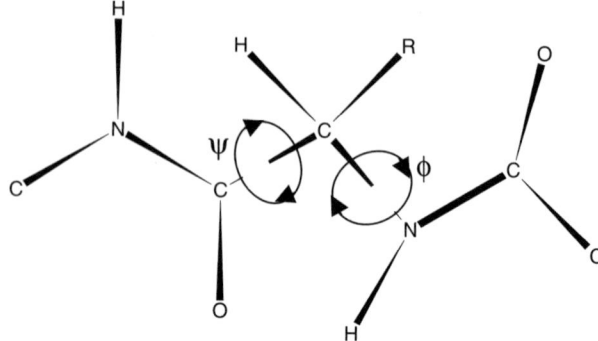

Figure 1.2 All amino acids have the same primitive structure and are connected with the same peptide linkage through C–C–N bonds
(O, N, C, H indicate oxygen, nitrogen, carbon and hydrogen atoms respectively. R is a pendant side-group which provides the amino acid with its identity, i.e. proline, glycine etc.)

side-groups and they are joined by two corresponding neutral counterparts that have a similar chemistry: aspartate, glutamate, asparagine, and glutamine.

The linkages between amino acids all have the same chemistry and basic geometry (Figure 1.2). The *peptide linkage* that connects all amino acids together consists of a carbon atom attached to a nitrogen atom through a single covalent bond. Although the chemistry of peptide linkages is fairly simple, to relate the primary sequence of amino acids to the resultant three dimensional structure in a protein is a daunting task and predominantly remains an unsolved problem. To describe protein structure in more detail it is useful to consider the motifs of secondary structure that occur in their morphology. The motifs include *alpha helices*, *beta sheets* and *beta barrels* (Figure 1.3). The full three dimensional *tertiary structure* of a protein typically takes the form of a compact globular morphology (the globular proteins) or a long extended conformation (fibrous proteins, Figures 1.4 and 1.5). Globular morphologies usually consist of a number of secondary motifs combined with more disordered regions of peptide.

Charge interactions are very important in determining of the conformation of biological polymers. The degree of charge on a polyacid or polybase (e.g. proteins, nucleic acids etc) is determined by the pH of a solution, i.e. the concentration of hydrogen ions. Water has the ability to dissociate into oppositely charged ions; this process depends on temperature

$$H_2O \rightleftarrows H^+ + OH^- \qquad (1.1)$$

PROTEINS

The product of the hydrogen and hydroxyl ion concentrations formed from the dissociation of water is a constant at equilibrium and at a fixed temperature (37 °C)

$$c_{H^+} c_{OH^-} = 1 \times 10^{-14} M^2 = K_w \quad (1.2)$$

where c_{H^+} and c_{OH^-} are the concentrations of hydrogen and hydroxyl ions respectively. Addition of acids and bases to a solution perturbs the equilibrium dissociation process of water, and the acid/base equilibrium

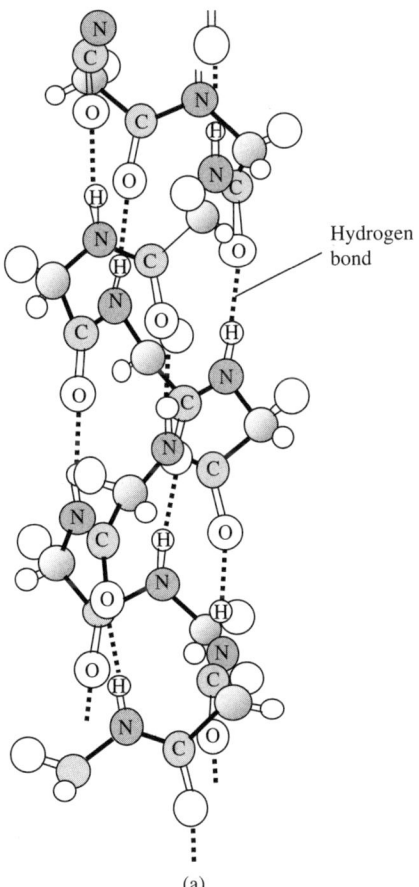

(a)

Figure 1.3 Simplified secondary structures of (a) an α-helix and (b) a β-sheet that commonly occur in proteins
(Hydrogen bonds are indicated by dotted lines.)

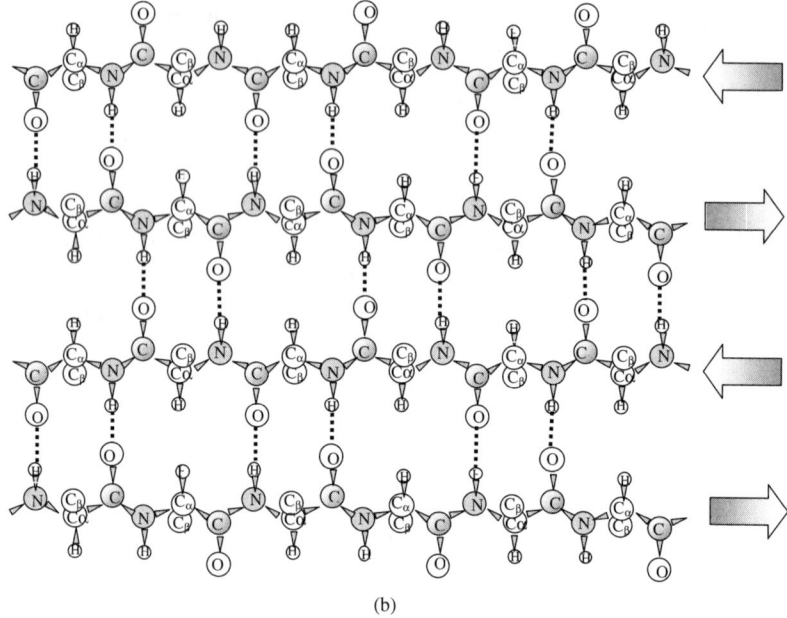

(b)

Figure 1.3 (*Continued*)

phenomena involved are a corner stone of the physical chemistry of solutions. Due to the vast range of possible hydrogen ion (H^+) concentrations typically encountered in aqueous solutions, it is normal to use a logarithmic scale (pH) to quantify them. The pH is defined as the

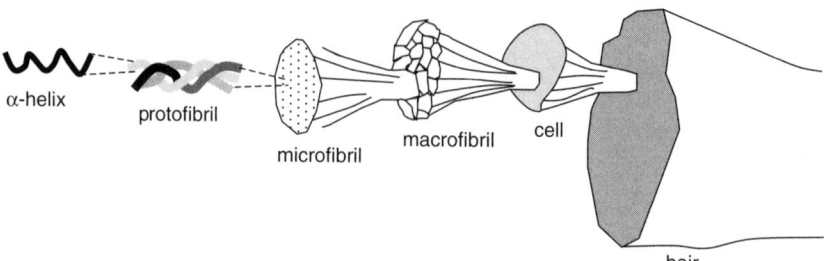

Figure 1.4 The complex hierarchical structures found in the keratins of hair (α-helices are combined in to protofibrils, then into microfibrils, macrofibrils, cells and finally in to a single hair fibre [*Reprinted with permission from J. Vincent, Structural Biomaterial, Copyright (1990) Princeton University Press*])

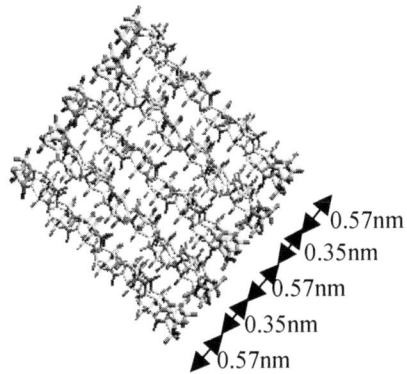

Figure 1.5 The packing of anti-parallel beta sheets found in silk proteins (Distances between the adjacent sheets are shown.)

negative logarithm (base 10!) of the hydrogen ion concentration

$$\text{pH} = -\log c_{H^+} \tag{1.3}$$

Typical values of pH range from 6.5 to 8 in physiological cellular conditions. Strong acids have a pH in the range 1–2 and strong bases have a pH in the range 12–13.

When an acid (HA) dissociates in solution it is possible to define an equilibrium constant (K_a) for the dissociation of its hydrogen ions (H^+)

$$\text{HA} \rightleftharpoons H^+ + A^- \qquad K_a = \frac{c_{H^+} c_{A^-}}{c_{HA}} \tag{1.4}$$

where c_{H^+}, c_{A^-} and c_{HA} are the concentrations of the hydrogen ions, acid ions, and acid molecules respectively. Since the hydrogen ion concentration follows a logarithmic scale, it is natural to also define the dissociation constant on a logarithmic scale (pK_a)

$$pK_a = -\log K_a \tag{1.5}$$

The logarithm of both sides of equation (1.4) can be taken to give a relationship between the pH and the pK_a value:

$$\text{pH} = pK_a + \log\left\{\frac{c_{\text{conjugate\_base}}}{c_{\text{acid}}}\right\} \tag{1.6}$$

where $c_{\text{conjugate\_base}}$ and c_{acid} are the concentrations of the conjugate base (e.g. A^-) and acid (e.g. HA) respectively. This equation enables the degree of dissociation of an acid (or base) to be calculated, and it is named after its inventors *Henderson and Hasselbalch*. Thus a knowledge of the pH of a solution and the pK_a value of an acidic or basic group allows the charge fraction on the molecular group to be calculated to a first approximation. The propensity of the amino acids to dissociate in water is illustrated in Table 1.1. In contradiction to what their name might imply, only amino acids with acidic or basic side groups are charged when incorporated into proteins. These charged amino acids are arginine, aspartic acid, cysteine, glutamic acid, histidine, lysine and tyrosine.

Another important interaction between amino acids, in addition to charge interactions, is their ability to form hydrogen bonds with surrounding water molecules; the degree to which this occurs varies. This amino acid hydrophobicity (the amount they dislike water) is an important driving force for the conformation of proteins. Crucially it leads to the compact conformation of globular proteins (most enzymes) as the hydrophobic groups are buried in the centre of the globules to avoid contact with the surrounding water.

Table 1.1 Fundamental physical properties of amino acids found in protein [*Ref.: Data adapted from C.K. Mathews and K.E. Van Holde, Biochemistry, 137*].

Name	pK$_a$ value of side chain	Mass of residue	Occurrence in natural proteins (%mol)
Alanine	—	71	9.0
Arginine	12.5	156	4.7
Asparagine	—	114	4.4
Apartic acid	3.9	115	5.5
Cysteine	8.3	103	2.8
Glutamine	—	128	3.9
Glutamic acid	4.2	129	6.2
Glycine	—	57	7.5
Histidine	6.0	137	2.1
Isoleucine	—	113	4.6
Leucine	—	113	7.5
Lysine	10.0	128	7.0
Methionine	—	131	1.7
Phenylalanine	—	147	3.5
Proline	—	97	4.6
Serine	—	87	7.1
Threonine	—	101	6.0
Tyrptophan	—	186	1.1
Tyrosine	10.1	163	3.5
Valine	—	99	6.9

PROTEINS

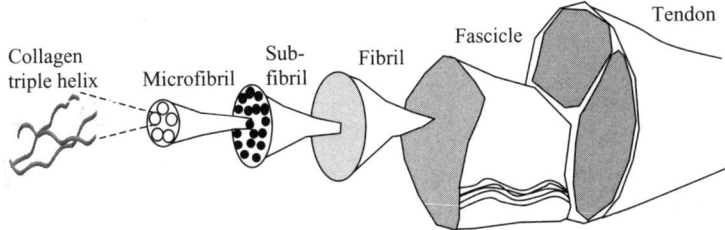

Figure 1.6 Hierarchical structure for the collagen triple helices in tendons (Collagen helices are combined into microfibrils, then into sub-fibrils, fibrils, fascicles and finally into tendons.)

Covalent interactions are possible between adjacent amino acids and can produce solid protein aggregates (Figures 1.4 and 1.6). For example, disulfide linkages are possible in proteins that contain cysteine, and these form the strong inter-protein linkages found in many fibrous proteins e.g. keratins in hair.

The internal secondary structures of protein chains (α helices and β sheets) are stabilised by hydrogen bonds between adjacent atoms in the peptide groups along the main chain. The important structural proteins such as keratins (Figure 1.4), collagens (Figure 1.6), silks (Figure 1.5), anthropod cuticle matrices, elastins (Figure 1.7), resilin

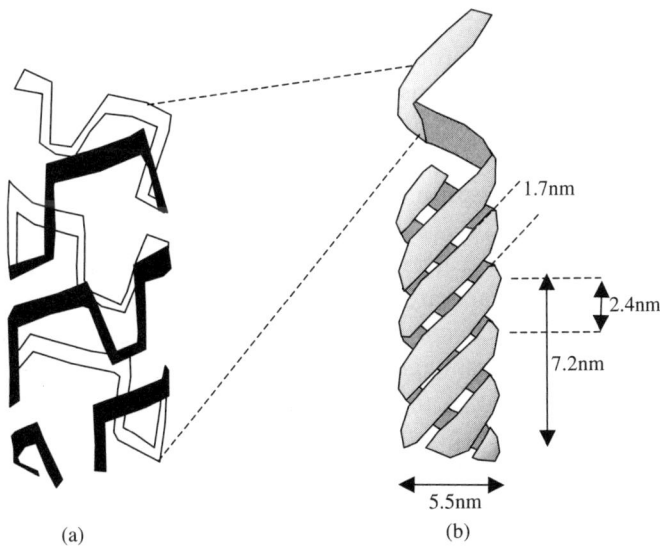

Figure 1.7 The β turns in elastin (a) form a secondary elastic helix which is subsequently assembled into a superhelical fibrous structure (b)

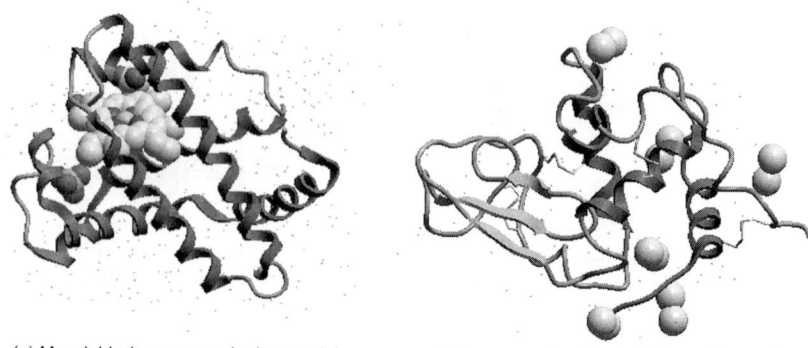

(a) Myoglobin (oxygen carrier in muscle)   (b) Lysozyme (antibacterial found in tears)

Figure 1.8 Two typical structures of globular proteins calculated using X-ray crystallography data

and abductin are formed from a combination of intermolecular disulfide and hydrogen bonds.

Some examples of the globular structures adopted by proteins are shown in Figure 1.8. Globular proteins can be denatured in a folding/unfolding transition through a number of mechanisms, e.g. an increase in the temperature, a change of pH, and the introduction of hydrogen bond breaking chaotropic solvents. Typically the complete denaturation transition is a first order thermodynamic phase change with an associated latent heat (the thermal energy absorbed during the transition). The unfolding process involves an extremely complex sequence of molecular origami transitions. There are a vast number of possible molecular configurations ($\sim 10^N$ for an N residue protein) that occur in the reverse process of protein folding, when the globular protein is constructed from its primary sequence by the cell, and thus frustrated structures could easily be formed during this process. Indeed, at first sight it appears a certainty that protein molecules will become trapped in an intermediate state and never reach their correctly folded form. This is called *Levinthal's paradox*, the process by which natural globular proteins manage to find their native state among the billions of possibilities in a finite time. The current explanation of protein folding that provides a resolution to this paradox, is that there is a funnel of energy states that guide the kinetics of folding across the complex energy landscape to the perfectly folded state (Figure 1.9).

There are two main types of inter-chain interaction between different proteins in solution; those in which the native state remains largely

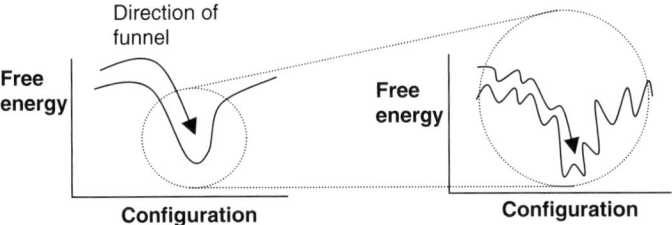

Figure 1.9 Schematic diagram indicating the funnel that guides the process of protein folding through the complex configuration space that contains many local minima. The funnel avoids the frustrated misfolded protein structures described in Levinthal's paradox

unperturbed in processes such as protein crystallisation and the formation of filaments in sheets and tapes, and those interactions that lead to a loss of conformation e.g. heat set gels (e.g. table jelly and boiled eggs) and amyloid fibres (e.g. Alzheimer's disease and Bovine Spongiform Encephalopathy).

1.2 LIPIDS

Cells are divided into a series of subsections or compartments by membranes which are formed predominantly from lipids. The other main role of lipids is as energy storage compounds, although the molecules play a role in countless other physiological processes. Lipids are amphiphilic, the head groups like water (and hate fat) and the tails like fat (and hate water). This amphiphilicity drives the spontaneous self-assembly of the molecules into membranous morphologies.

There are four principle families of lipids: fatty acids with one or two tails (including carboxylic acids of the form RCOOH where R is a long hydrocarbon chain), and steroids and phospholipids where two fatty acids are linked to a glycerol backbone (Figure 1.10). The type of polar head group differentiates the particular species of naturally occurring lipid. Cholesterol is a member of the steroid family and these compounds are often found in membrane structures. Glycolipids also occur in membranes and in these molecules the phosphate group on a phospholipid is replaced by a sugar residue. Glycolipids have important roles in cell signalling and the immune system. For example, these molecules are an important factor in determining the compatibility of blood cells after a blood transfusion, i.e. blood types A, B, O, etc.

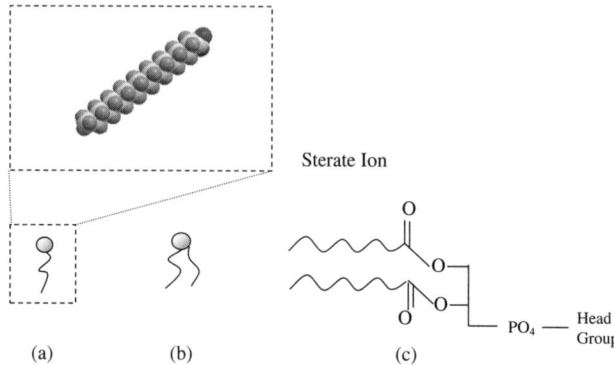

Figure 1.10 Range of lipid molecules typically encountered in biology (a) fatty acids with one tail; (b) steroids and fatty acids with two tails; (c) phospholipids

1.3 NUCLEIC ACIDS

The *'central dogma of biochemistry'* according to F.C.Crick is illustrated in Figure 1.11. DNA contains the basic blueprint for life that guides the construction of the vast majority of living organisms. To implement this blue print cells need to *transcribe* DNA to RNA, and this structural information is subsequently translated into proteins using specialised protein factories (the ribosomes). The resultant proteins can then be used to catalyse specific chemical reactions or be used as building materials to construct new cells.

This simple biochemical scheme for transferring information has powerful implications. DNA can now be altered systematically using *recombinant DNA technology* and then placed inside a living cell. The foreign DNA hijacks the cell's mechanisms for translation and the proteins that are subsequently formed can be tailor-made by the genetic engineer to fulfil a specific function, e.g. bacteria can be used to form biodegradable plastics from the fibrous proteins that are expressed.

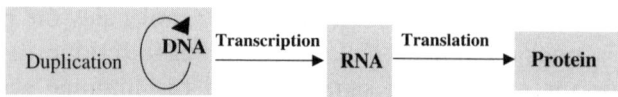

Figure 1.11 The central dogma of molecular biology considers the duplication and translation of DNA. DNA is duplicated from a DNA template. DNA is transcribed to form a RNA chain, and this information is translated into a protein sequence

NUCLEIC ACIDS

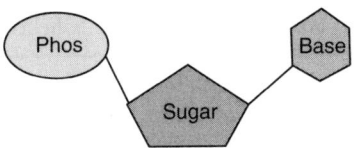

Figure 1.12 The chemical structure of the base of a nucleic acid consists of a phosphate group, a sugar and a base

The monomers of DNA are made of a sugar, an organic base and a phosphate group (Figure 1.12). There are only four organic bases that naturally occur in DNA, and these are thymine, cytosine, adenine and guanine (T,C,A,G). The sequence of bases in each strand along the backbone contains the genetic code. The base pairs in each strand of the double helical DNA are complementary, A has an afinity for T (they form two hydrogen bonds) and G for C (they form three hydrogen bonds). The interaction between the base pairs is driven by the geometry of the hydrogen bonding sites. Thus each strand of the DNA helix contains an identical copy of the genetic information to its complementary strand, and replication can occur by separation of the double helix and resynthesis of two additional chains on each of the two original double helical strands. The formation of helical secondary structures in DNA drastically increases the persistence length of each separate chain and is called a *helix-coil transition*.

There is a major groove and a minor groove on the biologically active A and B forms of the DNA double helix. The individual polynucleotide DNA chains have a sense of direction, in addition to their individuality (a complex nucleotide sequence). DNA replication in vivo is conducted by a combination of the DNA polymerases (I, II and III).

DNA in its double helical form can store torsional energy, since the monomers are not free to rotate (like a telephone cable). The ends of a DNA molecule can be joined together to form a compact supercoiled structure that often occurs in vivo in bacteria; this type of molecule presents a series of fascinating questions with regard to its statistical mechanics and topological analysis.

DNA has a wide variety of structural possibilities (Table 1.2, Figure 1.13). There are *3 standard types* of averaged double helical structure labelled A, B and Z, which occur ex vivo in the solid fibres used for X-ray structural determination. Typically DNA in solution has a structure that is intermediate between A and B, dependent on the chain sequence and the aqueous environment. An increase in the level of hydration tends to increase the number of B type base pairs in a double

THE BUILDING BLOCKS

Table 1.2 Structural parameters of polynucleotide helices

Property	A form	B form	Z-form
Direction of helix rotation	Right	Right	Left
Number of residues per turn	11	10	12
Rotation per residue	33°	36°	30°
Rise in helix per residue	0.255 nm	0.34 nm	0.37 nm
Pitch of helix	2.8 nm	3.4 nm	4.5 nm

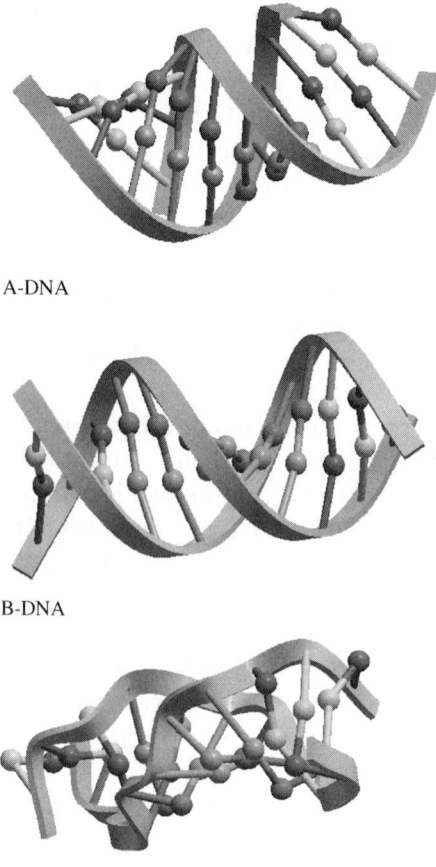

A-DNA

B-DNA

Z-DNA

Figure 1.13 Molecular models of A, B and Z type double helical structures of DNA (A and B type helical structures, and their intermediates typically occur in biological systems. Z-DNA helical structures crystallise under extreme non-physiological conditions.)

helix. Z-type DNA is favoured in some extreme non-physiological conditions.

There are a number of local structural modifications to the helical structure that are dependent on the specific chemistry of the individual DNA strands, and are in addition to the globally averaged A, B and Z classifications. The *kink* is a sudden bend in the axis of the double helix which is important for complexation in the nucleosome. The *loop* contains a rupture of hydrogen bonds over several base pairs, and the separation of two nucleotide chains produces loops of various sizes. In the process of DNA transcription RNA polymerase is bound to DNA to form a loop structure. In the process of *breathing* of a double helix, hydrogen bonds are temporarily broken by a rapid partial rotation of one base pair. The hydrogen atoms in the NH groups are therefore accessible and can be exchanged with neighbouring protons in the presence of a catalyst. The *cruciform structure* is formed in the presence of self-complementary palindromic sequences separated by several base pairs. Hydrophobic molecules (e.g. DNA active drugs) can be *intercalated* into the DNA structure, i.e. slipped between two base pairs. Helices that contain three or four nucleic acid strands are also possible with DNA, but do not occur naturally.

DNA has a number of interesting features with respect to its polymer physics. The persistence length (l_p) of DNA is in the order of 50 nm for *E. coli* (which depends on ionic strength), it can have millions of monomers in its sequence and a correspondingly gigantic contour length (L) (for humans L is ~ 1.5 m!). The large size of DNA has a number of important consequences; single fluorescently labelled DNA molecules are visible under an optical microscope, which proves very useful for high resolution experiments, and the cell has to solve a tricky packaging problem in vivo of how to fit the DNA inside the nucleus of a cell which is, at most, a few microns in diameter (it uses chromosomes).

1.4 CARBOHYDRATES

Historically, advances in carbohydrate research have been overshadowed by developments in protein science. This has in part been due to the difficulty of analysing of the structure of carbohydrates, and the extremely large variety of chemical structures that occur naturally. Carbohydrates play a vital role in a vast range of cellular processes that are still only partly understood.

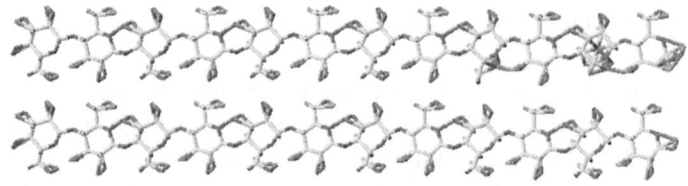

Figure 1.14 Sheet-like structures formed in cellulosic materials
(The $\beta(1 \to 4)$ linkages between glucose monomers induce extended structures, and the cellulose chains are linked together with hydrogen bonds.)

There are two important glucose polymers which occur in plants that are differentiated by the linkage between the monomers: cellulose and amylopectin. *Cellulose* is a very rigid polymer, and has both nematic and semi-crystalline phases. It is used widely in plants as a structural material. The straight chain formed by the $\beta(1 \to 4)$ linkage between glucose molecules is optimal for the construction of fibres, since it gives them a high tensile strength in the chain direction (Figures 1.14 and 1.15), and reasonable strength perpendicular to the chain due to the substantial intrachain hydrogen bonding in sheet-like structures. *Amylose* and its branched form, *amylopectin* (starch), are used in plants to store energy, and often amylopectin adopts smectic liquid crystalline phases

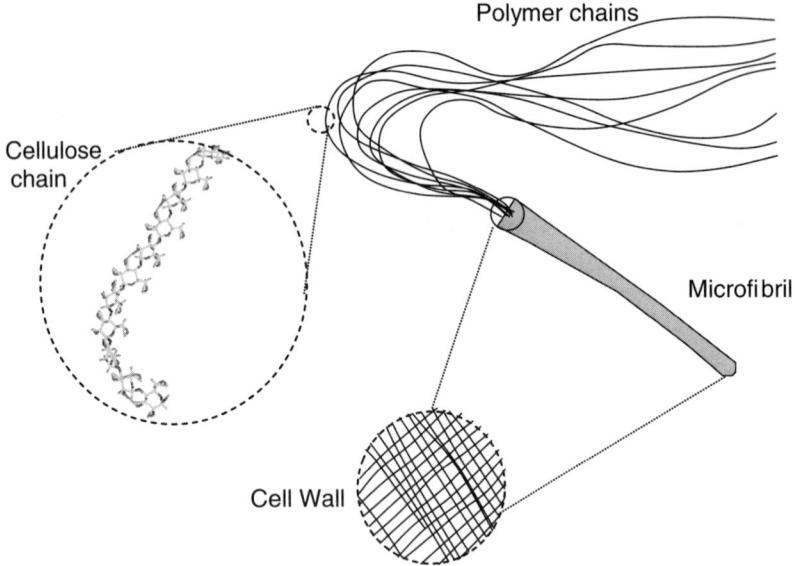

Figure 1.15 The hierarchical structure of cellulose found in plant cell walls
(Cellulose chains are combined into microfibrils that form the walls of plant cells [*Ref.: adapted from C.K. Mathews and K.E. Van Holde, Biochemistry, Benjamin Cummings*])

CARBOHYDRATES

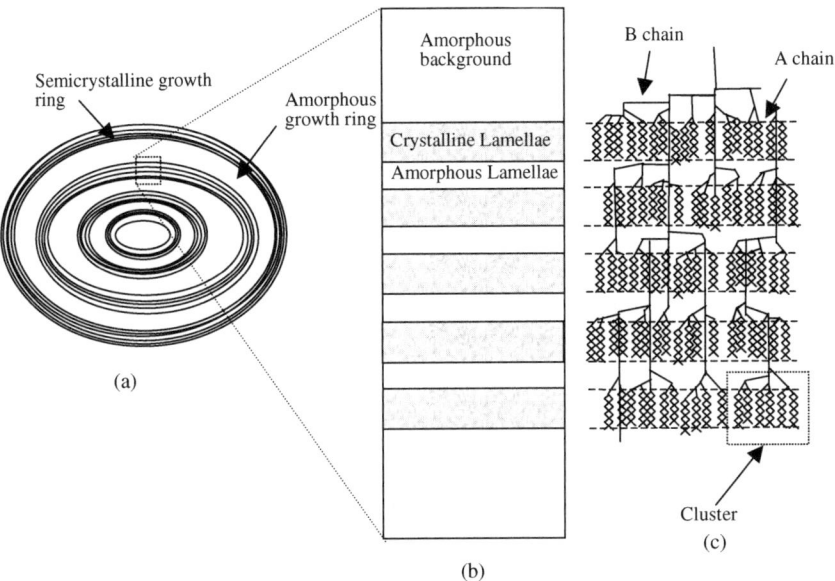

Figure 1.16 Four length scales are important in the hierarchical structure of starch; (a) the whole granule morphology ($\sim \mu$m), (b) the growth rings (~ 100 nms), (c) the crystalline and amorphous lamellae (~ 9 nm), and (d) the molecular structure of the amylopectin ($\sim$Å). [Ref.: T.A. Waigh, PhD thesis, University of Cambridge, 1996]

(Figure 1.16). Starch, an amylose/amylopectin composite, forms the principle component of mankind's food sources. In amylose the glucose molecules are connected together with an α $(1 \rightarrow 4)$ linkage. α-linkages between the glucose molecules are well suited to the formation of an accessible sugar store, since they are flexible and can be easily degraded by enzymes. Amylopectins are formed from amyloses with additional branched α $(1 \rightarrow 6)$ flexible linkages between glucose molecules (Figure 1.17). Glycogen is an amorphous hyperbranched glucose polymer analogous to amylopectin, and is used inside animal cells as an energy store.

Chitin is another structural polysaccharide; it forms the exoskeleton of crustaceans and insects. It is similar in its functionality to cellulose, it is a very rigid polymer and has a cholesteric liquid crystalline phase.

It must be emphasised that the increased complexity of linkages between sugar molecules, compared with nucleic acids or proteins, provides a high density mechanism for encoding information. A sugar molecule can be polymerised in a large number of ways, e.g. the six corners of a glucose molecule can each be polymerised to provide an additional N^6 arrangements for a carbohydrate compared with a protein

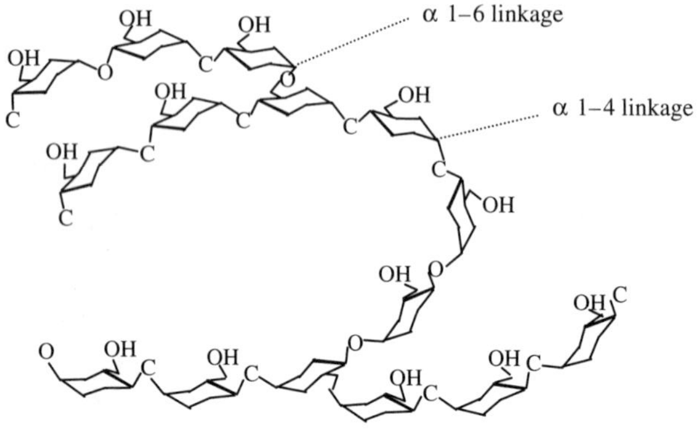

Figure 1.17 The branched primary structure found for amylopectin in starch (Both α(1→4) and α(1→6) flexible linkages occur between adjacent glucose monomers.)

of equivalent length (N). In proteins there is only one possible mechanism to connect amino acids, the peptide linkage. These additional possibilities for information storage with carbohydrates are used naturally in a range of immune response mechanisms.

Pectins are extra cellular plant polysaccharides forming gums (used in jams), and similarly *algins* can be extracted from sea weed. Both are widely used in the food industry. *Hyaluronic* acid is a long negatively charged semi-flexible polyelectrolyte and occurs in a number of roles in animals. For example it is found as a component of cartilage (a biological shock absorber) and as a lubricant in synovial joints.

1.5 WATER

Water is a unique polar solvent and its properties have a vast impact on the behaviour of biological molecules (Figure 1.18). Water has a high

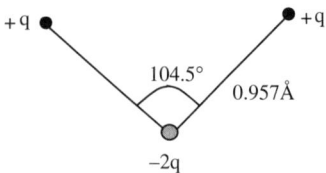

Figure 1.18 The geometry of a single water molecule
(The molecule tends to form a tetrahedral structure once hydrogen bonded in ice crystals (Figure 2.2).)

WATER

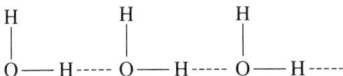

Figure 1.19 Schematic diagram of the network structure formed by water molecules (Dashed lines indicate hydrogen bonds. Such chains of hydrogen bonded water molecules occur over a wide range of angles for liquid water.)

dipole moment (P) of 6.11×10^{-30} Cm, a quadrupole moment of 1.87×10^{-39} Cm^2 and a mean polarisability of 1.44×10^{-30} m^3.

Water exists in a series of crystalline states at sub zero temperature or elevated pressures. The structure of ice formed in ambient conditions has unusual cavities in its structure due to the directional nature of hydrogen bonds, and it is consequently less dense than liquid water at its freezing point. The polarity of the O–H bonds formed in water allows it to associate into dimers, trimers etc (Figure 1.19), and produces a complex many body problem for the statistical description of water in both liquid and solid condensed phases.

Antifreeze proteins have been designed through evolution to impair the ability of the water that surrounds them in solution to crystallise at low temperatures. They have an alpha helical dipole moment that disrupts the hydrogen bonded network structure of water. These antifreeze molecules have a wide range of applications for organisms that exist in sub zero temperatures e.g. arctic fish and plants.

The imaging of biological processes is possible in vivo using the technique of nuclear magnetic resonance, which depends on the mobility of water to create the image. This powerful non-invasive method allows water to be viewed in a range of biological processes, e.g. cerebral activity.

Even at very low volume fractions water can act as a plasticiser that can switch solid biopolymers between glassy and non glassy states. The ingress of water can act as a switch that will trigger cellular activity in plant seeds, and such dehydrated cellular organisms can remain dormant for many thousands of years before being reactivated by the addition of water.

A wide range of time scales (10^{-18}–10^{3} s) of water are important to understand its biological function (Figure 1.20). The range of time scales includes such features as the elastic collisions of water at ultra fast times ($\sim 10^{-15}$ seconds) to the macroscopic hydrodynamic processes observed in blood flow at much slower times ($\sim$seconds).

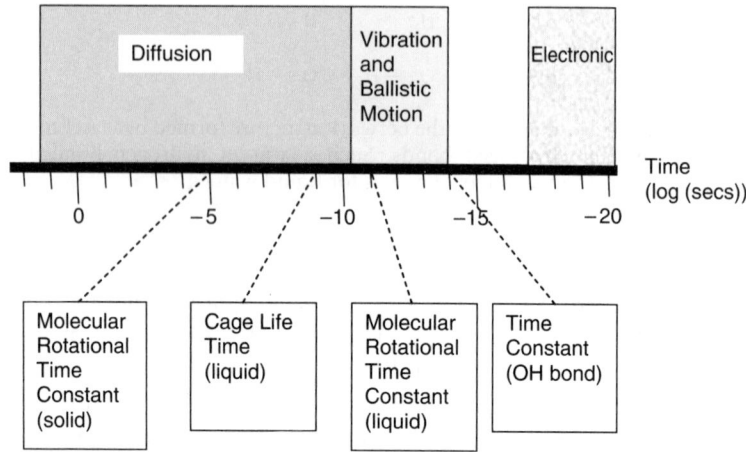

Figure 1.20 The range of time scales that determine the physical properties of water, shown on a logarithmic scale

1.6 PROTEOGLYCANS AND GLYCOPROTEINS

Proteoglycans (long carbohydrate molecules attached to short proteins) and glycoproteins (short carbohydrate molecules attached to relatively long proteins) are constructed from a mixture of protein and carbohydrate molecules (the glycosoaminoglycans). In common with carbohydrates, proteoglycans/glycoproteins exhibit extreme structural and chemical heterogeneity. Furthermore, the challenges presented to crystallography by their non-crystallinity means that a full picture of the biological function of these molecules is still not complete.

Many proteoglycans and glycoproteins used in the extracellular matrix have a bottle brush morphology (Figures 1.21 and 1.22). An

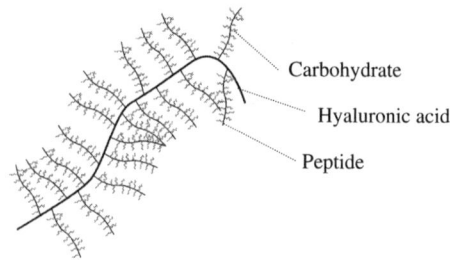

Figure 1.21 A schematic diagram of the aggrecan aggregate
(The aggrecan monomers (side brushes) consist of a core protein with highly charged carbohydrate side-chains. The bottle brushes are physically bound to the linear hyaluronic acid backbone chain to form a super bottle brush structure [*Ref.: A. Papagiannopoulos, T.A.Waigh, T. Hardingham and M. Heinrich, Biomacromolecules, 2006, 7, 2162–2172*])

CELLS (COMPLEX CONSTRUCTS OF BIOMOLECULES)

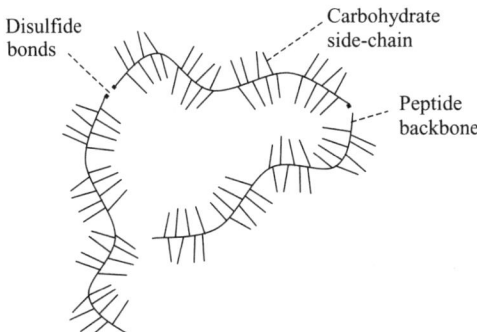

Figure 1.22 Porcine stomach mucin molecules contain a series of carbohydrate brush sections that are connected to a peptide backbone. The ends of the peptide are sticky, and these telechelic bottle brushes form thick viscoelastic gels at low pHs.

example of a sophisticated proteoglycan is aggrecan, a giant polymeric molecule that consists of a bottle-brush of bottle-brushes (Figure 1.21). These materials have a very large viscosity in solution, and are used to dissipate energy in collageneous cartilage composites and to reduce friction in synovial joints as boundary lubricants. An example of an extracellular glycoprotein is the mucins found in the stomach of mammals. These molecules experience telechelic (either end) associations to form thick viscoelastic gels that protect the stomach lining from autodigestion (Figure 1.22).

Other examples of glycoproteins occur in enzymes (Ribonuclease B), storage protein (egg white), blood clots (fibrin) and antibodies (Human IgG).

1.7 CELLS (COMPLEX CONSTRUCTS OF BIOMOLECULES)

Cells act co-operatively in multicellular organisms and are hierarchically arranged into tissues, organs and organ systems. Tissues contain both cells and other materials such as the extracellular matrix.

There are four distinct forms of mammalian *muscle cells*: skeletal and cardiac (which both form striated musclar tissues), smooth muscle (found in blood vessels and intestines) and myoepithlial cells (again present in intestines).

Nerve cells are used to send and receive signals. They are highly branched and this structure allows them to react to up to one

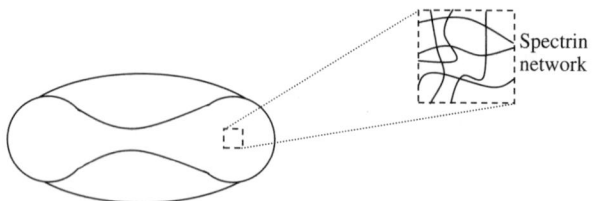

Figure 1.23 The cross-section through a squashed donut shaped blood cell (The spectrin network in the cell wall is a dominant factor for the determination of the morphology of the cell.)

hundred thousand inputs from other cells. The electrochemistry of nerve cells is a fascinating area; the efficiency and time response of these electrical circuits has been carefully optimised by evolution.

Blood cells have a squashed donut shape (Figure 1.23) which is related to the differential geometry of their cytoskeleton. Red blood cells carry oxygen and carbon dioxide, towards and away from the lungs. White blood cells play a role in the fight to remove infections from an organism.

Fibroblast cells are largely responsible for the secretion and regulation of the extracellular matrix, e.g. the production of molecules such as the collagens. *Epithelial cells* control the passage of material across the boundary of organs, e.g. in the interior of the intestinal tract.

1.8 VIRUSES (COMPLEX CONSTRUCTS OF BIOMOLECULES)

Viruses are intra-cellular parasites, biological entities that multiply through the invasion of cellular organisms. In addition to aspects related to their biological role in disease, viruses have attracted a great deal of attention from biophysicists for their physical properties. Viruses self-assemble into well defined monodisperse geometrical shapes (rods and polyhedra) (Figure 1.24) from their constituent components. Such materials have proven ideal model systems for the examination of the phase behaviour of charged colloids and lyotropic liquid crystals (Chapter 4), and allow the processes involved in their self-assembly to be investigated in detail (Chapter 6).

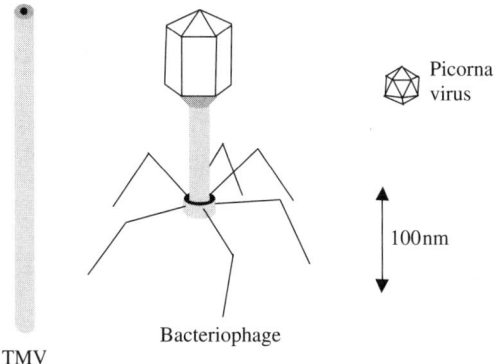

Figure 1.24 Schematic diagram of a range of virus structures (rod-like (TMV), asymmetric (bacteriophage), and icosohedral (picorna))

1.9 BACTERIA (COMPLEX CONSTRUCTS OF BIOMOLECULES)

Bacteria are small structurally simple cellular organisms. Only a minority of bacterial species have developed the ability to cause disease in humans. Bacteria take the form of spheres, rods and spirals. They will be encountered in terms of their mechanisms of molecular motility in Chapter 5 and Chapter 14.

1.10 OTHER MOLECULES

ADP and ATP are the 'currency of energy' in many biochemical processes. Energy is stored by the addition of the extra strongly charged phosphate group in the ATP molecule and can be released when it is metabolised into ADP. There are a vast range of other biomolecules that commonly occur in biology, and the reader should refer to a specialised biochemistry textbook for details.

FURTHER READING

For more exposure to the exquisite detail contained in molecular biophysics the student is directed to:
L. Stryer, *Biochemistry*, Freeman, 1995. Comprehensive coverage of basic biochemical processes.

B. Alberts, A. Johnson, J. Lewis *et al.*, *The Molecular Biology of the Cell*, Garland Science, 2002. A good introductory text to cellular biochemistry that is useful once the contents of Stryer have been fully digested.

D. Goodsell, *The machinery of life*, Springer, 1992. A simple discursive introduction to biochemistry with some attractive illustrations.

TUTORIAL QUESTIONS

1.1) A DNA chain has a molecular weight of 4×10^8 and the average monomer molecular weight of a nucleic acid subunit is 660 Da. For an A type helix there are 11 residues per helical pitch, and the translation per residue is 2.6 Å. For a B type helix there are 10 residues per helical pitch and the translation per residue is 3.4 Å. For a Z-type helix there are 12 residues per helical pitch and the translation per residue is 3.7 Å. What is the length in cm of a duplex DNA chain if it is in the A, B and Z helical forms? What is the average size of the nucleus of a mammalian cell? How does the cell manage to accommodate the DNA in its nucleus?

1.2) Suppose that you isolate a lipid micelle that contains a single protein that normally exists as a transmembrane molecule. How would you expect the lipid and protein to be arranged on the surface of the micelle?

1.3) Calculate the pH of a 0.2 M solution of the amino acid arginine if its pK_a value is 12.5.

1.4) Metals occur in a range of biological processes and form a key component of the structures of a number of biological molecules. Make a list of the biological molecules in which metal atoms occur.

2
Mesoscopic Forces

The reader should be familiar with some simple manifestations of the fundamental forces that drive the interactions between matter, such as electrostatics, gravity and magnetism. However, nature has used a subtle mixture of these forces in combination with geometric and dynamic effects to determine the interactions of biological molecules. These mesoscopic forces are not fundamental, but separation into the different contributions of the elementary components would be very time consuming and require extensive molecular dynamic simulations. Therefore, in this chapter a whole series of simple models for mesoscopic forces is studied and some generic methods to measure the forces experimentally reviewed. There is a rich variety of mesoscopic forces that have been identified. These include *Van der Waals, hydrogen bonding, screened electrostatics, steric forces, fluctuation forces, depletion forces* and *hydrodynamic interactions*.

2.1 COHESIVE FORCES

The predominant force of cohesion between matter is the *Van der Waals interaction*. Objects made of the same material always attract each other due to induced dipoles. The strength of Van der Waals bonds is relatively weak, with energies of the order of $\sim 1\,\text{kJmol}^{-1}$, but the forces act between all types of atom and molecule (even neutral ones).

A fundamental definition of the Van der Waals interaction is an attractive force of quantum mechanical origin that operates between any two molecules, and arises from the interaction between oscillating dipoles. Without overburdening the description with the detailed quantum mechanics, the potential ($V_{12}(r)$) which gives rise to the dispersive Van der Waals force between molecules 1 and 2 can be defined as:

$$V_{12}(r) = -\frac{1}{24(\pi\varepsilon_0)^2}\frac{1}{r^6}\sum_{n,k}\frac{|\langle n|\vec{m}|0\rangle_1|^2|\langle k|\vec{m}|0\rangle_2|^2}{(E_1^n - E_1^0) + (E_2^k - E_2^0)} \equiv -\frac{A_{12}}{r^6} \qquad (2.1)$$

where $\langle n|\vec{m}|0\rangle_1$ is the transition dipole moment from the quantised state n to 0 for molecule 1, r is the distance between the two molecules, and A_{12} is a constant. E_1^n and E_1^k are the quantum energies of state n and k for molecules 1 and 2 respectively. There is thus a characteristic $1/r^6$ decay of the potential between point-like molecules, with a single characteristic constant of proportionality (A_{12}, the *Hamaker constant*) that is dependent on the variety of molecule considered. Van der Waals forces are sometimes called a dispersion interaction, because the same quantities determine both the optical properties of the molecules (dispersion of light) and the forces between them. It is therefore possible to observe the effects of Van der Waals forces optically with micron sized colloidal particles in solution. If a material contains permanent dipoles, they can induce temporary dipoles in another material giving rise to further Van der Waals type interactions (Keesom or Debye forces).

The analysis of Van der Waals forces tends to be more complicated in practice than many of the fundamental interactions that may have been encountered previously in foundation physics courses. Van der Waals forces are *long range* and can be effective from large distances (> 10 nm) down to interatomic spacings (< 0.1 nm). The forces may be *repulsive* or *attractive*, and crucially in general, they do not follow a simple power law, as is illustrated in Figure 2.1 for four separate possible geometries. Van der Waals forces tend to both bring molecules together and mutually align or orientate them. Unlike gravitational and Coulomb forces, Van der Waals forces are *not generally additive*. At larger separations (> 10 nm) the effect of the finite speed of propagation (the speed of light, c) of the interaction also becomes important. This is the *retardation effect* and is observed experimentally in a r^{-7} dependence on separation (r) for point objects rather than r^{-6} at close distances, and for semi-infinite sheets it is r^{-3} rather than r^{-2} at close distances. The force laws illustrated in Figure 2.1 can be proved by careful summation of the contributions in equation (2.1) over an extended body.

COHESIVE FORCES

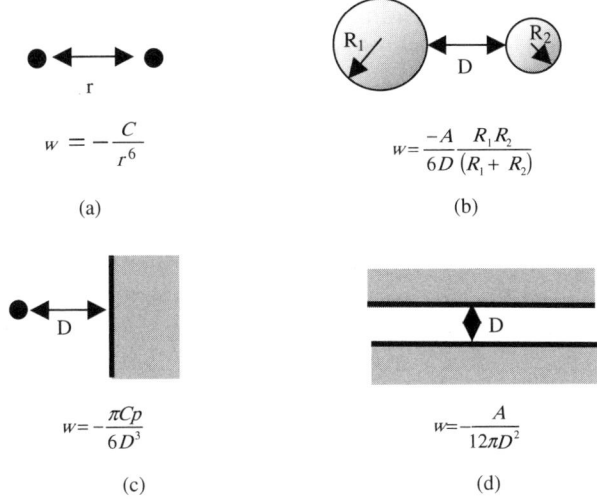

Figure 2.1 The leading term in the energy (w) of the Van der Waals interaction between surfaces depends on the geometry
(Four geometries are shown: (a) point atoms, (b) two spheres, (c) atom/plane, and (d) two plane surfaces.)

Question: Consider the strength of the Van der Waals forces on a fly stuck to the ceiling. There are three thousand hairs per foot and six feet per fly. What is the maximum mass (m) of the fly to maintain contact with the ceiling? Model the interaction as that due to two plane surfaces. The Hamaker constant (A) for the interaction is 10^{-19} J, the radius of curvature (d) of each of the fly's hairs is 200 nm and the separation distance (r) between the hair and the surface is 1 nm. The Van der Waals force is given by:

$$F_{vw} = -\frac{Ad}{6r^2} \qquad (2.2)$$

Answer: In equilibrium the fly's weight (mg) is balanced by the adhesive Van der Waals force (F_{vw}). The forces can thus be balanced:

$$mg = 6 \times 3000 \times 3.33 \times 10^{-9}\,\text{N} \qquad (2.3)$$

Therefore the maximum mass of the fly is 1.02 µg.

In molecular dynamics simulations the energy of interactions ($E(r)$) between biomolecules are often captured using the Lennard–Jones potential:

$$E(r) = \varepsilon\left[\left(\frac{r_0}{r}\right)^{12} - 2\left(\frac{r_0}{r}\right)^{6}\right] \quad (2.4)$$

where r_0 is the equilibrium separation between the particles, r is the distance between the molecules and ε is a characteristic energetic constant. The attractive (negative) term corresponds to the Van der Waals force for a point particle and the repulsive positive term is the hard sphere force (it originates from the Pauli exclusion principle – there is a large energy penalty when filled electronic orbitals overlap during the close approach of atoms).

2.2 HYDROGEN BONDING

Water exhibits an unusually strong interaction between adjacent molecules, which persists into the solid state (Figure 2.2). This unusual interaction is given a special name, *hydrogen bonding*, and is an important effect in a wide range of hydrogenated polar molecules and determines their different molecular geometries (Figure 2.3), e.g. chain structures, crystals, bifurcated associations and intramolecular bonds.

Hydrogen bonds are typically stronger than Van der Waals forces and have energies in the range 10–40 kJmol^{-1}, but are still weaker than ionic or covalent interactions by an order of magnitude. Hydrogen bonding

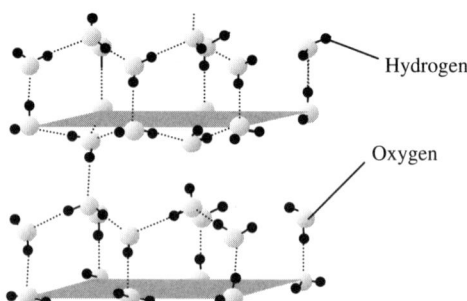

Figure 2.2 The molecular structure of crystalline water (ice I)
(The hydrogen bonds are indicated by dotted lines, and the covalent bonds by continuous lines [*Reprinted with permission from L. Pauling, Nature of Chemical Bond, Copyright (1960), Cornell University Press*])

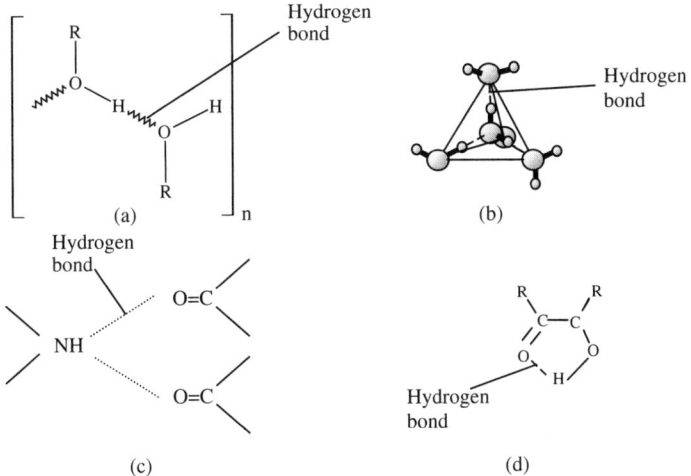

Figure 2.3 Examples of the range of possible geometries of hydrogen bonds encountered in organic molecules
((a) chain structure, (b) three dimensional structure, (c) bifurcated structure and (d) intramolecular bond)

plays a central role in molecular self-assembly processes such as micelle formation, biological membrane structure and the determination of protein conformation. Hydrogen bonds occur between a proton donor group (D), which is the strongly polar group in a molecule such as F–H, O–H, N–H, S–H, and a proton acceptor atom (A), which is a slightly electronegative atom such as fluorine, oxygen, nitrogen and sulfur.

Hydrogen bonding also has important consequences for apolar (nonpolar) biomolecules in aqueous solutions. Water molecules arrange themselves in *clathrate structures* around hydrophobic compounds, e.g. the hydrophobic tails of lipids (Figure 2.4). The clathrates are labile (the water molecules can exchange position with their neighbours), but the

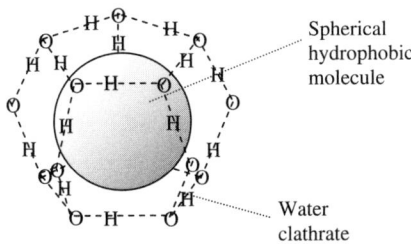

Figure 2.4 Schematic diagram of a clathrate of water molecules around a hydrophobic compound

water molecules are more ordered in the cages. Thus for apolar biomolecules (e.g. hydrocarbons) the free energy of transfer into an aqueous environment is proportional to the surface area of the molecules, since the entropy change is proportional to the area of the clathrate. This hydrophobic interaction is sometimes given the status of a separate mesoscopic force, since the reduction in free energy causes hydrophobic molecules to be driven together in aqueous solutions. The entropy of the associated water molecules plays a critical role in this case. However, to model such interactions care must be taken not to double count aqueous mesoscopic forces under both the 'hydrogen bonding' and 'hydrophobic' banners. Surface force apparatus experiments have provided evidence on the long range nature of the hydrophobic effect. It is still an active area of study, but the energy of repulsion (W) is thought to have the basic form:

$$W = W_0 e^{-r/\lambda} \qquad (2.5)$$

where λ is the decay length and is typically in the order of nanometers, W_0 is a constant (with units of energy) and r is the distance between the surfaces.

Ab initio computational methods to quantify the strength of hydrogen bonds remain at a rudimentary level. One stumbling block to the analysis is the ability of hydrogen bonds to bifurcate (e.g. a single oxygen atom can interact with two hydrogen molecules simultaneously); this leaves a would be modeller with a tricky multi-body problem. Another challenge is the wide spectrum of dynamic phenomena possible in hydrogen bonded solutions and care must be taken to determine the critical time window for the biological phenomena that need to be modelled (Section 1.5). A series of important experimental advances in the dynamics of hydrogen bonds have been recently made using pulsed femtosecond lasers. The lifetime of water molecules around solution state ions has been directly measured to be of the order of 10 picoseconds. It is hoped that such detailed experiments on the structure and dynamics of hydrogen bonds will allow tractable potentials to be refined.

2.3 ELECTROSTATICS

2.3.1 Unscreened Electrostatic Interactions

In principle, the electrostatic interaction between biomolecules can be calculated explicitly in a molecular dynamics simulation. Coulombs law,

ELECTROSTATICS

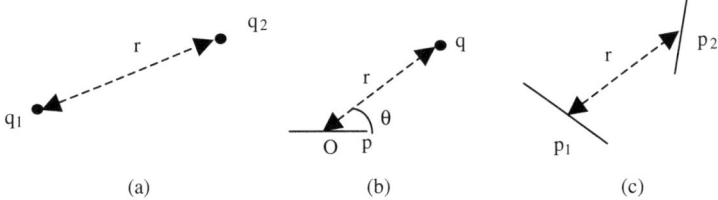

Figure 2.5 Geometry for the interaction between electrostatically charged ions and dipoles
(a) ion–ion (q_1, q_2), (b) ion–dipole (q, p), and (c) dipole–dipole (p_1, p_2))

ion–dipole and dipole–dipole interactions need to be treated, but rigorous quantitative potentials exist that can be calculated if the number of charged entities is sufficiently small (rarely the case with biological molecules, Figure 2.5). The direction of the interaction is paramount in the calculation of dipolar interactions, and also becomes important when Coulombic forces are experienced by extended objects, e.g. the parallel alignment of charged rods.

Coulomb's law for the interaction energy (E_c) between two point charges is given by:

$$E_C = \frac{q_1 q_2}{4\pi\varepsilon\varepsilon_0 r} \qquad (2.6)$$

where ε is the relative dielectric permittivity, ε_0 is the permittivity of free space, q_1 and q_2 are the magnitude of the two charges, and r is the distance between the charges. The next most important electrostatic interactions experienced by charged molecules are those between ions and dipoles. The energy of interaction (E_p) between a dipole (p) and a point charge (q) is given by:

$$E_p = -\frac{p^2 q^2}{(4\pi\varepsilon_0)^2 3kTr^2} \qquad (2.7)$$

where kT is the thermal energy. Similarly there is an interaction energy between two separate electric dipoles (E_{pp}) which is given by:

$$E_{pp} = \frac{p_1 p_2 K}{4\pi\varepsilon_0 r^3} \qquad (2.8)$$

where K is a constant. Higher order electrostatic interactions (quadrupolar etc) are also possible, but provide progressively smaller contributions to the force of interaction in most biological scenarios.

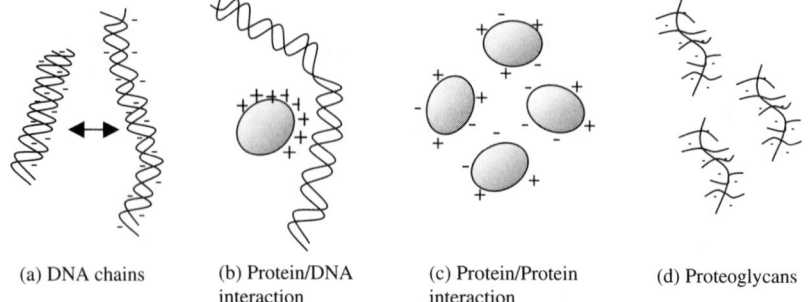

(a) DNA chains (b) Protein/DNA interaction (c) Protein/Protein interaction (d) Proteoglycans

Figure 2.6 Schematic diagram of some molecular systems in which electrostatic interaction dominates the intermolecular forces
((a) nucleic acids, (b) nucleic acids and proteins, (c) the aggregation of proteins and (d) proteoglycans)

Ionic bonds between molecules typically have a strength of the order of $\sim 500\,\text{kJmol}^{-1}$. For a large range of biological molecules, electrostatic forces are vitally important for their correct functioning (Figure 2.6) and provide the dominant long range interaction.

2.3.2 Screened Electrostatic Interactions

An electric 'double layer' forms around charged groups in aqueous solution (Figure 2.7) and the process by which the double layer screens the

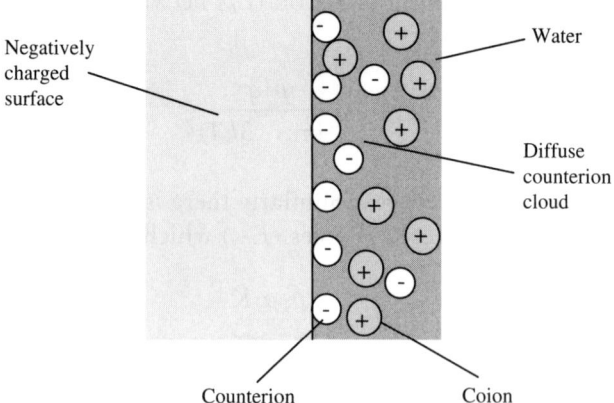

Figure 2.7 The distribution of counterions in an electric double layer around a negatively charged surface

Coulombic interaction is important in determining the resultant electrostatic forces. The concept of screening allows a complex many body problem involving two strongly charged objects immersed in an electrolyte of many billions of simple ions to be reduced to a simple two body problem with a modified potential between the two strongly interacting objects. The charging of a surface in a liquid can happen in two ways: by the dissociation of surface groups or by the adsorption of ions onto the surface. For example, surface carboxylic groups can be charged by dissociation:

$$-COOH \rightarrow -COO^- + H^+$$

which leaves behind a negatively charged surface and liberates approximately $14\,kT$ of energy. The adsorption of an ion from solution onto a previously uncharged or oppositely charged surface (e.g. binding Ca^{2+} onto a negatively charged protein) could charge the surface positively.

The chemical potential (μ, total free energy per molecule) for the electric double layer that surrounds a charged aqueous system is the sum of two terms:

$$\mu = ze\psi + kT \log \rho \qquad (2.9)$$

where ψ is the electric potential, ρ is the number density of counterions, kT is the thermal energy, z is the valence of the charged groups on the molecule, and e is the electronic charge. The first term is due to the electrostatic energy and the second is the contribution of the entropy of the counterions. The form of the chemical potential (equation 2.9) is consistent with the *Boltzmann distribution* for the density of the counterions (ρ) and can be reexpressed as:

$$\rho = \rho_0 e^{-ze\psi/kT} \qquad (2.10)$$

where ρ_0 is related to the chemical potential:

$$\rho_0 = e^{\mu/kT} \qquad (2.11)$$

A fundamental formula from electromagnetic theory is the *Poisson equation* for electrostatics. It relates the potential (ψ) to the free ion concentration (ρ_{freeion}) immersed in a dielectric at a distance x from a charged surface:

$$\varepsilon_r \varepsilon_0 \frac{d^2 \psi}{dx^2} = -\rho_{\text{freeion}} \qquad (2.12)$$

where ε_0 is the permittivity of free space and ε_r is the relative permittivity of the dielectic (e.g. water) in which the ions are embedded. The one dimensional version of the Poisson equation, dependent only on the perpendicular distance from the surface (x), is quoted for simplicity. The Poisson equation for electrostatics can be combined with the Boltzmann distribution for the thermal distribution of ion energies and gives the *Poisson–Boltzmann (PB) equation*:

$$\frac{d^2\psi}{dx^2} = -\frac{ze\rho}{\varepsilon_r\varepsilon_0}e^{-ze\psi/kT} \qquad (2.13)$$

The PB equation can be solved to give the potential (ψ), the electric field ($E = \partial\psi/\partial x$) and the counterion density (ρ) at any point in the gap between two planar surfaces. The density of counterions and coions from a planar surface can therefore be calculated as shown schematically in Figure 2.8.

There are some limitations on the validity of the PB equation at short separations which include: ion correlation effects (electronic orbitals become correlated), finite ion effects (ions are not point-like), image forces (sharp boundaries between dielectrics affect the solutions of the electromagnetism equations), discreteness of surface charges (the surface charge is not smeared out smoothly), and solvation forces (interaction of water molecules with the charges). Some of these important questions will be analysed in more detail in Chapter 9.

The pressure (P) between two charged surfaces in water can often be calculated using the *contact value theorem*. It relates the force between two surfaces to the density of contacts (or ions in this case) at the midpoint ($\rho_s(r)$):

$$P(r) = kT[\rho_s(r) - \rho_s(\infty)] \qquad (2.14)$$

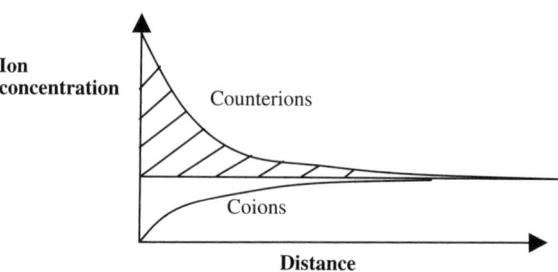

Figure 2.8 The concentration of counterions and coions as a function of distance from a charged planar surface in water

where $\rho_s(\infty)$ is the ion concentration at infinity (e.g. bulk salt concentration), kT is the thermal energy and r is the separation between the surfaces. The force between charged surfaces is discussed in more detail with respect to the physics of cartilage in Chapter 15.

With screened electrostatic interactions the intersurface pressure is given by the increase in the ion concentration at the surfaces as they approach each other. The theorem is valid as long as there is no specific interaction between the counterions and the surfaces. The contact value theorem also functions well in other calculations of mesoscopic forces such as those of solvation interactions, polymer associated steric and depletion interactions, undulation and protrusion forces.

Example: The solution to the PB equation at a distance x from a charged membrane surface is:

$$\psi = \left(\frac{kT}{ze}\right) \log(\cos^2 \kappa x) \quad (2.15)$$

and the characteristic length scale (κ^{-1}) that defines the screening (Section 9.2) is given by:

$$\kappa^2 = \frac{(ze)^2 \rho_0}{2\varepsilon_r \varepsilon_0 kT} \quad (2.16)$$

where ρ_0 is the charge on the surfaces, z is the valence of the counterions, kT is the thermal energy, ε_r is the relative dielectric and ε_0 is the dielectric of free space.

If two surfaces with charge density (σ) of 0.4 Cm^{-2} are placed at a separation (D) of 2 nm and the inverse Debye screening length (K) is 1.34×10^9 m^{-1} what is the repulsive pressure (P) between them?

Answer: From the contact value theorem the pressure (P) can be calculated directly using equation (2.16):

$$P = kT\rho_0 = 2\varepsilon_r \varepsilon_0 \left(\frac{kT}{ze}\right)^2 K^2 = 1.68 \times 10^6 \, Nm^{-2} \quad (2.17)$$

There is thus a large pressure between the surfaces equivalent to 16.6 atmospheres.

2.3.3 The Force Between Charged Spheres in Solution

A surprisingly successful theory for forces between colloidal particles is that due to Derjaguin, Landau, Verwey and Overbeek (DLVO). It has received confirmation from a wide range of experimental techniques such as optical tweezers, light scattering, neutron/X-ray scattering, coagulation studies and surface force apparatus. The theory described the competition between attractive van der Waals and repulsive double layer forces, which is thought to determine the stability of many colloidal systems. The DLVO potential includes both Van der Waals and electrostatic terms for spherical particles, and assumes they are additive (Figure 2.9). Algebraically the potential ($V(r)$ as a function of particle separation (r)) is given by

$$\frac{V(r)}{B} = -\frac{A_{121}}{12\pi r^2} + \frac{64kTc_0^*\Gamma_0^2}{\kappa}e^{-\kappa r} \qquad (2.18)$$

where A_{121} is the Hamaker constant for the Van der Waals force, κ^{-1} is the Debye screening length, c_0^* is the bulk salt concentration, Γ_0 is defined as $\tan(zq\psi/4kT)$, z is the valence of the particles, q is the electronic charge, kT is the thermal energy, B is the surface area of the particle, and ψ is the surface potential. The first component on the right hand side of equation (2.18) is due to the Van der Waals interaction and the second is from the screened electrostatic potential. The agreement between the DLVO model and experiment is excellent in a wide range of systems (Figure 2.10).

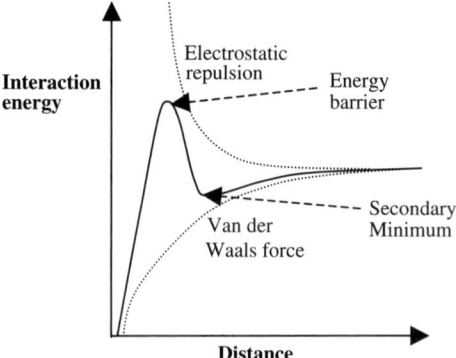

Figure 2.9 Schematic diagram of the DLVO potential between two colloidal particles (The interaction energy is shown as a function of the distance of separation. The secondary minimum is due to an interplay between electrostatic and Van der Waals forces.)

ELECTROSTATICS

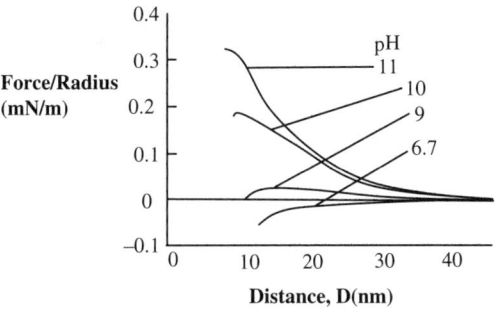

Figure 2.10 The agreement between the predictions of the DLVO theory with SFA experiments for the mesoscopic forces between two planar sapphire surfaces is almost perfect. Lines of best fit are shown
[Reprinted with permission from R.G. Horn, D.R. Clarke and M.T. Clarkson, J. Materials Res., 3, 413–416, Copyright (1998) Materials Research Society]

For a number of colloidal materials it is found that the critical coagulation concentration (ρ_∞) varies as the inverse sixth power of the valency (z) of the electrolyte counterions i.e. $\rho_\infty \propto 1/z^6$. The total DLVO interaction potential between two spherical particles that interact at constant surface potential is:

$$W(r) = \left(\frac{64\pi kTR\rho_\infty \gamma^2}{\kappa^2}\right) e^{-\kappa r} - \frac{AR}{6r} \quad (2.19)$$

where r is the interparticle distance, A is the Hamaker constant, kT is the thermal energy, κ^{-1} is the Debye screening length, γ is the surface potential and R is the radius of the colloid. The critical coagulation concentration occurs when both the potential ($W = 0$) and the force ($dW/dr = 0$) are equal to zero. The condition of zero potential ($W = 0$) upon substitution in equation (2.19) leads to:

$$\frac{\kappa^2}{\rho_\infty} = 384\pi kTr\gamma^2 \frac{e^{-\kappa r}}{A} \quad (2.20)$$

The second condition on dW/dr leads to the result that $\kappa r = 1$, which shows that the potential maximum occurs at $r = \kappa^{-1}$. This expression can be inserted into equation (2.20) to provide the relationship:

$$\frac{\kappa^3}{\rho_\infty} = \frac{768\pi kT\gamma^2}{eA} \quad (2.21)$$

From the definition of the Debye screening length, equation (2.16), it is known that:

$$\kappa^2 \propto \frac{\rho_\infty z^2}{\varepsilon T} \quad (2.22)$$

The surface potential (γ) is known to be constant at high surface potentials ($= 1$) and equation (2.21) can be squared and substituted in the cube of equation (2.22). This provides an expression for the dependence of the critical concentration on the valence:

$$z^6 \rho_\infty \propto \varepsilon^3 T^5 \frac{\gamma^4}{A^2} \quad (2.23)$$

Therefore the critical coagulation concentration follows the expression $\rho_\infty \propto 1/z^6$ which proves the experimentally observed relationship. The origin of this macroscopic coagulation phenomena can be found in the interparticle DLVO forces.

2.4 STERIC AND FLUCTUATION FORCES

The packing constraints on solvents in confined geometries produces oscillatory force/distance curves with a period determined by the solvent size. These forces are most readily measured experimentally between two smooth hard surfaces (Figure 2.11). For example the *packing force* (F_{pack}) due to identical hard sphere solvent molecules confined between

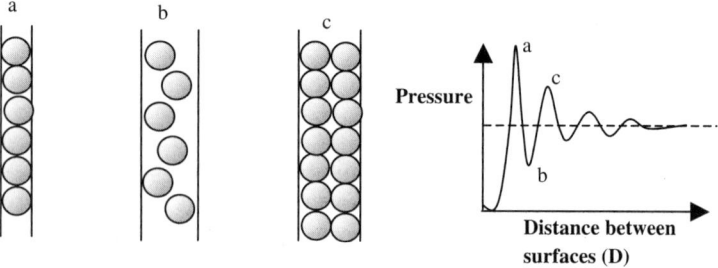

Figure 2.11 The pressure between two planar surfaces as a function of the distance of separation (D). The force is mediated by the excluded volume of the spherical molecules trapped between the surfaces – a depletion potential

STERIC AND FLUCTUATION FORCES

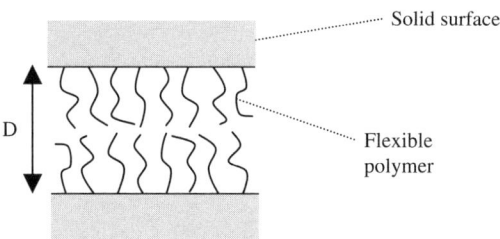

Figure 2.12 The steric forces between two surfaces are produced by the entropic contribution of the grafted flexible polymer chains to the free energy

two hard surfaces is given by:

$$F_{\text{pack}}(r) = A\cos\left(\frac{2\pi r}{\lambda}\right)e^{-r\lambda} \quad (2.24)$$

where r is the separation of the surfaces, A is a constant and λ is the diameter of the confined molecules.

Polymers at surfaces can also give rise to *steric entropic forces* due to the conformational entropy of the chains. Biological realisations of this phenomena of entropic stabilisation include flexible proteins on the surface of interacting membranes, friction reducing polymers in synovial joints, and DNA chains absorbed on to histones. For a polymer to be effective at colloidal stabilisation due to a steric mechanism the polymer must adopt an extended conformation, i.e. good solvent conditions must apply for its configurational statistics. The range of interaction of steric forces is governed by the distance from the surface that the polymer chains extend (Figure 2.12). Typically in biophysical examples, the polymer chains that provide the steric stabilisation for a surface are attached by absorption from solution or are grafted on to the surface by specialist enzymes.

For *polymerically stabilised systems* the repulsive energy per unit area between the surfaces is roughly exponential:

$$W(r) \approx 36kTe^{-r/R_g} \quad (2.25)$$

where R_g is the unperturbed radius of gyration of the polymer chains and r is the separation between the surfaces. The steric force as a function of the distance between two mica surfaces with surface grafted polymer chains in a good solvent is shown in Figure 2.13.

Interacting membranes experience a repulsive steric force due to the fluctuations of membrane structures which are called membrane forces

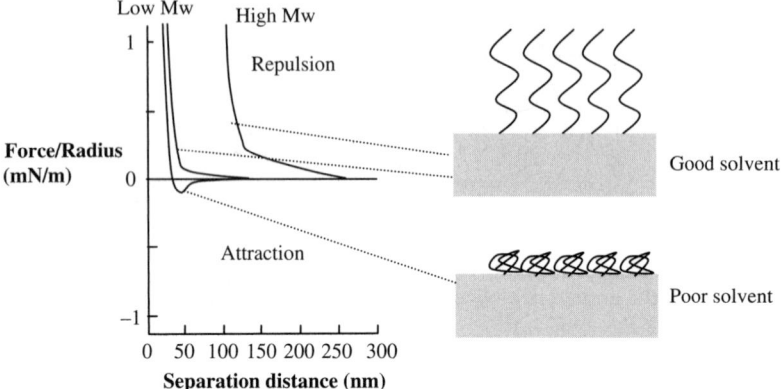

Figure 2.13 The force between two surfaces with surface adsorbed polymers in a good or poor solvent
(The length of the interaction increases with the molecular weight (Mw) of the adsorbed polymer. This experimental data is from SFA measurements [*Reprinted with permission from G.Hadzioannu, G. Patel, S. Granick and M. Tirrell, J. Am. Chem. Soc, 108, 2869–2876, Copyright (1986) American Chemical Society*])

(Figure 2.14). The entropic force per unit area (the pressure $P(r)$) between two surfaces is given by the contact value theorem:

$$P(r) = kT[\rho(r) - \rho(\infty)] \tag{2.26}$$

where $\rho(r)$ is the volume density of molecular contacts, kT is the thermal energy and $\rho(\infty)$ is the number of molecular contacts at infinity

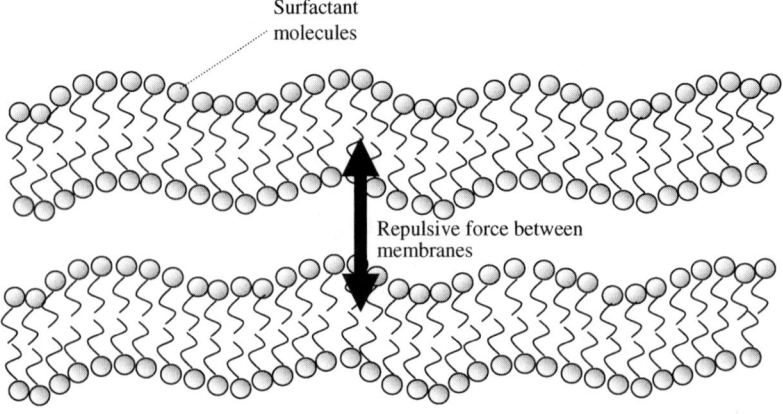

Figure 2.14 Repulsive undulation forces occur between flexible membranes due to the thermally driven collisions

STERIC AND FLUCTUATION FORCES

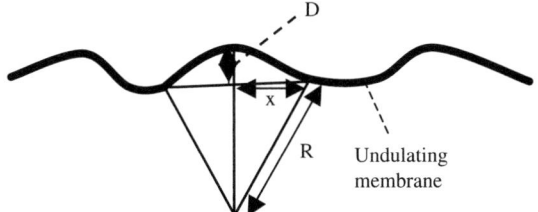

Figure 2.15 Geometry for calculating the magnitude of undulation forces experienced by a membrane

(Figure 2.15). A simple scaling calculation can be made for the magnitude of the steric *undulation forces* between two membranes. The density of contacts ($\rho(r)$) at a certain height (r) above a single membrane is equal to the inverse volume of a single mode:

$$\rho(r) = 1/(\text{volume of mode}) = 1/\pi x^2 r \quad (2.27)$$

where x is equal to the radius of the contacts. The density of the contacts at infinite separation ($\rho(\infty)$) of the membranes must be zero:

$$\rho(\infty) = 0 \quad (2.28)$$

From a simple continuum elasticity model the energy of a bending mode (E_b per unit area) is:

$$E_b = \frac{2k_b}{R^2} \quad (2.29)$$

where k_b is the bending rigidity and R is the curvature of a membrane. Each bending mode occupies an area that depends on its wavelength (πx^2). The bending energy can be equated with the thermal energy (kT) in thermal equilibrium and gives:

$$kT \approx \frac{2\pi x^2 k_b}{R^2} \quad (2.30)$$

The 'Chord theorem' for the geometry of the membrane is:

$$x^2 \approx 2Rr \quad (2.31)$$

This allows equation (2.30) to be reexpressed as:

$$kT \approx \frac{4\pi r k_b}{R} \qquad (2.32)$$

The entropic force per unit area ($P(r)$) between the membranes can now be constructed from the contact value theorem (equation (2.26)) and is:

$$P(r) = \frac{kT}{\pi x^2 r} \approx \frac{kT}{2\pi R r^2} = \frac{(kT)^2}{k_b r^3} \qquad (2.33)$$

This r^3 dependence of the pressure between membranes has been verified experimentally.

Membranes also experience *peristaltic forces* (hydrodynamics causes the membranes to be pulled together at close separations – a similar effect can occur at the macroscopic level between the hulls of boats that is driven by surface water waves) and *protrusion forces* (the detailed nature of the excluded volume of the membranes is important for very close molecular overlap). These intermembrane forces are normally weaker than the undulation forces, but can still be significant.

2.5 DEPLETION FORCES

Depletion forces are another mesoscopic interaction that are formed by a subtle range of more fundamental forces. An illustrative example is when colloidal spheres and polymers are mixed in aqueous solution. The colloids can experience an effective attractive interaction when the polymers are excluded from the volume between the approaching colloidal spheres (Figure 2.16). Such phenomena were originally verified through macroscopic measurements on the phase separation of polymer/colloid mixtures. Recent experiments with dual trap optical tweezers have provided direct evidence for the depletion potential between two colloidal probes in DNA solutions (Figure 2.17).

The depletion force can be understood from an analysis of the thermodynamics. The addition of the polymer lowers the solvent's chemical potential, and creates a depletion force that drives the colloidal surfaces together. This is a trick often used to promote protein crystallisation for structural studies (Section 3.4).

DEPLETION FORCES

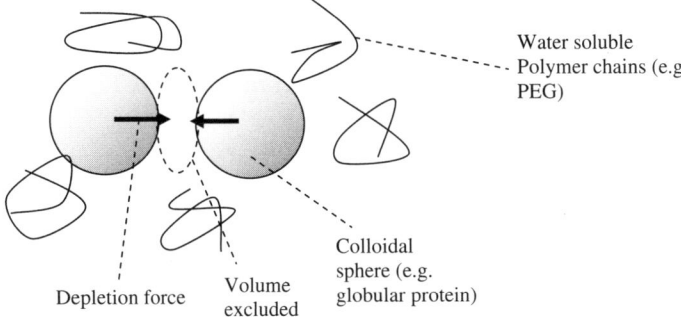

Figure 2.16 Depletion forces between two colloids in a solution of water soluble polymer chains, e.g. polyethylene glycol

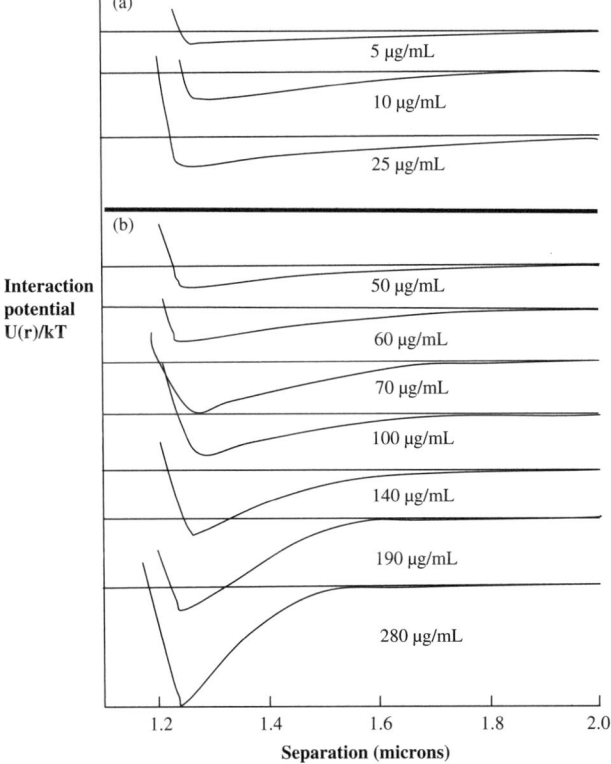

Figure 2.17 Depletion potential between two 1.25 µm silica spheres as a function of DNA concentration in (a) dilute and (b) semi-dilute DNA solutions measured with dual trap optical tweezers
[*Reprinted with permission from R. Verma, J.C. Crocker, T. C. Lubensky and A.G. Yodh, Macromolecules, 33, 177–186, Copyright (2000) American Chemical Society*]

For a dilute concentration of polymers in a colloidal solution the osmotic pressure (Π) is proportional to the number density of polymer chains (N/V), with each chain contributing kT to the osmotic pressure:

$$\Pi = \frac{N}{V} kT \tag{2.34}$$

This pressure is analogous to that of an ideal gas, that gives rise to the Van der Waals equation ($P = NRT/V$, where R is the ideal gas constant). The depletion force (F_{dep}) is approximately the product of the osmotic pressure with the volume from which the chains are depleted (V_{dep}):

$$F_{\text{dep}} = -\Pi V_{\text{dep}} = -\Pi \tfrac{4}{3} \pi R_g^3 \tag{2.35}$$

where R_g is the radius of gyration of the polymer molecules. Thus a high molecular weight (large R_g) and a high polymer concentration (large Π) are required for a strong depletion force. Equation (2.35) was first verified experimentally by the measurement of the force between two interacting bilayer surfaces in a concentrated dextran solution. In a general biological context the naturally occurring intracellular environment is extremely crowded and an intricate hierarchy of excluded volume depletion interactions can occur.

2.6 HYDRODYNAMIC INTERACTIONS

Each of the mesoscopic forces discussed in the previous section have a time scale associated with their action, since they can not occur instantaneously. Therefore the dynamics of the components of each system (e.g. solvents, counterion clouds and tethered polymers) needs to be understood to realistically gauge the strength of the interaction potentials. More advanced treatments of mesoscopic forces often consider how to evaluate the dependence of the forces on time.

2.7 DIRECT EXPERIMENTAL MEASUREMENTS OF INTERMOLECULAR AND SURFACE FORCES

There are a large number of experimental probes for intermolecular forces, which operate using a small range of physical principles

DIRECT EXPERIMENTAL MEASUREMENTS

(Figure 2.18). Some of the most important principles are briefly reviewed here (see also Chapter 13).

The *thermodynamic properties* of gases, liquids and solids (pressure–volume–temperature phase diagrams, boiling points, latent heats of vaporisation and lattice energies) provide important information on short range interparticle forces. Similarly, adsorption isotherms provide information on the interactions of molecules with surfaces.

A range of *direct physical techniques* on gases, liquids and solids (e.g. molecular beam scattering, viscosity, diffusion, compressibility, NMR, x-ray and neutron scattering experiments) can provide information on short range interactions of molecules, with particular emphasis on their

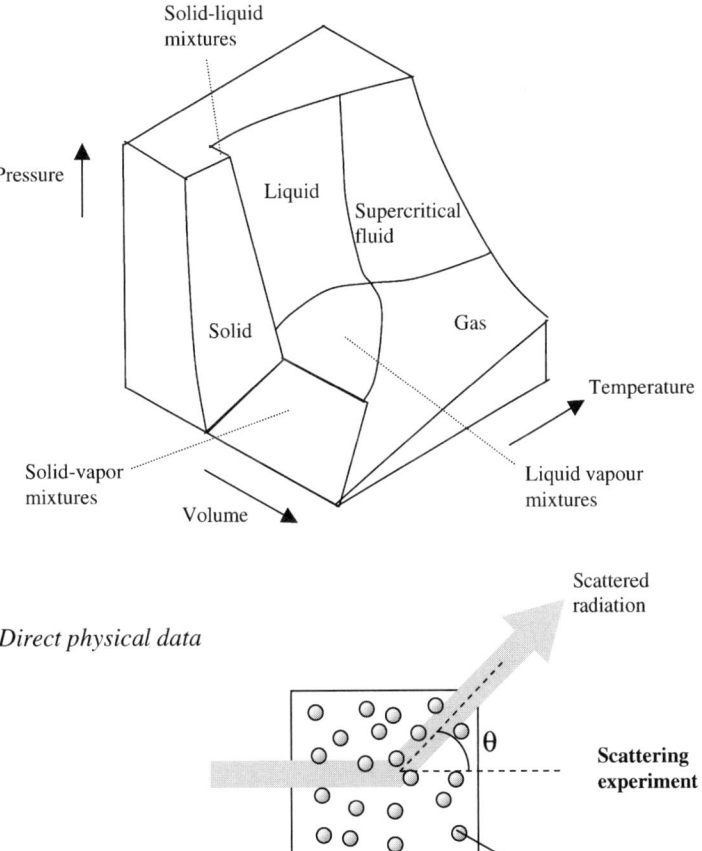

Figure 2.18 Range of techniques for the measurement of intermolecular forces

Thermodynamic data on solutions (Phase stability)

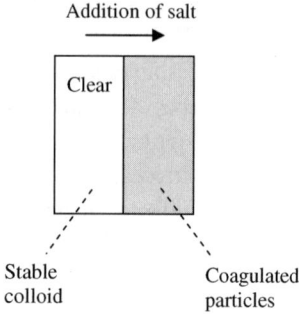

Adhesion experiments

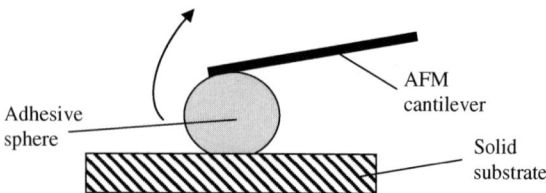

Direct force measurement (Optical tweezers)

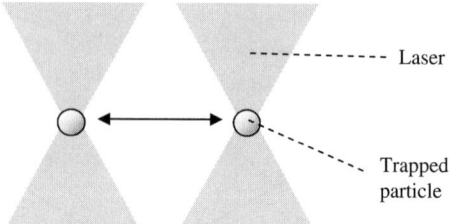

Surface studies
Reflectivity

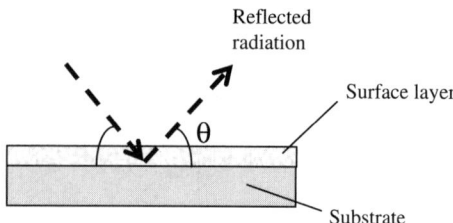

Contact angle

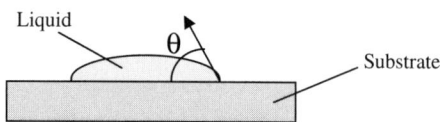

Figure 2.18 (*Continued*)

repulsive forces. A sensitive method for the characterisation of hydrated molecules is to measure the separation of the molecules (using X-ray scattering) in parallel with osmotic pressure measurements. This provides a non-invasive piconewton measurement of intermolecular forces.

Thermodynamic data on solutions (phase diagrams, solubility, partitioning, miscibility and osmotic pressure) provide information on short range solute–solvent and solute–solute interactions. With colloidal dispersions coagulation studies as a function of the salt concentration, pH or temperature yield useful information on interparticle forces.

Adhesion experiments provide information on particle/surface adhesion forces and the adhesion energies of solid surfaces in contact.

Direct force measurement between two macroscopic/microscopic surfaces as a function of surface separation can characterise forces of interaction in great detail, e.g. surface force apparatus (SFA), optical/magnetic tweezers and atomic force microscopes (AFM).

Surface studies such as surface tension and contact angle measurements can elucidate liquid–liquid and solid–liquid adhesion energies. Similarly the reflectivity of radiation (neutrons, X-rays and light) from surfaces can provide invaluable information on surface energies. With *film balances*, the thicknesses of free soap films and liquid films absorbed on surfaces can be measured using optical interferometry as a function of salt concentration or vapour pressure, and this again provides a direct measurement of intersurface potentials.

Hydrodynamic studies of liquids can be made using nuclear magnetic resonance spectroscopy (NMR), and elastic/inelastic scattering of light, X-rays and neutrons. These methods are particularly useful for the measurement of hydrodynamic effects that lead to time dependent mesoscale forces.

FURTHER READING

J. Israelachvili, *Intermolecular and Surface Forces*, Academic Press, 1992. A classic text on mesoscopic forces.

S.F. Evans and H. Wennerstrom, *The colloidal domain*, John Wiley & Sons Ltd, 1994. Useful account of forces in colloidal systems.

D. Tabor, *Solids, liquids and gases*, Cambridge University Press, 1991. Simple refined presentation of the basic phenomena in condensed matter.

J.L. Barrat and J.P. Hansen, *Basic concepts for simple and complex fluids*, Cambridge University Press, 2003. Clear mathematical approach to soft matter physics.

TUTORIAL QUESTIONS

2.1) Two adjacent atoms experience a Lennard–Jones potential that determines their atomic spacing:

$$\xi(r) = \varepsilon\left\{\left(\frac{r_0}{r}\right)^{12} - 2\left(\frac{r_0}{r}\right)^{6}\right\}$$

The energy constant for the interaction (ε) is 0.8×10^{-18} J and the equilibrium distance (r_0) is 0.33 nm. What force would the atoms experience if they were compressed to half of the equilibrium distance?

2.2) What is the Debye screening length for solutions of sodium chloride at concentrations of 0, 0.001, 0.01, 0.1 and 1 M? Water has an intrinsic dissociation constant of around 10^{-14}. What is the Debye screening length for a divalent salt solution (e.g. magnesium chloride ($MgCl_2$)) at the same concentration? What is the Debye screening length in standard physiological conditions (0.1 M salt)?

2.3) A charged polymer can adopt both globular and extended linear conformations. It is assumed that the charge is conserved during the change in conformation and there is a negligible amount of salt in the solution and no charge screening. With which geometry does the potential decrease most rapidly with the distance from the chain? If the charge is smeared out on a planar surface how does the form of the potential compare with the linear and globular morphologies?

2.4) Charged spherical viruses have polymer chains attached to their surfaces, e.g. they are 'PEGelated' in the language of a synthetic chemist. At what distance of separation does the entropic force due to the chains become significant compared with that due to the intervirus electrostatic repulsion?

3
Phase Transitions

Everyone has experience of a range of phase transitions in everyday life. These could be boiling a kettle (a liquid–gas phase transition) or melting a wax candle (a solid–liquid phase transition). In terms of the molecular arrangement, phase transitions almost always involve a change from a more ordered state to a less well ordered state. They can be in stable equilibrium or non-equilibrium if their form evolves with time.

3.1 THE BASICS

A material can adopt a number of phases simultaneously at equilibrium. Thus *phase diagrams* are required to calculate the relative importance of the coexisting substates. The standard everyday states of matter are the crystalline (a regular incompressible lattice), liquid (an irregular incompressible lattice) and gas (an irregular compressible lattice) phases. However, there are also other less conventional phases that occur in nature, such as amorphous solids, rubbers and glasses, and our intuition concerning their mechanical behaviour and microstructure needs careful consideration. Liquid crystals and gels also commonly occur in biology, and present additional possibilities for the thermodynamic state of a material. Other more exotic examples of phase changes studied in this book include wetting (e.g. water drops rolling off the ultra hydrophobic surfaces of lotus leaves) and complexation (e.g. DNA compaction in to chromosomes); the range of possible thermodynamic phases of a biological molecule is vast.

Some of the basic phenomena associated with phase transitions are recapped here before they are applied to a range of biophysical problems. Hopefully many of these concepts will be familiar from introductory thermodynamics studies. The two states between which a *first order* phase transition occurs are distinct, and occur at separate regions of the thermodynamic configuration space. *First order* phase transformations experience a discontinuous change in all the dependent thermodynamic variables except the free energy. In contrast, the states between which a *second order* phase transition occur are contiguous in the thermodynamic configuration space. In these continuous phase transitions (2nd, 3rd, 4th etc), the dependent variables such as the heat capacity, compressibility and surface tension, diverge or vanish as the independent variables approach a critical value, e.g. a critical temperature (T_c). This divergence occurs in a characteristic manner around the critical point for continuous phase transitions; for example if the heat capacity (C_V, the ability of the material to store thermal energy) diverges as a function of temperature near to the critical temperature for a phase transition (T_c), its functional form is $C_v \sim (T - T_c)^{-\alpha}$. The exponent α is found to be universal, dependent on the class and symmetry of the phase transition, but not on the exact details of the molecular components of the system.

The differences between first and second order phase transitions, can be illustrated through the consideration of the heat capacity (C_v) and the enthalpy (H) as a function of temperature (Figure 3.1). The discontinuous nature of the first order phase transition is clearly evident in this example.

The *Gibbs phase rule* provides the number of parameters that can be varied independently in a system that is in phase equilibrium. Consider an ideal gas whose properties depend on the volume (V), the number of particles (n), the pressure (P) and the temperature (T). The equation of state (F), is a unique function dependent on the four variables that describe the system, and is given mathematically by:

$$F(V, n, P, T) = 0 \tag{3.1}$$

And with an ideal gas it is well known that the equation of state takes the form of the Van der Waals law:

$$\frac{PV}{nRT} - 1 = 0 \tag{3.2}$$

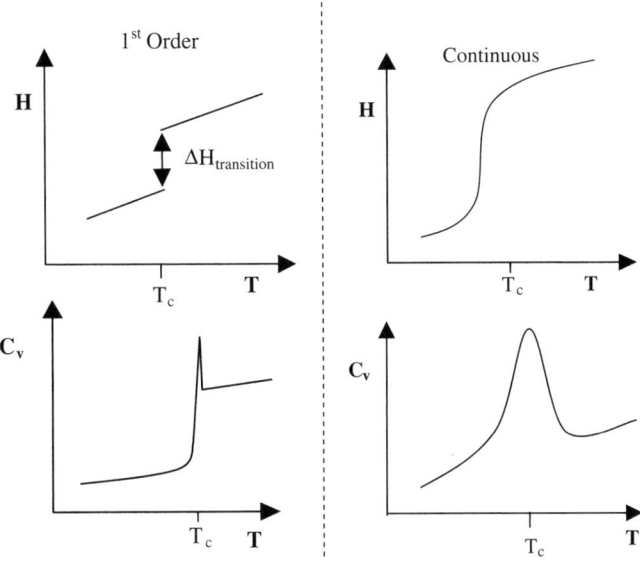

Figure 3.1 Comparison of the enthalpy (H) and heat capacity (C_v) for first order and continuous phase transitions (2nd etc)

where R is the ideal gas constant. Only three independent variables are thus needed to define the system, the fourth variable is dependent. In the general case the Gibbs phase rule states that the number of degrees of freedom in a system (f, the number of independent intensive variables that remain after all the possible constraints have been taken into account), the number of simultaneously existing phases (p), and the number of components (c) are related by:

$$f + p = c + 2 \qquad (3.3)$$

This rule is very useful to predict the phase behaviour for multicomponent colloidal systems, and can be proved from general thermodynamic principles.

To describe a phase transition in more detail it is useful to define an *order parameter*. This order parameter takes a zero value in the disordered phase, and a finite value in the ordered phase. How the order parameter varies with temperature (or other independent variable such as pressure, volume etc.) describes the nature of the transition. Acceptable order parameters for biological phase transitions include the density of a sample, the sample volume and the degree of molecular orientation. Orientational order parameters are crucially important for an understanding of liquid crystalline phases and will be examined in Chapter 4.

The behaviour of *continuous phase transitions* was extensively developed during the second half of the twentieth century. Universal behaviour was found for materials that experience continuous phase transitions near to the critical point, dependent on only the symmetry of the constituent phases. Continuous phase transitions are thus characterised by the values of critical exponents needed to describe the divergence of the order parameters at the phase transition temperature.

For a continuous phase transition a limited number of thermodynamic variables diverge or become zero at the critical point. For example the heat capacity (C_v) near the critical temperature (T_C, Figure 3.1) is given by:

$$C_V = \begin{cases} A(T - T_c)^{-\alpha} & T > T_c \\ A(T_c - T)^{-\alpha} & T < T_c \end{cases} \quad (3.4)$$

where A is a constant, T is the temperature and α is the critical exponent. The same critical exponent (α) applies to either side of the transition point.

For a gas–liquid system that approaches the critical temperature (T_C) the density (ρ) also diverges with a characteristic exponent (β):

$$\rho(\text{liquid}) - \rho(\text{gas}) = B(T_c - T)^{\beta} \quad (3.5)$$

where B is a constant and T is the temperature. This can be pictured experimentally with a transparent kettle full of a subcritical boiling liquid close to a second order gas–liquid transition. It is assumed that all the dissolved gas is removed (which obscures the effect), and the liquid will become milky at the transition temperature. The milkiness is due to large fluctuations in the density (order parameter) that occur at the critical point (the boiling point of the fluid). The density fluctuations can be quantified and are related to the compressibility (κ) of the fluid (how much the volume (V) of a material changes in response to a change in pressure (P)), which diverges at the critical point:

$$\kappa = -\frac{1}{V}\left(\frac{\partial V}{\partial P}\right) = E(T - T_c)^{-\nu} \quad (3.6)$$

where E is a constant and ν is the critical exponent. The expression for the compressibility is valid for temperatures above the critical temperature ($T > T_c$). As explained previously these large density fluctuations can be viewed experimentally in the form of *critical opalescence*; liquids become cloudy due to large fluctuations of density, which scatter incident light as the material approachs a phase transition. The correlation length for the physical size of the density fluctuations also diverges at the critical point.

Another example of a continuous phase transition relates to the thermodynamics of surfaces. The surface tension (γ) between two liquid phases approaches zero at the critical point temperature (T_c), although the width of the interface diverges:

$$\gamma = \gamma_0 \left(1 - \frac{T}{T_c}\right)^{\mu} \qquad (3.7)$$

where μ is another critical exponent and γ_0 is the average surface tension. Thus phase transitions are not limited to bulk three dimensional systems.

Other sources of continuous phase transitions include: systems in which finite size effects play a dominant role (the formation of small lipid micelles leads to a broad critical micelle concentration), systems that contains impurities or inhomogeneities which broadens the phase transition, and systems in which equilibrium times are long compared to observation times, e.g. glass transition of polymers (Section 12.3). In this chapter a range of phase transitions that are important for molecular biophysics are considered in detail: the *helix–coil transition*, *globule–coil transition*, *crystallisation* and *liquid–liquid phase separation*.

3.2 HELIX–COIL TRANSITION

Helix–coil transitions occur in a wide variety of biological situations with an immense range of biological molecules, e.g. carbohydrates, proteins and nucleic acids. Reversible thermodynamic double and single helix–coil transitions both commonly occur in nature (Figure 3.2). The chains can be transferred from the helix to the coil state when a series of parameters such as the temperature, the quality of the solvent or the pH of the solution are changed.

The *Zipper model* is the simplest method for a quantitative description of the thermodynamics of the helix–coil transition (Figure 3.3). Let s be

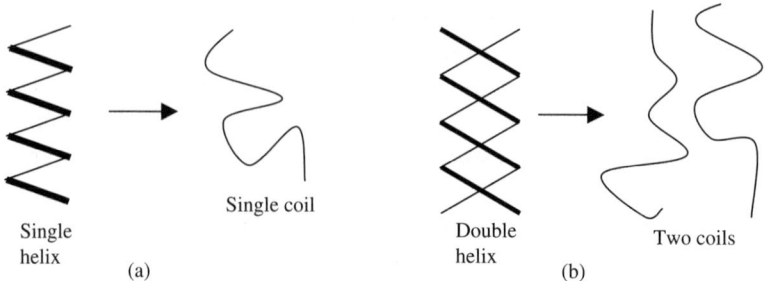

Figure 3.2 Schematic diagram of the helix–coil transition in single (a) and double (b) helices

defined as the equilibrium constant for making a new helical unit at the end of a helical sequence. This can be written as:

$$s = \frac{\ldots cchh\underline{h}cc\ldots}{\ldots cchh\underline{c}cc\ldots} \quad (3.8)$$

where h is the helical state of a polymer chain, c is the coil state of a polymer chain and the underlining corresponds to the chain link considered. The equilibrium constant (s) is called the *propagation* step and provides the statistical weight for the growth of a helical section of a polymer chain provided the nucleation step has occurred. σ is defined as the equilibrium constant for the *nucleation* step ($\sigma \ll 1$); the formation of a helical unit on a coil from the flexible chain state:

$$\sigma = \frac{(\ldots cc\underline{h}cc\ldots)}{(\ldots cc\underline{c}cc\ldots)} \quad (3.9)$$

For a chain of n units the partition function (Z) for the helix–coil transition can be constructed (see the Box):

$$Z = 1 + \sum_{k=1}^{n} \Omega_k \sigma s^k \quad (3.10)$$

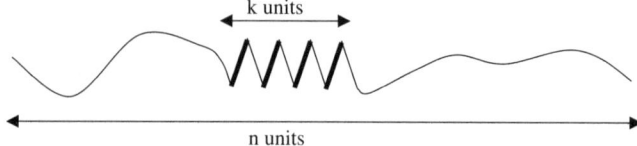

Figure 3.3 The Zipper model for an alpha helix consists of a stretch of k helical units arranged along a chain of n units. $(n - k)$ units are in the coil state

where Ω_k is the number of ways that k helical units can be put on a chain of n units. The number of distinct permutations for Ω_k can be counted and is equal to:

$$\Omega_k = (n - k + 1) \qquad (3.13)$$

> The partition function (Z) is of paramount importance in equilibrium statistical mechanics.
>
> There are a number of useful relationships using this partition function:
>
> 1) The partition function can be directly related to the free energy (F) and the thermal energy $-F/kT = \ln Z$, where kT is the thermal energy.
> 2) To find the probability $p(E_i)$ that an energy state E_i is occupied the partition function can be used:
>
> $$p(E_i) = \frac{e^{-E_i/kT}}{Z} \qquad (3.11)$$
>
> 3) Average quantities can then be constructed from the probabilities, e.g.:
>
> $$\langle E_i \rangle = \sum p(E_i) E_i = \frac{\sum e^{-E_i/kT} E_i}{Z} \qquad (3.12)$$

The fractional helicity (θ) of a sample is a parameter that can be easily measured experimentally by such means as polarimetry, X-ray diffraction or NMR. The helicity (θ) is simply defined as the number of monomers in the helical state (k) divided by the total number of monomers (n):

$$\theta = \frac{k}{n} \qquad (3.14)$$

The helicity predicted by the Zipper model can be calculated from the partition function defined in equation (3.10) using:

$$\theta = \frac{1}{n} \frac{\partial \ln Z}{\partial \ln s} \qquad (3.15)$$

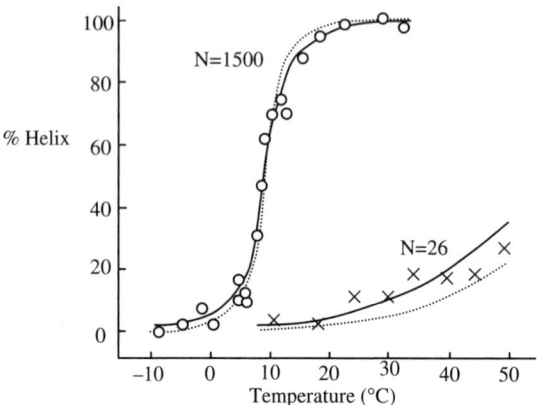

Figure 3.4 Agreement of the Zipper model with data for the helicity of a polypeptide of two separate lengths $N = 1500$ and $N = 26$
(The polypeptide forms an alpha helical structure at high temperatures. [*Reprinted with permission from B.H. Zimm and J.K. Bragg, J. Chem. Phys, 31, 526, Copyright (1959) American Institute of Physics*])

The zipper model is found to be in good agreement with experimental data for the behaviour of short alpha helical chains (Figure 3.4). However, the zipper model breaks down for long chains, since thermodynamic fluctuations can lead to sections of helix interspersed with parts of random coil. The possibility of such fluctuations is not included in the zipper partition function and is thus badly described by the model.

The *Zimm–Bragg* (Ising) model provides a more sophisticated description of the helix–coil transition that includes fluctuations of the helicity along the chain. As before, due to the co-operative nature of the helical conformations, links are assumed to exist in either of two discrete states: the helix and the coil. The junction between the helical and coil sections carries a large positive free energy (Δf) penalty that encourages long lengths of helix. There are two Zimm–Bragg parameters introduced for the model, the statistical weights of the states s and σ:

$$s \equiv \exp(-\Delta f/T) \qquad \sigma \equiv \exp(-2\Delta f_s/T) \qquad (3.16)$$

where Δf is the free energy change upon the addition of an extra helical section, Δf_s is the free energy change for nucleation of a new section of helix, and T is the temperature. For naturally occurring biopolymers the statistical weight for nucleation (σ) is typically very small, of the order of $\sim 10^{-3}$–10^{-4}. The Zimm–Bragg model constitutes a simple method for coarse graining the complex network of

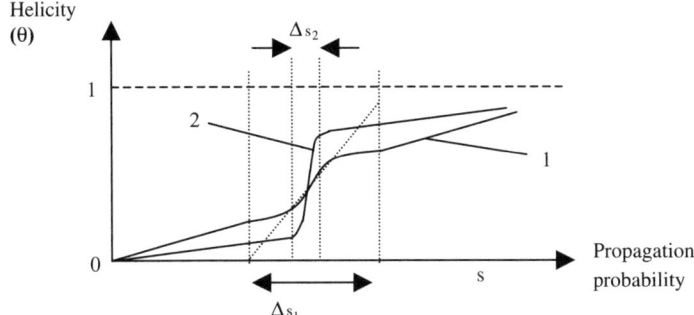

Figure 3.5 Dependence of the helicity (θ) on the strength of the propagation step (s) for the Zimm–Bragg model of the helix–coil transition
(Δs_1 and Δs_2 are the width of the transitions for low and high cooperativities respectively)

hydrogen bonds required for helix formation (including their three dimensional geometry), and describes the thermodynamic behaviour of the phase transition very well.

The partition function for the Zimm–Bragg model becomes more complicated than with the zipper model and only numerical solutions to the behaviour are examined (Figures 3.5 and 3.6). It is found that the helix–coil transition in a single stranded homopolymer occurs over a very narrow temperature interval, which becomes narrower as the co-operativity parameter decreases (σ). The mean lengths of helical and coil sections are finite and independent of the total chain length even as the polymer length (N) tends to infinity.

The helicity (θ) defined in equation (3.14) can be calculated for the Zimm–Bragg model as a function of the degree of co-operativity ($\sigma = 1$, no co-operativity and $\sigma \ll 1$, strong co-operativity), and is shown in Figure 3.5. Curve 1 indicates a small degree of co-operativity and curve 2 shows strong co-operativity. Biological helices normally demonstrate a high degree of co-operativity.

The complete phase diagram predicted by the Zimm–Bragg model for a helical chain with regions of disorder is shown in Figure 3.6. A range of distinct phases are possible; these include random chains, chains with alternating random and helical sections, coexisting random chains and chains with single helices, and single helices with occasional disorder at the ends. The phase diagram is very rich, even for such a simple idealised system. Single molecule experiments are able to investigate each of these scenarios on a molecule by molecule basis (Chapter 13) and are in reasonable agreement with the theory.

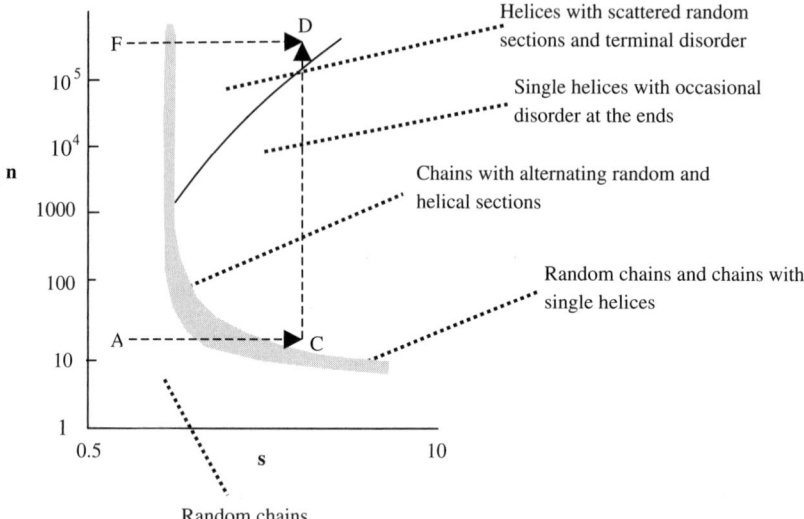

Figure 3.6 Phase diagram for the states of a helical Zimm–Bragg chain with $\sigma = 10^{-4}$ (n is the length of the chain and s is the propagation step. [*Reprinted with permission from B.H. Zimm and J.K. Bragg, J. Chem. Phys, 31, 526, Copyright (1959) American Institute of Physics*])

Although frequently regarded as a melting of helices (the process is accompanied by an endotherm that can be quantified by differential scanning calorimetry, which is a sensitive technique for measuring the amount of heat required to change a samples temperature), the single helix–coil transformation should not be considered a true thermodynamic phase transition. In a one dimensional system the equilibrium coexistence of macroscopic phases is prohibited due to a theorem by Landau. The *Landau theorem* is deduced from the exceedingly small energy associated with phase separation in one-dimensional systems; it is impossible for a true phase transition to occur in one dimension. For a short chain the width of the helix–coil transition is thus anomalously large due to the dominance of end effects.

A helix–coil transition in charged polymers can be initiated by a change in the pH of the medium, e.g. nucleic acids. In this case the transition is accompanied by a sharp change in the average charge of the helical molecule, which provides another experimental method to study the phase behaviour. However, there are a number of subtleties for the analysis of counterion condensation with charged polymers (the interaction of the polymers with their counterion clouds) that will be returned to in Chapter 9, which need to be considered in detail to quantitatively understand the phenomena involved.

The analysis of phase transitions with double helices is slightly different to the case of single helices. With the *double helix–coil transition*, internal coil sections of the double chain are loops. The strong entropic disadvantage of long loops results in increased co-operativity of the phase transition when compared to that for single helices. The loop factor leads to an abrupt sharpening of the helix–coil transition, and calculation shows that it can be considered a true phase transition.

Other rearrangements of secondary structure can be described using modified Zimm–Bragg type models. The formation of beta sheets can be satisfactory described with such a model, but again charge effects need to be carefully considered.

In heteropolymers the helicity constants that correspond to the monomers of each separate chemistry are different, and the character of the helix–coil transition therefore depends on the primary structure. Such a model is thus more realistic for naturally occurring nucleic acids (there are four varieties of monomer for the DNA heteropolymer) and proteins (twenty amino acids could be involved). In a real heteropolymer, the helix–coil transition proceeds by consecutive melting of definite helical sections, whose primary structures possess a higher than average concentration of low-melting temperature links. More sophisticated Zimm–Bragg models describe this behaviour and again there is good agreement with experiment. The statistical description of the helix–coil and beta sheet–coil transition are thus a success story of molecular biophysics.

3.3 GLOBULE–COIL TRANSITION

There is a wide range of experimental evidence for the globule–coil transition for polymeric chain molecules. For example, the temperature can be reduced for a variety of extended polymers in good solvents. This changes the quality of the solvent for the chains and causes them to shrink into dense spherical globules (Figure 3.7). Some of the most

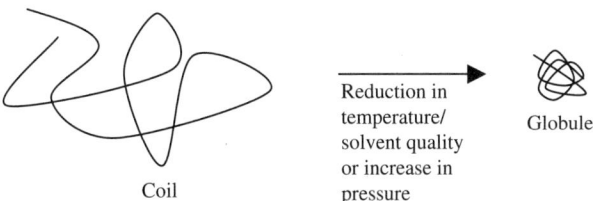

Figure 3.7 Schematic diagram of the coil to globule transition for a single chain induced by a reduction in temperature/solvent quality or an increase in pressure

illustrative data on the globule–coil transition have been measured using fluorescence microscopy with DNA molecules. Under an optical microscope fluorescently tagged DNA can be clearly seen to expand or contract in an abrupt transition when an attractive interaction is introduced between the monomers, e.g. through the introduction of positive multivalent counterions. Furthermore, neutron, X-ray and light scattering, dynamic light scattering, differential scanning calorimetry, electron microscopy and atomic force microscopy all clearly point to the existence of a globularisation phase transition with macromolecules. The globule–coil transition is closely associated with the phenomena involved during both the folding of proteins and the compaction of nuclear DNA, which make it of large biological significance.

A swelling coefficient (α) is often used as the order parameter for the globule–coil transition and is defined as the ratio of the radius of gyration (R) of the chain to the unperturbed radius of gyration (R_0) of the coil, $\alpha = R/R_0$. The radius of gyration will be discussed in detail in Chapter 8. When the molecule shrinks α is less than one ($\alpha < 1$, a bad solvent) and when it swells α is greater than one ($\alpha > 1$, a good solvent). The globule–coil transition is found to be a first order phase transition; there is a sharp change in α as a function of temperature, solvent quality or pressure.

The entropic contribution to the free energy (F_{ent}) when a polymer coil is stretched by a factor α is given by:

$$F_{ent} = -TS(\alpha) \qquad (3.17)$$

where T is the temperature, and $S(\alpha)$ is the entropy associated with the swelling of the chain. Remember that the total free energy (F) for a system is related to the internal energy (U), temperature (T) and entropy (S) by the expression $F = U - TS$, a result from elementary thermodynamics. The energy of self-interactions as a function of the degree of expansion can be defined as $U(\alpha)$. The standard form of the free energy of a globule–coil transition is the sum of the entropy (F_{ent}) and the self-interaction terms (U):

$$F(\alpha) = F_{ent}(\alpha) + U(\alpha) \qquad (3.18)$$

The internal energy (U) can be expanded in powers of the monomer density n (similar to the approximation for a Van der Waals gas leading to $PV = nRT$):

$$U = VkT[n^2 B + n^3 C + ..] \qquad (3.19)$$

GLOBULE–COIL TRANSITION

where V is the volume of the coil and kT is the thermal energy. B and C are the second (two body collision) and third (three body collision) virial coefficients respectively, which describe the strength of the intersegment attraction. Negative B implies an attractive intrachain potential, positive B implies a repulsive intrachain potential, and a negligible B ($B \sim 0$ in a theta solvent) causes C to dominate the behaviour. The monomer density (n) is of the order of the degree of polymerisation of a chain (N) divided by the chain size (R), N/R^3, so equation (3.19) can be written:

$$U \sim R^3 kT \left[B\left(\frac{N}{R^3}\right)^2 + C\left(\frac{N}{R^3}\right)^3 \right] \tag{3.20}$$

By definition $\alpha = R/R_0$, and the unperturbed radius is $R_0 = lN^{\frac{1}{2}}$ (Section 8.1), where l is the size of a monomer. The internal energy can therefore be reexpressed as:

$$U(\alpha) = kT \left[\frac{BN^{\frac{1}{2}}}{\alpha^3 l^3} + \frac{C}{\alpha^6 l^6} \right] \tag{3.21}$$

Where N is the number of monomers in the chains, B and C are the virial constants, and kT is the thermal energy.

The entropic contribution to the free energy can also be calculated using a simple statistical model and is given by:

$$F_{\text{ent}}(\alpha) \sim kT(\alpha^2 + \alpha^{-2}) \tag{3.22}$$

Through minimisation of the total free energy (equation 3.18) of the globule–coil transition using equations (3.22) and (3.20) for the entropic and internal free energies respectively, it is possible to construct a phase diagram for the polymer chain (Figure 3.8). The coil condenses onto itself when the second virial coefficient is sufficiently negative. There is a jump in the molecular size (parameterised by α) as this process of chain condensation proceeds and the polymer chain experiences a first order phase transition.

Quasielastic light scattering studies show clear non-invasive evidence for the globule–coil transition in biopolymers as the temperature is reduced. However, such experiments with large numbers of polymer chains are very sensitive to the total polymer concentration. Low

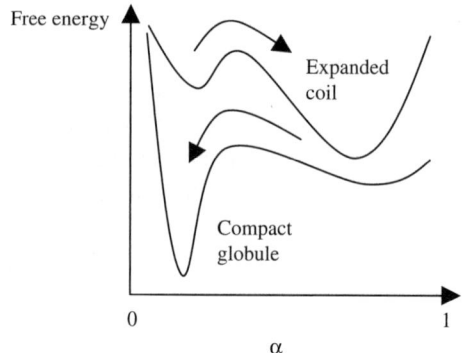

Figure 3.8 Schematic diagram of the free energy of a flexible polymeric globule as a function of the degree of chain extension (α)
(Minima for the expanded coil and compact globule states are shown. At low temperatures the chain adopts a globular conformation whereas at high temperature it expands up into a flexible extended chain conformation.)

concentrations of biopolymer are necessary to avoid aggregation of neighbouring globules in poor solvent conditions due to the strong chain–chain attraction, since this would confuse the interpretation of the results. The use of highly charged biopolymers can circumvent the experimental problems with aggregation, but additional terms must be added to the free energy (equation (3.18)) to describe both the Coulombic repulsion between monomers and the entropy of the counterions associated with the chains, which significantly complicates the theoretical analysis.

Detailed theoretical studies show that large neutral globules consist of a dense homogeneous nucleus and a relatively thin surface layer, a fringe of less dense material. In equilibrium the size of the globule adjusts itself so that the osmotic pressure of the polymer in the globule's nucleus equals zero. As the θ point (where B, the second virial coefficient, is zero) of a polymer/solvent mixture is approached from poor solvent conditions, a globule gradually swells and its size becomes closer to that of a coil. The transition becomes continuous, a second-order phase transition, as the second virial coefficient approaches zero. The width of the globule–coil transition in a chain with N monomers is proportional to $N^{-1/2}$ and becomes infinitely sharp as N tends to infinity.

The character of a globule–coil transition also sensitively depends on the stiffness of the polymeric chains considered. For stiff chains (e.g. DNA, helical proteins and highly charged biomacromolecules), the transition is very sharp and closely resembles a first order phase transition.

GLOBULE–COIL TRANSITION

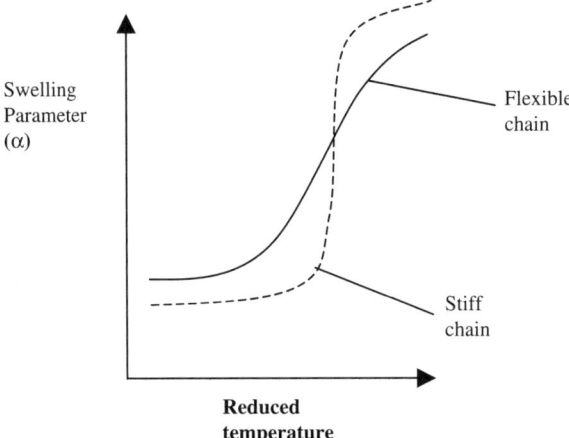

Figure 3.9 Dependence of the swelling parameter (α) on the reduced temperature for flexible and rigid polymers

For flexible chains it is much smoother and is a second order phase transition (Figure 3.9).

The globule-coil condensation of a macromolecule of moderate length and significant stiffness can result in the creation of distinctive small globule morphologies. When the size of a small globule is comparable with the persistence length of the chain, the structure of the globule sensitively depends on the flexibility mechanism. Experimentally for a semi-flexible chain (e.g. DNA), a small globule is found to take the shape of a donut (a torus, Figure 3.10). Furthermore these torroidal shapes often have a liquid crystalline internal structure. DNA chains are rigid

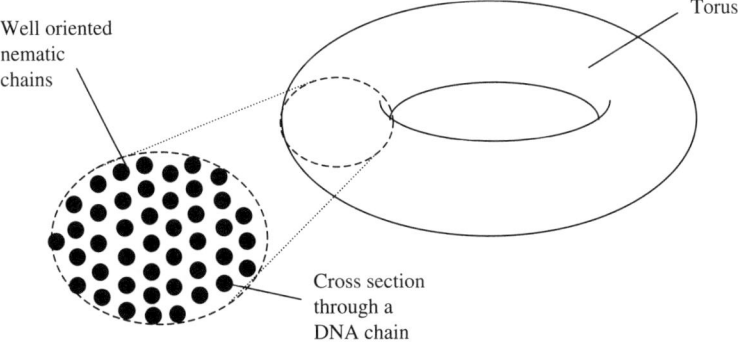

Figure 3.10 Schematic diagram of a DNA donut (torus) that shows the liquid crystalline internal structure

and charged, but compaction can be induced by the addition of a flexible polymer to change the chemical potential of the surrounding water molecules (osmotically stressing the solution) or the addition of multivalent counterions e.g. Mg^{2+} or spermidine to induce an attractive electrostatic intrachain force.

Giant DNA molecules naturally exist in a very complex compacted state, often combined with proteins. The structure and phase transitions of these chromosomes will be covered in more detail in Chapter 16.

The native structure of a *globular protein* is characterised by a precise series of amino acids which possess the property of self-assembly. The compactness of a protein globule is maintained primarily by the hydrophobic effect. Hydrophobic groups are predominantly located inside the globule and hydrophilic ones on the surface. The protein globule is a system of rigid blocks of secondary structure and their surfaces bristle with the side groups of amino acids; van der Waals interactions between side groups of neighbouring blocks fix the details of the tertiary structure. Electrostatic interactions typically play an essential role in the folding of globular proteins only far from the isoelectric point, and charged links are often only located on the surface of the globule. The charged links provide electrostatic stabilisation of the globule against aggregation with other globules. Together with coil and native globular states, the phase diagram of a protein molecule also includes a molten globular state, and it is this state which is qualitatively explained by simple globule–coil theories such as those described by the Flory equations (3.17)–(3.22).

The *self-assembly* of the tertiary structure of a globular protein is found to proceed in two stages: a rapid globule–coil transition driven by hydrophobicity and electrostatics, which is followed by the slow formation of the native structure inside the globule. This complicated molecular origami is exceedingly subtle and reference should be made to the literature of a specific protein for exact details. Often additional proteins (chaperones) are required along the folding pathway to produce the biologically active native structure.

3.4 CRYSTALLISATION

Protein crystallisation in three dimensions is of small relevance to the functioning of living organisms (except for a few unusual examples in seed storage proteins and extracellular proteins), but it is of central importance to structural biology. Only through the production of large high-quality defect-free crystals can the structure of proteins be obtained

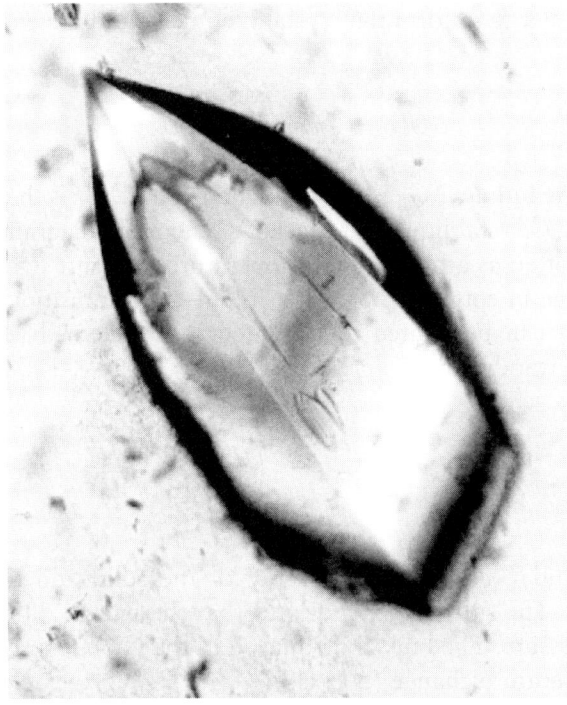

Figure 3.11 Optical photograph of a crystal of a globular protein prepared for crystallographic studies

using diffraction techniques (Figure 3.11). A billion pound question is thus posed by the biotechnology industry; how to form large high quality protein crystals for structure determination to understand structure/function relationships.

The *liquid–solid transition* is much more complicated than the liquid–liquid transition (Section 3.5), since an infinite number of order parameters are required to completely describe the resultant crystalline structure; typically the Fourier components of the density are chosen for this role. The liquid–solid transition is invariably a first order phase transition.

The process of crystallisation is often induced by reducing the temperature of an aqueous suspension of particles. At the melting temperature (T_m), a liquid material never freezes, because it costs energy to form an interface. Without impurities the sample must be undercooled to initiate crystallisation (homogeneous nucleation conditions). The free energy change $(\Delta G(r))$ upon crystallisation can be constructed as the sum of the energy to form the crystalline nuclei and the energy to form the surfaces:

$$\Delta G(r) = \tfrac{4}{3}\pi r^3 \Delta G_b + 4\pi r^2 \gamma_{sl} \qquad (3.23)$$

where r is the radius, ΔG_b is the bulk energy, and γ_{SL} is the surface free energy of the crystal/liquid interface. A schematic diagram of the free energy is plotted as a function of temperature in Figure 3.12.

The change in entropy (ΔS_m) in a liquid–solid transition at constant pressure (P) can be related to the latent heat released (ΔH) using a standard thermodynamic expression:

$$\Delta S_m = \left(\frac{\partial G_s}{\partial T}\right)_p - \left(\frac{\partial G_1}{\partial T}\right)_p = \frac{\Delta H}{T_m} \qquad (3.24)$$

where G_s and G_l are the free energy of the solid and liquid phases respectively. The subscript P indicates that the partial differential is at constant pressure. The bulk contribution to the Gibbs energy associated with a temperature change (ΔT) is:

$$\Delta G_b \approx -\frac{\Delta H_m}{T_m}\Delta T \qquad (3.25)$$

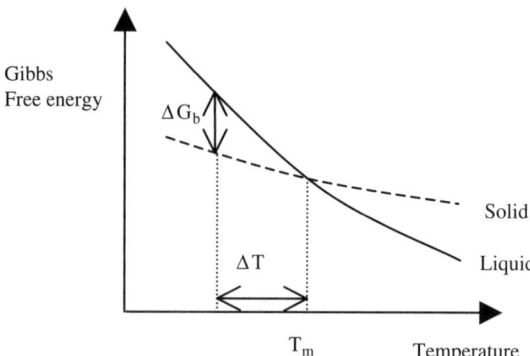

Figure 3.12 Schematic diagram of the Gibbs free energy for the solid and liquid phases of a material as a function of the temperature
(T_m is the melting temperature. ΔG_b is the difference in free energy between the solid and liquid phases.)

CRYSTALLISATION

Thus the total energy from equation (3.23) to form a crystal is:

$$\Delta G(r) = -\tfrac{4}{3}\pi r^3 \frac{\Delta H}{T_m}\Delta T + 4\pi r^2 \gamma_{sl} \quad (3.26)$$

$G(r)$ can be differentiated with respect to the crystal radius, and this analysis shows that the free energy has a maximum at a critical radius (r^*) given by:

$$r^* = \frac{2\gamma_{sl} T_m}{\Delta H_m \Delta T} \quad (3.27)$$

A free energy barrier is thus associated with the formation of stable nuclei due to the surface energy and is given by:

$$\Delta G^* = \frac{16\pi}{3}\gamma_{sl}^3 \left(\frac{T_m}{\Delta H_m}\right)^2 \frac{1}{\Delta T^2} \quad (3.28)$$

The functional form of the free energy with respect to the crystallite radius is shown in Figure 3.13. Thus crystallites must spontaneously nucleate into crystallites with radii above this critical size in order for crystallisation to take place.

Dynamic effects are also important in understanding crystallite growth. Arrhenius dynamics are often used to model the kinetics of crystallisation with an activation energy given by ΔG^*, and the probability that a crystal is nucleated is found to be proportional (Section 5.3) to:

$$\exp(-\Delta G^*/kT) \quad (3.29)$$

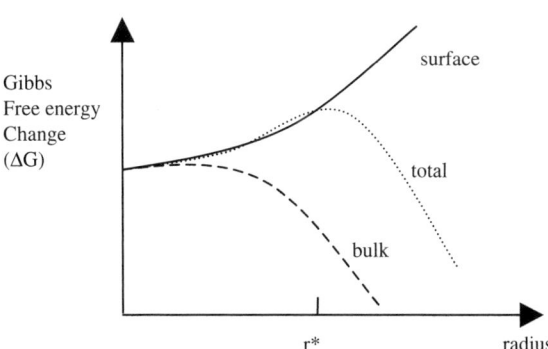

Figure 3.13 The Gibbs free energy for the crystallisation of a solid as a function of the crystallite radius
(The crystals have a critical radius (r^*) dependent of the interplay between the bulk and surface energies for the creation of stable crystallites.)

Equations (3.28) and (3.29) can be combined and it is observed that the surface energy (γ_{SL}) governs the growth kinetics of the crystals. For colloidal crystals the rate of nucleation (Γ) can be written as a product of two factors:

$$\Gamma = e^{-\Delta G^*/kT} \nu \qquad (3.30)$$

where ν is a measure of the rate at which critical nuclei, once formed, transform into larger crystallites. A current challenge is to relate the phase behaviour (e.g. through the activation energy ΔG and the rate constant v) to the inter-protein potential and this is still a hot area of research. Advances have been made that map the behaviour of globular protein crystallisation on to that observed for spherical colloids that contain a number of sticky adhesive patches on their surfaces. The electrostatic and Van der Waals forces must also be considered in combination with these adhesive forces.

Once small protein crystals are formed, the surface free energy continues to play a role in the development of crystalline morphologies as described by equation (3.23). Small crystals are absorbed by large crystals as their surface free energy is minimised in a process called *Ostwald ripening*. It is found to be an important effect in the production of ice cream, as anyone who has eaten melted and subsequently refrozen ice cream will testify.

Naturally occurring solid proteins often adopt fibrous semi-crystalline morphologies, and the kinetics and morphologies produced by the crystallisation of such materials are much more complicated than those found with globular proteins. Furthermore many of these materials adopt intermediate liquid crystalline mesophases due to their extended molecular structures; this behaviour will be examined in Chapter 4.

3.5 LIQUID–LIQUID DEMIXING (PHASE SEPARATION)

Another common process in biological systems is *liquid–liquid phase separation* (Figure 3.14). Examples include the production of food gels, aggregation of ocular proteins in the eye and the partitioning of intracellular ionic species. Liquid–liquid phase separation can also happen in lower dimensional systems, e.g. lipid rafts on the surface of cell membranes can experience a process of two dimensional liquid–liquid phase separation that determine their morphology.

LIQUID–LIQUID DEMIXING

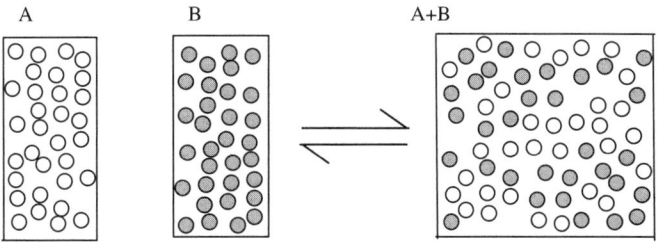

Figure 3.14 Two phase separated systems A and B are involved in a reversible mixing transition

A useful simple reference system, before more complicated biological molecules are considered, is a model for the free energy of mixing of simple molecular liquids. It is an important reference model that needs to be understood, before the more sophisticated phenomena involved in the phase separation of colloids, surfactants and polymers can be approached.

The change in free energy upon mixing (F_{mix}, Figure 3.15) of two simple fluids of type A and B is the difference in free energies before (F_{A+B}) and after ($F_A + F_B$) phase separation:

$$F_{mix} = F_{A+B} - (F_A + F_B) \quad (3.31)$$

The Boltzmann formula for the entropy of mixing (S) of the fluids on a lattice is given by:

$$S = -k_B \sum_i p_i \ln p_i \quad (3.32)$$

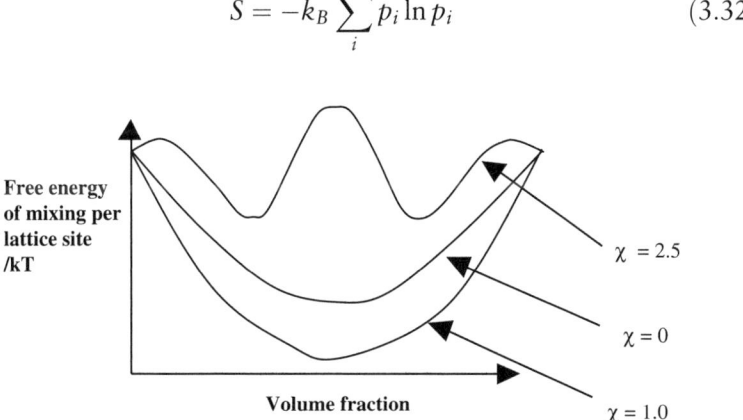

Figure 3.15 The free energy of mixing of liquids A and B as a function of the volume fraction
(χ is the interaction parameter. For $\chi = 0$, $\chi = 1$ the phases are mixed, whereas for $\chi = 2.5$ phase separation occurs.)

where p_i is the probability of occupation of state i of the system by a liquid molecule of type A, and k_B is the Boltzmann constant. The internal degrees of freedom of the two fluids are neglected in this model. The molecules (As and Bs) are assumed to interact in a pairwise additive manner, i.e. only the energies of the nearest neighbours are added. ϕ_A and ϕ_B are the volume fractions of the A and B molecules respectively. The mean field assumption (i.e. when concentration fluctuations are neglected) is that a given site has $z\phi_A$ A neighbours and $z\phi_B$ B neighbours. The interaction energy per site is:

$$\frac{z}{2}\left(\phi_A^2 \varepsilon_{AA} + \phi_B^2 \varepsilon_{BB} + 2\phi_A\phi_B\varepsilon_{AB}\right) \quad (3.33)$$

where ε_{AA}, ε_{BB}, and ε_{AB} are the binary interaction energies between AA, BB and AB molecules respectively. The energy of the unmixed state is:

$$\frac{z}{2}(\phi_A\varepsilon_{AA} + \phi_B\varepsilon_{BB}) \quad (3.34)$$

The difference in the two interaction energies (U_{mix}) associated with the process of mixing is therefore:

$$U_{mix} = \frac{z}{2}\left[(\phi_A^2 - \phi_A)\varepsilon_{AA} + (\phi_B^2 - \phi_B)\varepsilon_{BB} + 2\phi_A\phi_B\varepsilon_{AB}\right] \quad (3.35)$$

If every site is occupied there is a further condition on the sum of the two volume fractions:

$$\phi_A + \phi_B = 1 \quad (3.36)$$

The mathematics are simplified through the definition of an interaction parameter (χ):

$$\chi = \frac{z}{2kT}(2\varepsilon_{AB} - \varepsilon_{AA} - \varepsilon_{BB}) \quad (3.37)$$

The mixing energy, equation (3.35), can therefore be expressed as:

$$U_{mix} = \chi\phi_A\phi_B \quad (3.38)$$

LIQUID–LIQUID DEMIXING

The standard thermodynamic equation for the free energy is $F = U - TS$, and the total free energy of the phase separating mixture (F_{mix}) can therefore be constructed as:

$$\frac{F_{mix}}{kT} = \phi_A \ln \phi_A + \phi_B \ln \phi_B + \chi \phi_A \phi_B \qquad (3.39)$$

If the mixed phase separates into two distinct coexisting phases, the total free energy of the separated mixture is:

$$F_{sep} = \frac{\phi_0 - \phi_2}{\phi_1 - \phi_2} F_{mix}(\phi_1) + \frac{\phi_1 - \phi_0}{\phi_1 - \phi_2} F_{mix}(\phi_2) \qquad (3.40)$$

Coexisting compositions ϕ_1 and ϕ_2 are formed, since the separated free energy is smaller than the homogeneous mixture ($F_{sep} < F_0$) and the system seeks to minimise its free energy in thermal equilibrium (Figure 3.16). Whether the system is stable to fluctuations also helps determine the phase behaviour (Figure 3.17). The stability of the system depends on the second derivative of the free energy ($d^2F/d\phi^2$), which defines the spinoidal line for the process of phase separation. It is interesting that purely repulsive interactions can promote the formation of an ordered phase in this model, and this behaviour has been demonstrated experimentally with model hard sphere colloidal systems.

A good example of phase separation in biocolloids is demonstrated by proteins from the eye, e.g. the simple gamma crystalline protein/water system. The compressibility and correlation length for the phase separation of the gamma crystallins are shown in Figure 3.18. At the point of phase separation the compressibility and correlation length measured

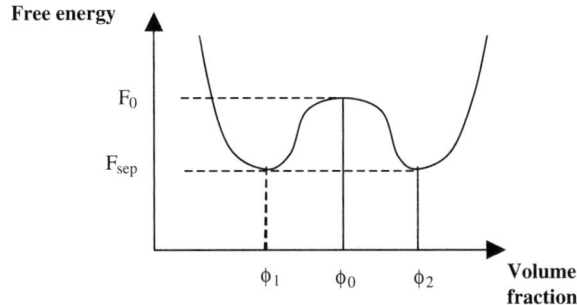

Figure 3.16 Graphical construction for the free energy of separation of liquid A and B (The separated free energy (F_{sep}) is lower that the mixed free energy (F_0) which causes the mixture to separate into two volume fractions ϕ_1 and ϕ_2.)

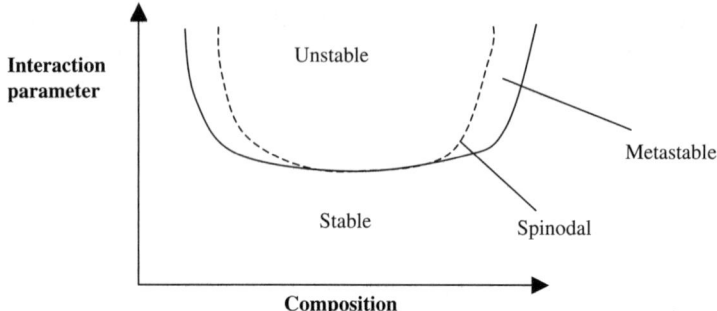

Figure 3.17 The composition (i.e. volume fraction) of a binary mixture of liquids as a function of the interaction parameter (χ). Both the spinoidal line and the metastable region are shown

with light scattering diverges with a power scaling law, with a characteristic exponent for this continuous phase transition as described in Section 3.1. Such phenomena are thought to be associated with the formation of cataracts.

Chemically different neutral polymers mix very poorly in solution and the slight repulsion between the monomeric links is often sufficient to separate the mixture into two virtually pure phases. A small degree of monomeric repulsion is magnified by a high degree of chain polymerisation and can lead to macroscopically observable effects. The separation of

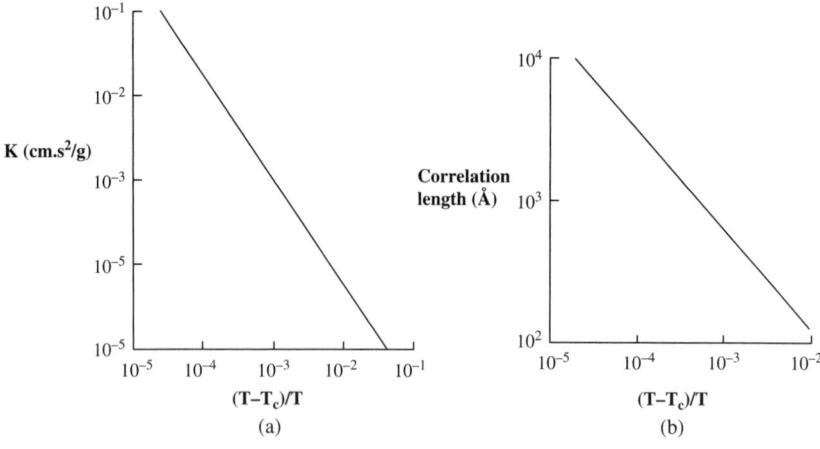

Figure 3.18 Both the compressibility (a) and correlation length (b) measured with light scattering diverge for ocular proteins as the critical temperature for phase separation (T_c) is approached. The line is a best fit from linear regression
[Reprinted with permission from P. Schurtenberger, R.A. Chamberlin, G.M. Thurston, et al., Physical Review Letters, 63, 19, 2064–2067, Copyright (1989) American Physical Society]

the two polymers is a process of phase transition, and it can be realised by either spinoidal decomposition or nucleation and growth (as with the model of the liquid–liquid transition discussed previously). Mixtures of polymers occur in countless biological systems, so it is an important effect to study for its impact in vivo, e.g. aggrecan/collagen mixtures in cartilage. Furthermore, the morphologies formed upon phase separation are sensitively dependent on the dynamics of the constituent molecules. Large degrees of dynamic asymmetry (e.g. long slow-moving polymers mixed with fast moving colloidal particles) lead to a range of novel time dependent morphologies. The phase separated morphology that can occur in biopolymer food gels during their preparation is shown in Figure 3.19.

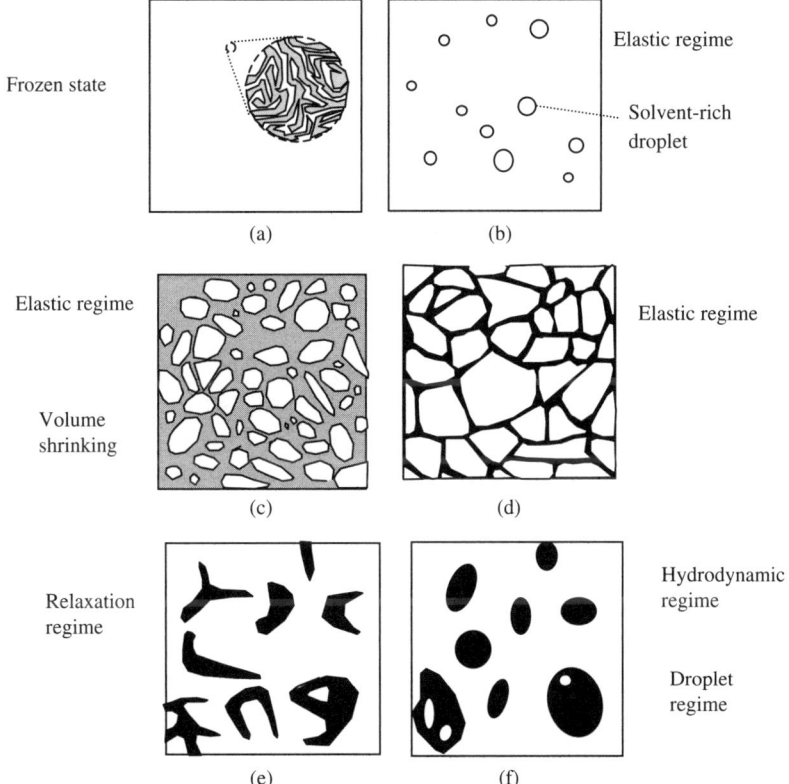

Figure 3.19 The viscoelastic phase separation of food biopolymers has a complex series of evolutionary steps
(A frozen state (a) is followed by an elastic regime (b, c and d), followed by a relaxation regime (e), followed by a hydrodynamic regime (f). The unusual morphologies are driven by the dynamic asymmetry of the phase separating components [*Reprinted with permission from H. Tanaka, J. Phys: Condensed Matter, 12, R207–264, Copyright (2000) IOP Publishing*])

The compatibility of two polymers improves substantially if one of the components becomes weakly charged. This is due to the contribution of the entropy of the counterions to the free energy. Thus polyelectrolyte mixtures tend to resist liquid–liquid phase separation to some degree due to this mechanism.

During the early stages of tissue differentiation and morphogenesis cells undergo a sorting process that resembles liquid–liquid phase separation. Here the Flory interaction parameter (χ) is equivalent to the energy of adhesion between the cells. Such phenomena are vitally important to life in multicellular organisms, and a coherent picture of this extremely complicated process is only slowly starting to evolve.

FURTHER READING

A.Y. Grosberg and A.R. Khoklov, *Statistical physics of macromolecules*, Academic Press, 1994. A good pedagogic account of polymer physics.

M. Rubinstein and R.H. Colby, *Polymer physics*, Oxford University Press, 2004. Similar in level to Grosberg's book and includes a great range of tutorial exercises.

T. Lubensky and P. Chaikin, *Principles of condensed matter physics*, Cambridge University Press, 1995. Mathematically sophisticated coverage of soft matter physics.

M. Kleman and O. P. Lavrentovich, *Soft matter physics*, Springer, 2003. Similar in level and scope to Chaikin and Lubensky.

R.A.L. Jones, *Soft condensed matter*, Oxford University Press, 2002. Reasonably simple introductory treatment of soft condensed matter physics.

H.B. Callen, *Thermodynamics and an introduction to thermostatics*, John Wiley & Sons Ltd, 1985. Classic text on thermodynamics.

TUTORIAL QUESTIONS

3.1) The DSC endotherm for a helix–coil transition for an α helical polypeptide becomes narrower as the length of the peptide is increased. Can you explain the phenomena in terms of the thermodynamics of the transition?

3.2) Peptides in low water conditions are only partially plasticised and can demonstrate glassy (non-ergodic) behaviour at room temperature. How might this alter the behaviour of the globule–coil transition of a long peptide chain?

3.3) What is the free energy barrier for the nucleation of a lysozyme crystal if the critical crystallite size is 50 nm, the surface free energy is 1.2 mJm^{-2}, the melting temperature is 50 °C, and the temperature is reduced by 1 °C below the melting temperature.

4
Liquid Crystallinity

Rod-like molecules can spontaneously align in solution to form anisotropic fluids of reduced viscosity if the concentration of the molecules is increased beyond a critical value (*lyotropic liquid crystals*) or if the correct temperature range is chosen (*thermotropic liquid crystals*). Mankind has recently developed synthetic examples of such materials in a wide range of roles, such as the displays in television screens, bullet proof jackets and soap powders, but nature has already been using the rich variety of phenomena associated with liquid crystals in a range of biological processes over millions of years.

4.1 THE BASICS

Liquid crystals are an intermediate state of matter (a mesophase) between a liquid and a solid. They are characterised by the orientational ordering of the molecules (solid-like behaviour), which maintain an ability to flow (liquid-like behaviour).

Small angle neutron scattering data from a novel example of a biological liquid-crystal, purified mucin from the stomach of a pig, is shown in Figure 4.1. Above a critical concentration the long axes of the molecules spontaneously align themselves at the $\sim 10\,nm$ length scale. This is observed experimentally in the ellipsoidal scattering patterns determined by small-angle neutron scattering, endotherms from calorimetry experiments, and birefringence under an optical microscope. The long axes of the molecules are aligned perpendicular to the major axis of the ellipsoid

---

Applied Biophysics: A Molecular Approach for Physical Scientists   Tom A. Waigh
© 2007 John Wiley & Sons, Ltd

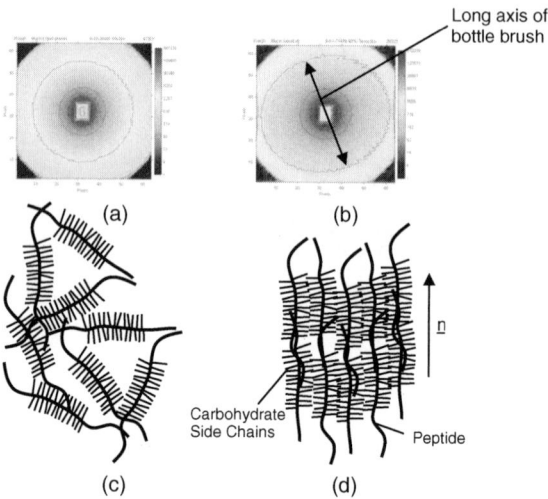

Figure 4.1 Small-angle neutron scattering patterns from (a) isotropic (low concentration) and (b) nematic (high concentration) phases of mucin molecules. (c) and (d) show schematic diagrams of the orientation of comb mucin molecules in the isotropic and nematic phases respectively deduced from (a) and (b)
[*Ref.*: T.A. Waigh, A.P. Papagiannopoulos, A. Voice et al., *Langmuir*, 2002, 18, 7188–7195]

observed in the neutron scattering experiment (Figure 4.1). This alignment of the mucin molecules radically changes their viscoelasticity, and has important implications for their role in the stomach as a barrier against autodigestion.

There are a wide range of biological molecules which form liquid crystalline phases. These include lipids, nucleic acids, proteins, carbohydrates and proteoglycans. There is therefore a correspondingly wide range of liquid crystalline phenomena that are biologically important. For example, cell membranes are maintained in liquid crystalline phases that are used to compartmentalise the cell while still allowing the transfer of important molecules (Figure 4.2(a)), slugs move on nematic trails of proteoglycan molecules whose viscoelasticity is intimately connected with their chosen form of locomotion (Figure 4.2(b)), starch assembles in to smectic structures as a high density energy store in plant storage organs (see Question 4.1), spider silk has a low viscosity liquid crystalline phase as it is extruded from the spider's spinneret to form the super tough materials used to make its web (Figure 15.14), protocollagen forms nematic phases during the construction of the tough viscoelastic collagen networks in skin, and cellulose microfibrils form chiral nematic phases in

THE BASICS

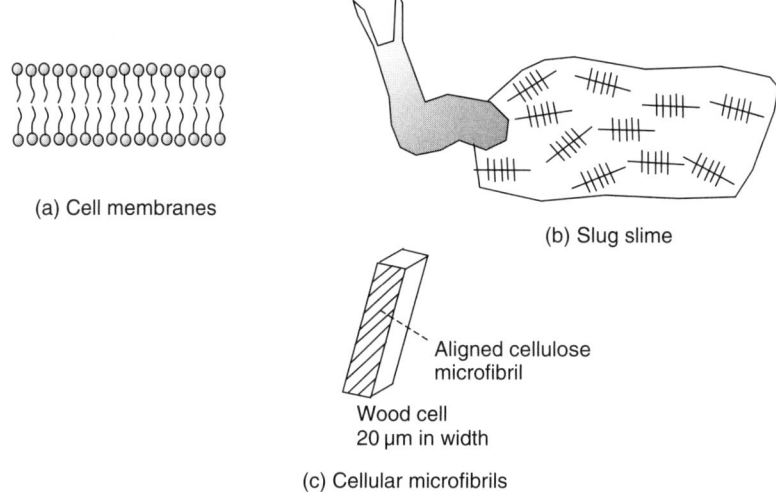

Figure 4.2 Schematic diagrams of naturally occurring examples of liquid crystalline materials

plant cell walls that provide trees with their strength and resilience (Figure 4.2(c)).

The principle *phases* formed by soft condensed matter are due to an interplay between *positional*, *orientational* and *conformational disorder* (Figure 4.3). There are a wide variety of different mesophases that can be further refined within the broad category of liquid crystalline materials, i.e. materials with orientational and conformational ordering (Table 4.1, Figure 4.4). In addition to liquid crystalline phases, the possibility for internal conformational ordering of molecules combined with positional ordering (with a related lattice) leads to the phases of condis crystals and plastic crystals. A further mesophase sub-classification is possible upon the inclusion of *molecular chirality*, i.e. the molecules have a well defined handedness. Many biological molecules are chiral (e.g. DNA is normally left handed) and their mesophase structure reflects the chiral interaction between subunits. The principal chiral liquid crystalline mesophases are the *cholesterics* (chiral nematics) and *tilted smectics* (chiral smectics). Chirality also has a large impact on the defect textures that liquid crystalline molecules adopt, and thus on their macroscopic properties.

To detect *liquid crystalline phase transitions* a wide selection of experimental techniques is typically used. These include differential scanning calorimetry to study the latent heat absorbed, polarising microscopy to view the strength and variety of defect textures, X-ray and

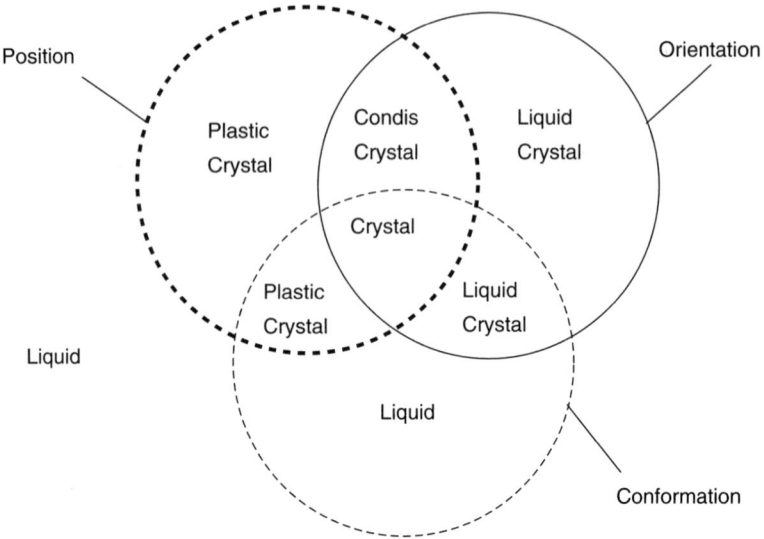

Figure 4.3 Types of phase found in condensed matter from a mixture of conformational, positional and orientational ordering
[*Reprinted with permission from C. Viney in Protein Based Materials, Eds K. McGrath and D. Kaplan, Birkhauser, 281–311, Copyright (1997) Birkhauser Boston Inc.*]

neutron scattering to measure orientational and lattice order parameters, and atomic force microscopy and ellipsometry to measure terraces on the surfaces of the samples.

Liquid crystals are often first detected through their optical textures using crossed polars under an optical microscope, since it is a cheap

Table 4.1 The range of mesophases commonly encountered with biological molecules is primarily determined by a combination of the positional and orientational order parameters

Phase	Positional Order	Orientational Order	Possible Mesophase Order Parameter
Liquid Figure Aa	None	None	None
Nematic Figure Ab	None	Yes	Legendre polynomials
Smectic Figure B	One dimensional	Yes	Fourier components in one dimension
Columnar Figure C	Two dimensional	Yes	Fourier components in two dimensions
Crystalline Figure D	Three dimensional	Yes	Infinite number of Fourier components in three dimensions

THE BASICS 81

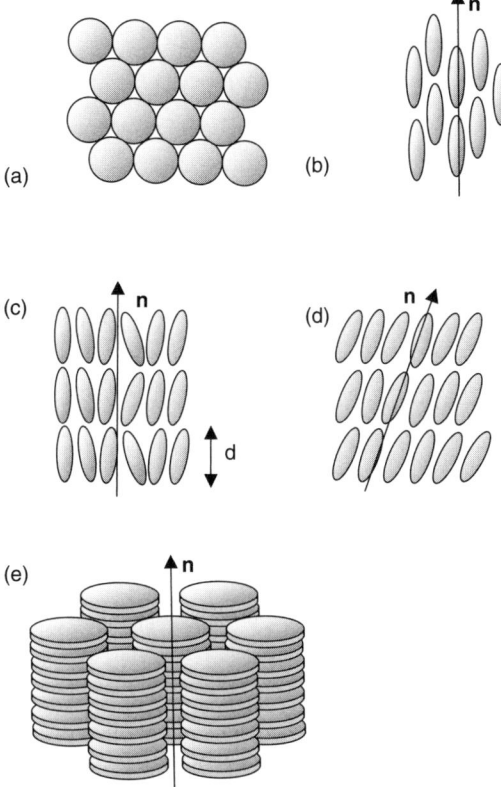

Figure 4.4 (a) Perfect crystalline solid formed from spherical colloidal particles, (b) nematic phase formed from rod-like mesogens, (c) smectic A (perpendicular) phase formed from rod-like mesogens, (d) smectic C (tilted) phase formed from chiral rod-like mesogens and (e) hexatic phases are a rarer liquid crystalline phase in biology formed from disk-like mesogens. Hexatic phases have been observed with nucleosome particles

readily available technique. These *defect textures* can often enable the exact liquid crystalline phase to be categorised. The quantitative evaluation of these defect textures in biological liquid crystals will be considered in Section 4.2.

Differential scanning calorimetry reveals the presence of liquid crystalline phase transitions in a material through the detection of the associated enthalpy changes. The isotropic–nematic phase transition is clearly demonstrated to be a true thermodynamic event with an associated endotherm in such experiments (Figure 4.5). Usually a liquid crystalline material will experience a sequence of thermodynamic phase transitions as a function of temperature e.g. a crystal transforms into a smectic liquid crystal, then into a nematic liquid crystal, and finally it is

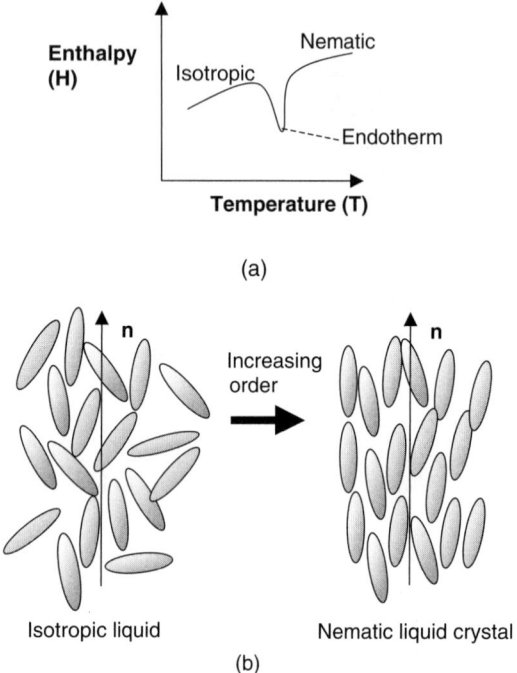

Figure 4.5 (a) Differential scanning calorimetry can measure the enthalpy as a function of temperature for a liquid crystalline material, and shows the endotherm due to an isotropic-nematic phase change. (b) Schematic diagram that indicates the increased orientational ordering upon an isotropic-nematic phase transition for small rigid molecules

converted into an isotropic liquid as a function of increasing temperature.

Following the discussion of phase changes in Chapter 3 it is useful to consider the relevant *order parameters* for a liquid crystalline phase transition. The three most primitive phases of liquid crystals are *nematics* (a single direction of preferred orientation), *cholesterics* (nematics with the orientational direction twisting along a helix) and *smectics* (with long range order in one or two dimensions).

With *nematic liquid crystals* it is conventional to use the second order Legendre polynomial (P_2), which is based on a spherical coordinate system (θ, ϕ, r), to quantify the degree of orientational alignment of the molecules:

$$S = \langle P_2(\cos\theta) \rangle = \left\langle \frac{3}{2}\cos^2\theta - \frac{1}{2} \right\rangle \qquad (4.1)$$

THE BASICS

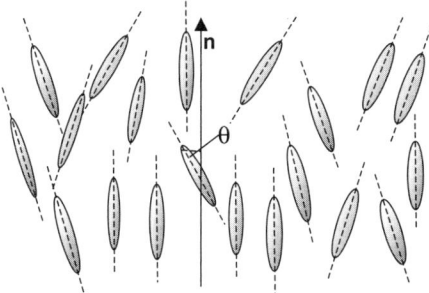

Figure 4.6 The long axes of the mesogens in a nematic phase of biological molecules become ordered along a single axis
(n is the direction of the average value of the director field. θ is the angle the rods make with the director.)

where θ is the angle that the long axes of the molecules make with the nematic director and $\langle\rangle$ implies that the mean value is to be calculated (Figure 4.6). The director is a unit vector that indicates the average direction of alignment of the molecules. The calculation for the nematic order parameter can be pictured by the calculation of a $\frac{3}{2}\cos^2\theta - \frac{1}{2}$ term for each molecule in the solution, and then to average them over all the molecules. This nematic order parameter (S) is equal to one for a perfectly aligned nematic parallel to the director, $-\frac{1}{2}$ for a perfect perpendicular alignment and zero for an isotropic liquid. A typical plot for the variation of S as a function of temperature during the isotropic–nematic phase change is shown in Figure 4.7. The nematic order parameter (S) decreases with increasing temperature and is lost abruptly at T_c, the temperature of the first order phase transition.

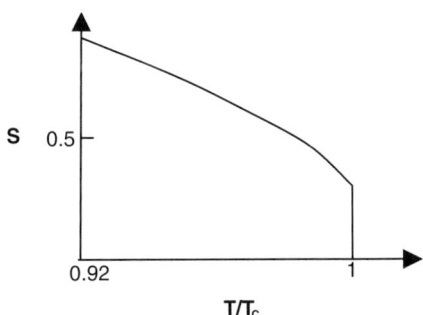

Figure 4.7 Temperature dependence of the nematic order parameter (S)
(Above T_c the material forms an isotropic phase whereas below T_c it is a nematic liquid crystal.)

An additional order parameter in combination with that for nematicity is required to describe the lamellar ordering found with smectic liquid crystals, i.e. the one dimensional lattice structure needs to be described. A *lamellar order parameter* is defined through the expansion of the electron density of the periodic smectic stack as a Fourier series in the direction perpendicular to the stack (z). The first cosine term is then kept as a good first approximation for the stack density. The order parameter (ψ) is defined as the amplitude of this cosine function:

$$\rho(z) = \rho_0 \left(1 + \psi \cos\left(\frac{2\pi z}{d}\right)\right) \quad (4.2)$$

where d is the spacing of the layers, ρ_0 is a constant electron density and $\rho(z)$ is the electron density of the stack as a function of z. The lamellar order parameter (ψ) can be measured using X-ray, atomic force microscopy (AFM), neutron and light scattering techniques (Figure 4.8). It is possible to theoretically predict the behaviour of the smectic and nematic order parameter described by equations (4.1) and (4.2) near to a critical point of a phase transition using a model due to Landau (Section 4.2). A further chiral order parameter is needed for a discussion of the phase behaviour of cholesterics and is introduced with an analysis of the elasticity of these twisted nematic phases.

The *isotropic–nematic phase transition* for small rigid biological molecules in solution is now fairly well understood. Both analytic models and simulations quantitatively predict the onset of a nematic phase and are in good agreement with experimental systems in which intermolecular potentials approximate to a hard sphere interaction. For short rigid rods in solution, Lars Onsager analytically determined the phase diagram

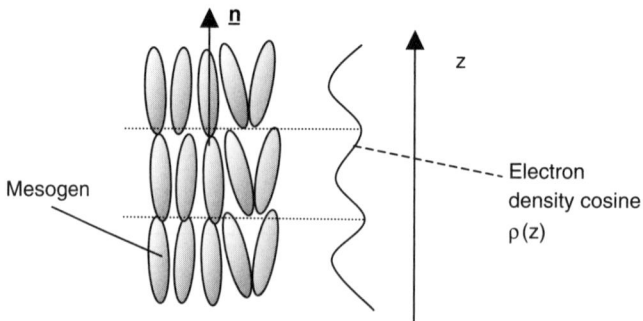

Figure 4.8 The smectic order parameter is typically defined as the amplitude of a sinuisoidal expansion of the electron density of the lamellar stack

THE BASICS

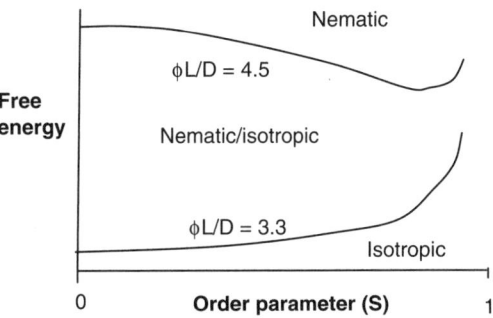

Figure 4.9 The free energy diagram of a solution of hard rods as a function of the nematic order parameter (S) shows the stable regions of nematic, nematic/isotropic, and isotropic phases. The critical parameter that determines the phase behaviour is the volume fraction (ϕ) multiplied by the aspect ratio (L/D)

for nematic liquid crystals and found that it could be simply predicted as a function of the aspect ratio of the molecules (L/D, length (L) and diameter (D)) and the volume fraction (ϕ) (Figure 4.9). Below a critical value of the product $\phi L/D$, the solution adopts an isotropic liquid structure:

$$\text{Isotropic ordering } \frac{\phi L}{D} < 3.34 \qquad (4.3)$$

whereas above an upper bound the rods adopt a perfect nematic ordering:

$$\text{Nematic ordering } \frac{\phi L}{D} > 4.49 \qquad (4.4)$$

There is coexistence of the isotropic and nematic phases between the two critical values (Figure 4.10) described by equations (4.3) and (4.4). This process of liquid crystalline ordering is a first order phase transition.

Liquid crystalline phases can also be adopted by long *semi-flexible polymeric molecules* such as DNA, collagen and carrageenan. The behaviour is more complicated than that displayed with simple rod-like molecules, because the internal degrees of freedom (conformations) of the polymer must be considered. The adoption of liquid-crystalline phases in polymers is therefore related to their persistence lengths

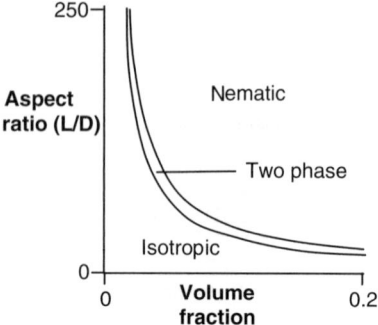

Figure 4.10 Phase diagram for a solution of hard rods as a function of the aspect ratio (L/D) and the volume fraction
(The phase behaviour is determined from the free energy shown in Figure 4.9.)

(Section 8.1). The isotropic/nematic phase diagram is qualitatively similar to that with small molecules (Figure 4.10) when the chains are semi-flexible, with the phase boundaries renormalised by the magnitude of the persistence length. As the orientational ordering grows in a solution of semi-flexible chains so does the mean size of the chains along the ordering axis. The phase diagram of fD virus solutions in shown in Figure 4.11. These virus molecules are an ideal experimental system to examine semi-flexible liquid crystals. fD viruses form an optically observable microscope smectic phase (Figure 4.12) and are perfectly monodisperse, since their protein sequence is genetically determined.

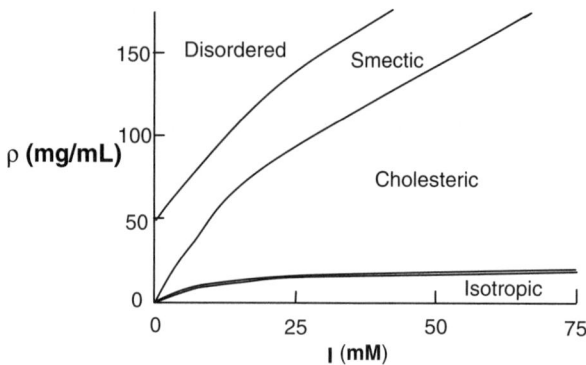

Figure 4.11 Experimentally determined phase diagram for fD virus solutions (chiral semi-flexible molecules) as a function of density (ρ) and ionic strength (I)
[*Reprinted with permission from S. Fraden in 'Observation, Prediction and Simulation of Phase Transitions in Complex Fluids', Eds M. Baus, L.F. Rull and J.P. Ryckaert, Kluwer Acdaemic Press, NATO ASI Series C, Vol. 460, Copyright (1995) Springer Science and Business Media*]

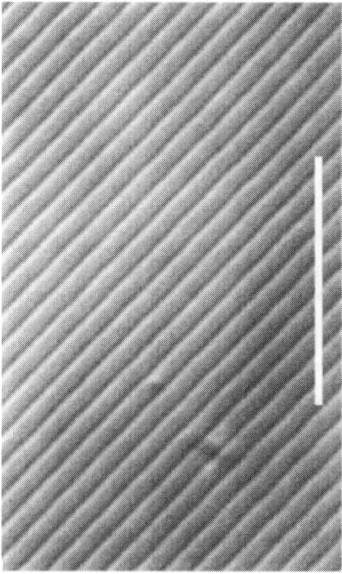

Figure 4.12 Differential interference contrast microscopy image from smectic phases of fD viruses whose phase diagram is shown in Figure 4.11 (The white scale bar is 10 μm and the smectic periodicity is 0.92 μm [*Reprinted with permission from Z. Dogic and S. Fraden, Physical Review Letters, 78, 12, 2417–2420, Copyright (1997) American Physical Society*])

For an elastic Hookean spring in one dimension the energy (E) stored in the system as it is extended beyond its equilibrium length is given by the familiar expression:

$$E = \tfrac{1}{2}Kx^2 \qquad (4.5)$$

where x is the extension and K is the spring constant (Figure 4.13). The factor of a half is included so that upon differentiation the restoring force on the spring is given by Kx (Hooke's Law). In three dimensions (Chapter 11) the elasticity of an arbitrarily chosen material is much more complicated and a compliance tensor (a matrix of 81 numbers) is needed in place of K. Fortunately, liquid crystalline materials have an elasticity which is dependent on only the orientation of the director and the distortion of the director field. Thus simple nematic liquid crystals in three dimensions only require three elastic constants K_1 (*splay*), K_2 (*twist*) and K_3 (*bend*) for a complete description of their elasticity. The elastic energy (or free energy) of a nematic is constructed using symmetry relations. A fair amount of mathematical effort is required for the

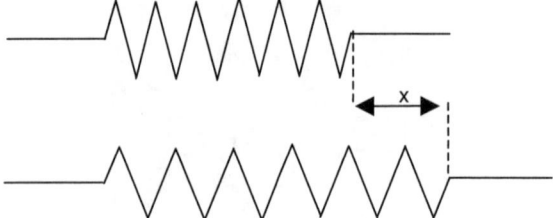

Figure 4.13 Schematic diagram of the extension of a spring in one dimension (The spring is extended by a distance x and it stores elastic energy.)

derivation, but the free energy per unit volume (F_v) of a non-chiral nematic liquid crystal is found to be:

$$F_v = \tfrac{1}{2}K_1[\nabla . \underline{n}]^2 + \tfrac{1}{2}K_2[\underline{n}.(\nabla \times \underline{n})]^2 + \tfrac{1}{2}K_3[\underline{n} \times (\nabla \times \underline{n})]^2 \qquad (4.6)$$

where K_1, K_2, and K_3 are the spring constants with units Joules m^{-3}, and $\underline{n}$ is the unit vector describing the direction of the orientation of the rod-like molecules (the director). This nematic free energy (equation (4.6)) has the same form as that of a single elastic spring (equation (4.5)), except three elastic constants are now required to describe the material. A typical value of K_3 for a short molecular liquid crystal is 10^{-11} Nm^{-2} and it is normally two to three times bigger than K_1 and K_2.

To physically understand the formula for the free energy (F_v) it is useful to simplify the vector calculus of equation (4.6). If the director ($\underline{n}$) is along the z direction all of the derivatives of the z component of the director (n_z) are equal to zero:

$$\frac{\partial n_z}{\partial x} = \frac{\partial n_z}{\partial y} = \frac{\partial n_z}{\partial z} = 0 \qquad (4.7)$$

The individual terms in equation (4.6) can then be calculated using equation (4.7)

$$[\nabla . n]^2 = \left[\left(\frac{\partial n_x}{\partial x}\right)_{y,z} + \left(\frac{\partial n_y}{\partial y}\right)_{x,z}\right]^2 \qquad (4.8)$$

$$[n \times (\nabla \times n)]^2 = \left(\frac{\partial n_y}{\partial z}\right)^2_{x,y} + \left(\frac{\partial n_y}{\partial z}\right)^2_{x,y} \qquad (4.9)$$

$$[n.(\nabla \times n)]^2 = \left[\left(\frac{\partial n_y}{\partial x}\right)_{y,z} - \left(\frac{\partial n_x}{\partial y}\right)_{x,z}\right]^2 \qquad (4.10)$$

THE BASICS 89

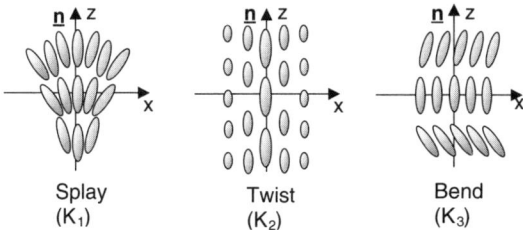

Figure 4.14 Visualisation of the splay, twist, and bend constants for the director field of a nematic liquid crystal aligned along the z-axis

The director field when only the first term on the right hand side of equations (4.8), (4.9) and (4.10) is not zero is shown in Figure 4.14. This figure allows a useful intuitive picture to be developed for the meaning of splay, twist and bend distortions.

The Frank's free energy (equation (4.6)) can be extended to describe a cholesteric liquid crystal. To model a chiral nematic (cholesteric) phase aligned perpendicular to the x-axis the director needs to follow a helical path (Figure 4.15) given by:

$$n_x = 0 \qquad (4.11)$$

$$n_y = -\sin\left(\frac{2\pi x}{P}\right) \qquad (4.12)$$

$$n_z = \cos\left(\frac{2\pi x}{P}\right) \qquad (4.13)$$

The cholesteric director thus twists around the x-axis with a characteristic length scale that describes the axial repeat called the pitch (P). The div and curl of the director are required to construct the free energy in equation (4.6) and they are:

$$\nabla . v = 0 \qquad (4.14)$$

$$(\nabla \times n)_y = \frac{2\pi}{P}\sin\frac{2\pi x}{P} \qquad (4.15)$$

$$(\nabla \times n)_z = -\frac{2\pi}{P}\cos\frac{2\pi x}{P} \qquad (4.16)$$

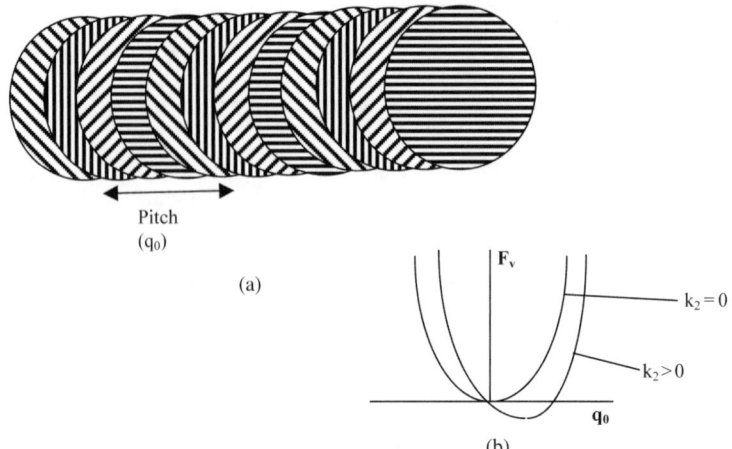

Figure 4.15 (a) The helical pitch (q_0) of a cholesteric liquid crystal. It is the separation distance over which the directors are aligned. (b) The free energy density (F_v) for non-chiral ($k_2=0$) and chiral liquid crystals ($k_2>0$) as a function of the helical pitch (q_0)

Therefore, each of the three terms on the right hand side of equation (4.6) are:

$$[\nabla.n]^2 = 0 \qquad (4.17)$$

$$[n.(\nabla \times n)]^2 = \left(\frac{2\pi}{P}\right)^2 \qquad (4.18)$$

$$[n \times (\nabla \times n)]^2 = 0 \qquad (4.19)$$

and the total free energy only has a contribution from the twist term:

$$F_v = \frac{1}{2}K_2\left(\frac{2\pi}{P}\right)^2 = \frac{1}{2}K_2 a^2 \qquad (4.20)$$

where $a = 2\pi/P$. Twist is the only distortion present in this simple example of a cholesteric liquid crystal. In a nematic liquid crystal (with no chirality) a helical distortion is therefore unstable and a twisted nematic will relax the twist distortion to minimise its free energy. Upon minimisation of equation (4.20), a is zero and the pitch becomes infinite. The free energy for a chiral nematic thus needs a linear term to be added to equation (4.20) to stabilise its free energy:

$$K_4[n.(\nabla \times n)] \qquad (4.21)$$

THE BASICS

where K_4 is a new elastic constant, which measures the degree of chirality of the system. K_4 corresponds to an intrinsic chirality of the mesogens which is common facet of biological molecules (e.g. DNA, helices in proteins and carbohydrates, beta sheets in proteins, and lipids can all have an intrinsic chirality) and in principle can be calculated from the molecular details of the mesogens. The helical director field equations (4.11–4.13) can be substituted into the Frank's free energy with the additional term (equation (4.21)) to give:

$$P = \frac{2\pi K_2}{K_4} \quad (4.22)$$

$$a_0 = \frac{K_4}{K_2} \quad (4.23)$$

The pitch (P) is related to the ratio of the two elastic constants (a_0), the chiral term divided by the twist term. The free energy per unit volume is given by:

$$F_v = -\frac{K_4^2}{2K_2} \quad (4.24)$$

The free energy as a function of cholesteric pitch is shown in Figure 4.15(b). The pitch relaxes to zero if there is no molecular chirality and the corresponding elastic constant (K_4) will be zero.

The Franks free energy equation (4.6) can be used to calculate the free energy of a nematic liquid crystal around a defect. Let the director field parallel to the z-axis be:

$$n_x = \cos[\theta(x, y)] \quad (4.25)$$

$$n_y = \sin[\theta(x, y)] \quad (4.26)$$

$$n_z = 0 \quad (4.27)$$

where $\theta(x, y)$ is the radial director of the mesogens. It is assumed that all the elastic constants are equal ($K = K_1 = K_2 = K_3$), and the free energy for an *axial disclination* can be calculated as:

$$F_v = \frac{1}{2} K \frac{m^2}{\rho^2} \quad (4.28)$$

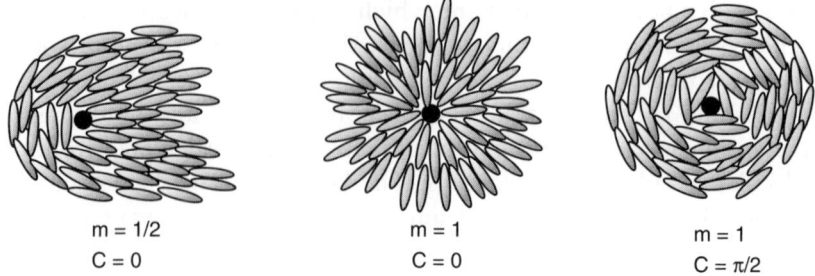

Figure 4.16 Schematic diagram of some of the defect textures encountered with nematic liquid crystals

where m is the strength of the disclination ($m = \pm\frac{1}{2}, \pm 1, \pm\frac{3}{2}\ldots$) and ρ is the radius of the core of isotropic material at the centre of the defect. The free energy diverges as the radius of the disclination reduces to zero ($\rho \to 0$) (Figure 4.16), and thus the nematic ordering becomes frustrated at the centre of such an axial orientational defect.

4.2 LIQUID–NEMATIC–SMECTIC TRANSITIONS

The *Onsager theory* allows the shape of the phase diagram for an isotropic–nematic transition to be calculated as a function of the aspect ratio of the molecules and the volume fraction of the solution. The theory allows both the form of the isotropic–nematic–smectic phase diagram to be motivated and provides quantitative predictions for the value of the order parameters near to the critical point of the phase transition.

The *Landau theory* for the isotropic–nematic phase transition develops on the ideas introduced by Onsager (Figure 4.17). The theory gives accurate information on the scaling of the phase behaviour near the critical points. The Landau theory assumes that the nematic order parameter (S) is small for the nematic phase in the vicinity of the isotropic–nematic transition and the difference between the free energy per unit volume of the two phases ($G(S, T)$) can be expanded in powers of the nematic order parameter (S):

$$G(S, T) = G_{\text{iso}} + \tfrac{1}{2}A(T)S^2 + \tfrac{1}{3}BS^3 + \tfrac{1}{4}CS^4 + \quad (4.29)$$

where G_{iso} is the free energy change for the isotropic material and A, B and C are the expansion coefficients. $A(T)$ is the most important

LIQUID–NEMATIC–SMECTIC TRANSITIONS

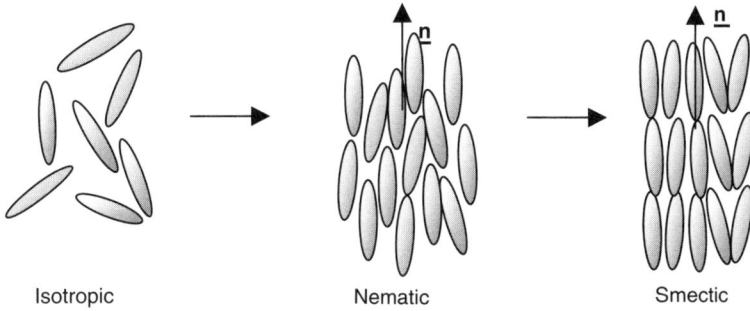

Isotropic   Nematic   Smectic

Figure 4.17 Schematic diagram of the isotropic–nematic and nematic–smectic phase transitions in a molecular liquid crystal, e.g. these could be induced by an increase in temperature

parameter for the determination of the free energy change during the phase transition and can be given a simple form (Figure 4.18) to a first (good) approximation:

$$A(T) = A_0(T - T^*) \qquad (4.30)$$

where T^* is the critical temperature for the transition and A_0 is a constant. It is then possible to study the stability of the free energy using this functional form. The solutions of equation (4.29) are given by:

$$S = 0 \text{ (isotropic)} \qquad (4.31)$$

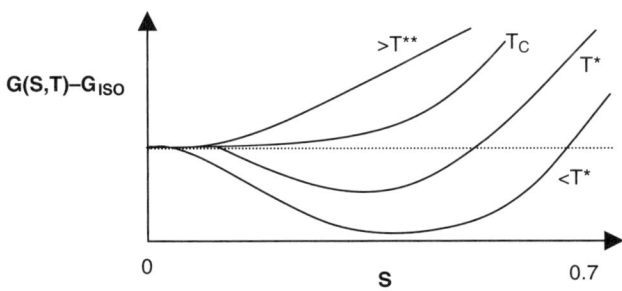

Figure 4.18 Schematic diagram of the free energy change $(G - G_{iso})$ for a nematic phase as a function of the order parameter (S) for the Landau model
(The temperatures shown are T^* for a isotropic phase, and T^{**} for a nematic phase. T_c is the critical temperature for the nematic phase transition.)

And when the free energy change is minimised as a function of the orientation:

$$\frac{\partial G}{\partial S} = A(T)S + BS^2 + CS^3 = 0 \qquad (4.32)$$

with the solution:

$$S = \frac{-B \pm \sqrt{B^2 - 4AC}}{2C} \qquad (4.33)$$

for the local maximum and minimum of G as a function of S. Such theories for the behaviour of the orientational parameter as a function of temperature are in reasonable agreement with experiment (Figure 4.19).

For the *nematic–smectic transition* it is also possible to construct a similar theory for the free energy change during the process (Figure 4.17). The order parameter (ψ) is now the amplitude of the density of the layered structure (equation (4.2)). From the translational symmetry of the layered stack the Landau free energy change can be constructed as:

$$G(|\psi|, T) = G_{\text{nem}} + \tfrac{1}{2}\alpha(T)|\psi|^2 + \tfrac{1}{4}\beta|\psi|^4 + \tfrac{1}{6}\gamma|\psi|^6 \qquad (4.34)$$

where G_{nem} is the free energy change per unit volume of the nematic phase, and α, β, and γ are characteristic constants. Using this free energy a phase diagram for the nematic–smectic transition of a material can be motivated and is in reasonable agreement with experiment.

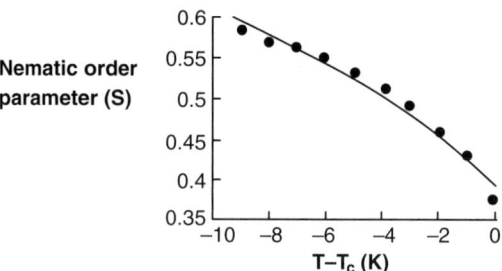

Figure 4.19 Experimental order parameter determined near the isotropic–nematic phase transition as a function of temperature compared with a fit of the Landau de Gennes theory
[*Reprinted with permission from P.J. Collings and M.J. Hird, Introduction to Liquid Crystals, Copyright (1997) Taylor and Francis*]

4.3 DEFECTS

Defects are an important facet of the structure of both solid and liquid crystalline biological materials. Theory and experiment aim to explain a whole series of complex phenomena, e.g. how helices pack together in soft solids, the dynamics of phase transitions and how chromosomes are constructed.

In solid materials which exhibit *lamellar* (a one dimensional lattice) and *columnar* (a two dimensional lattice) ordering, defect structures always occur. The *Landau–Pierels theorem* states that the geometry of one and two dimensional lattices is unable to constrain fluctuations in the positions of the molecules and they must display defect structures on large length scales, e.g. lamellar and columnar solids must be semi-crystalline or liquid crystalline. Another important biological example of solid defect structures is the ordering of helical molecules on a hexagonal lattice, which is commonly the case with solid biopolymers. Each helix often has six identical neighbours and it is not possible for the helices to align with all of their neighbours simultaneously unless the helices have a perfect intrinsic sixfold symmetry, which is rarely the case. There must be frustration in these helical crystals and they must contain helical screw defects (Figure 4.20).

Defects in solids consist of two main categories. *Point imperfections* are vacancy interstitial defects that involve an atom or molecule taken from a surface and inserted in an interior site not normally occupied. *Line imperfections* are defects localised along a continuous curve that

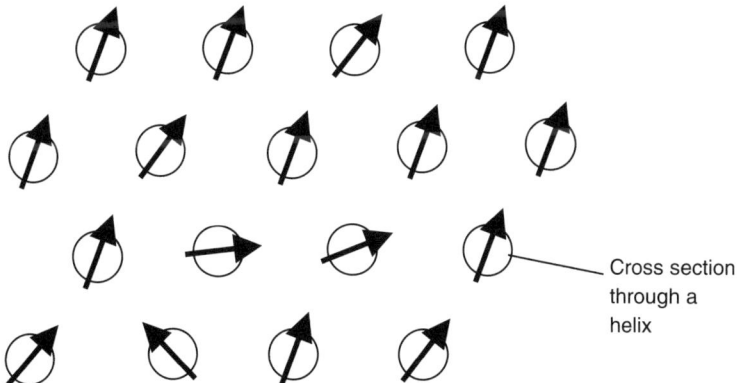

Figure 4.20 Defect structures in the packing of helices on a hexagonal lattice (The arrows indicate the orientation of the inter-helix potential in the horizontal cross section. The inter-helix potential is frustrated by the symmetry of the lattice.)

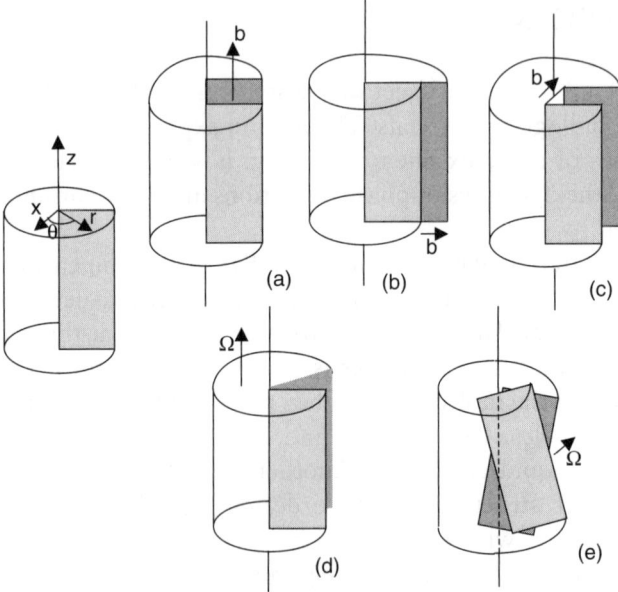

Figure 4.21 Dislocations and disclinations in a cylindrical section of an elastic media ((a) Screw dislocations, (b) and (c) edge dislocations, (d) wedge disclination, and (e) twist disclinations can occur.)

passes through the ordered medium. The Volterra process is a geometrical method for the creation of dislocations in solids and is shown in Figure 4.21. This process is useful for classifying dislocation structures.

There are two distinct categories of line defects; *dislocation line defects*, which involve translation of one part of a crystal with respect to another part, and *disclination line defects* that involve rotation of one point of the material relative to another part (Figure 4.21). The energy of disclinations is provided by the elastic energy associated with long range distortions of the director field and can be calculated using generalised theories of elasticity, such as those discussed earlier (equation (4.26)). The strength of a disclination is determined by tracing a closed path that surrounds the disclination core, which tracks the orientation of the director field (Figure 4.22). The disclination strength (m) is defined as the normalised total angle or director reorientation in a complete circuit around the defect:

$$m = \frac{1}{2\pi}\oint \frac{d\phi}{d\theta}d\theta = \frac{\phi_{\text{total}}}{2\pi} \qquad (4.35)$$

DEFECTS 97

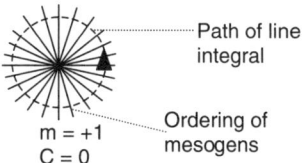

Figure 4.22 To calculate the strength (m) of a defect a line integral is used to add up the changes in direction. A hedgehog disclination is shown

where $\phi(r)$ is the angle of the director at a position r, ϕ_{total} is the total angle in a complete circuit and θ is the angle that r makes with the positive horizontal axis. $\oint$ is the line integral around a complete circuit.

An illustration of a hedgehog disclination of strength one found in a polymeric biological liquid crystal is shown in Figure 4.23. The diffraction patterns from small micron sized elements in the biopolymer indicate the direction of the helical mesogens and are mapped across a single starch granule (~ 60 μm) using a scanning X-ray microdiffraction technique. This diffraction technique is a direct molecular probe of the structure of micron sized disclinations in the material. Another method to characterise disclinations in a liquid crystalline material is with polarised optical microscopy (Figure 4.24). The number of brushes that emanate from a defect (N) allows the strength of the defect to be

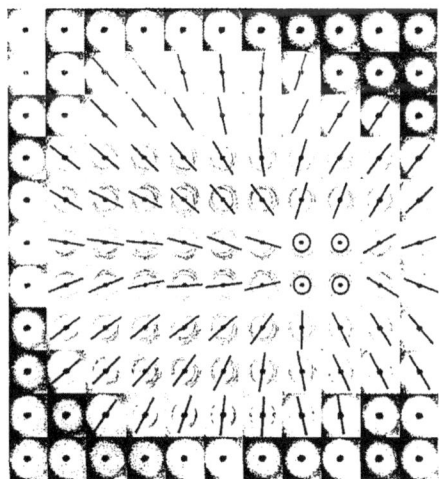

Figure A

'Disclination' map of a single potato starch granule based on the orientation of the (100) interhelix reflections. Each box samples a 5 μm by 5 μm region. The directors (lines) tend to point towards an eccentric region within the granule with no orientation, believed to be the hilum (circles). This map is physically analogous to a hedgehog disclination in liquid crystals, of strength $s = +1$.

Figure 4.23 Single hedgehog defects naturally occur for the orientation of helical mesogens in carbohydrate granules
[Ref.: T.A. Waigh, K.L. Kato, A.M. Donald et al., Starch, 2000, 52, 12, 2000]

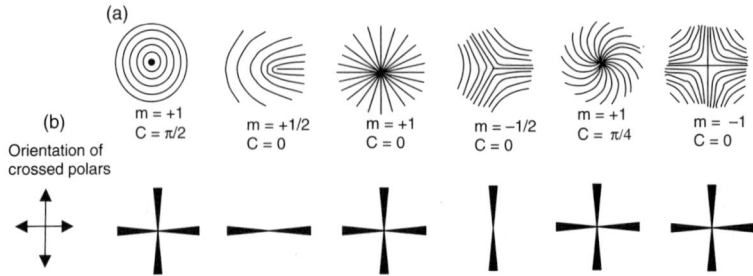

Figure 4.24 (a) Examples of the orientation of the disclination fields of liquid crystals. (b) The defect textures that correspond to (a) when the liquid crystals are observed under a polarising microscope with crossed polars

calculated ($m = N/2$). The sign of the defect ($\pm$) can be determined from the direction of rotation of the brushes when one of the polarisers in the microscope is rotated.

From the generalised theory of elasticity the elastic energy per unit length (E) of a solid disclination located along the central line of a cylinder of radius R can be calculated as:

$$E = 2\pi K m^2 \ln\left(\frac{R}{r_c}\right) + E_{\text{core}} \qquad (4.36)$$

where K is the average elastic constant of the material, m is the strength of the defect, r_c is the radius of the core and E_{core} is the core energy. The core of the line defect is in the order of the molecular size and it is assumed to contain isotropic randomly oriented material. The force (f_{12}) between a pair of straight parallel disclinations with strength m_1 and m_2 separated by a distance r_{12} is:

$$f_{12} = -2\pi K m_1 m_2 \frac{r_{12}}{(r_{12})^2} \qquad (4.37)$$

When the strength of the two defects is equal, but of opposite sign, f_{12} is attractive and the defects can combine and cancel each other out.

For smectic materials step-like terraced defects are observed under optical and atomic force microscopes (Grandjean terraces). Optical polarising microscopy indicates that there are large scale defect patterns specific to smectic liquid crystals, such as Duplin cyclides and focal conic domains (Figure 4.25).

DEFECTS

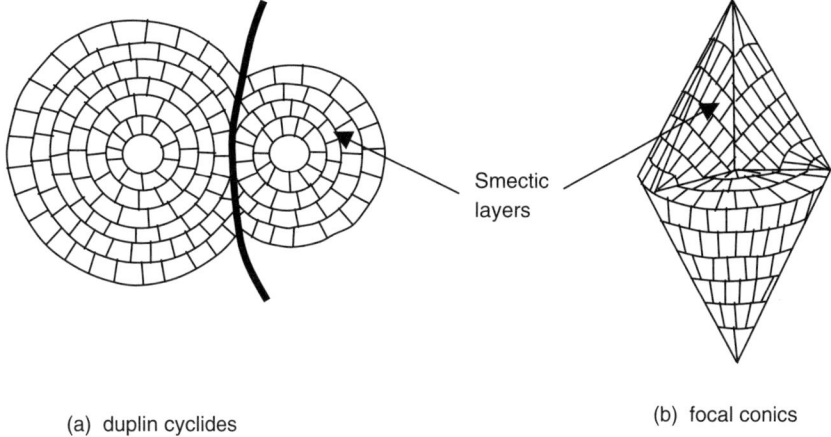

(a) duplin cyclides

(b) focal conics

Figure 4.25 Defect textures observed with smectic liquid crystals

X-ray scattering is a standard technique for characterising smectic liquid crystalline materials (Figure 4.26). Owing to the Landau–Peierls instability the Bragg peaks that normally occur with crystalline materials are broadened into power law cusps with X-ray scattering from smectics. Measurement of the functional form of these cusps allows the bending rigidity of smectic stacks to be accurately calculated, which is intimately connected with the defect texture demonstrated by the material.

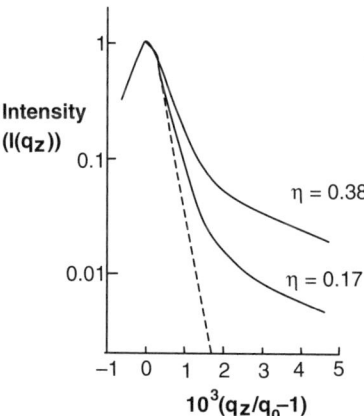

Figure 4.26 Bragg peaks are broadened into power law cusps by the thermal fluctuations of the smectic layers
(The intensity is shown as a function of the reduced momentum transfer perpendicular to the smectic stack. The dashed line is a fit to $I \sim |q_z - q_0|^{-2+\eta}$ that can provide the bending rigidity of the layers.)

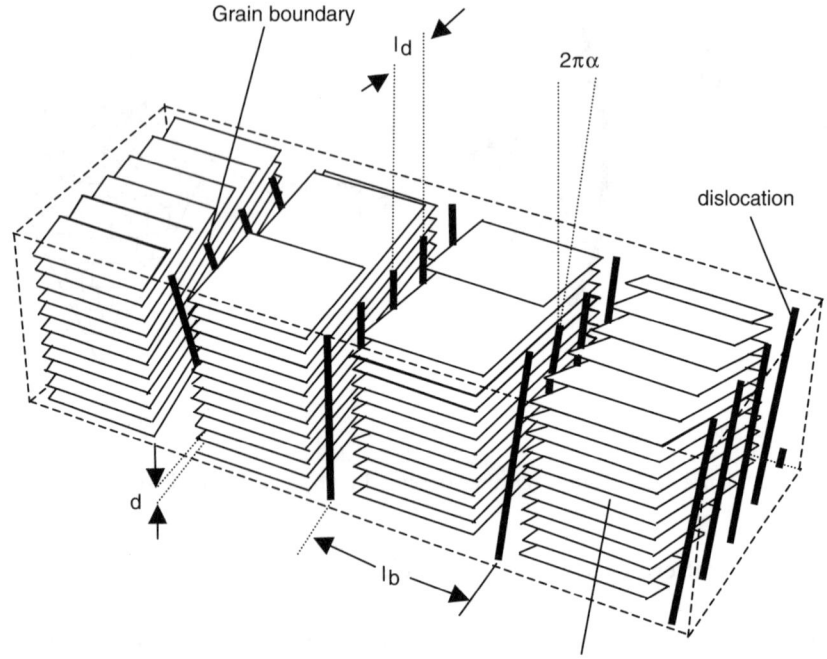

Figure 4.27 Schematic diagram of a twisted grain boundary phase in a smectic liquid crystalline material. The chirality of the mesogens causes this unusual defect phase [*Reprinted with permission from J. Goodby, M.A. Waugh, S.M. Stein et al., Nature, 337, 449, Copyright (1988) Macmillan Publishers Ltd*]

Twisted grain boundary defects in smectic–cholesteric materials such as DNA (Figure 4.27) have recently experienced a large amount of theoretical research, since there is an analogy with the phase behaviour of superconductors using the Landau equation (4.33) for the nematic-smectic transition. Many other liquid crystalline defect textures have been observed in naturally occurring 'soft solid' biological materials (Figure 4.28). Another sophisticated example of a biological defect structure is the blue phase in collagen (Figure 4.29), where the chirality of the collagen molecules is again closely related to the texture observed.

4.4 MORE EXOTIC POSSIBILITIES FOR LIQUID CRYSTALLINE PHASES

Biological polymeric liquid crystals are often induced by the increased persistence length produced by a *helix–coil transition* (e.g. DNA, and

MORE EXOTIC POSSIBILITIES FOR LIQUID CRYSTALLINE PHASES 101

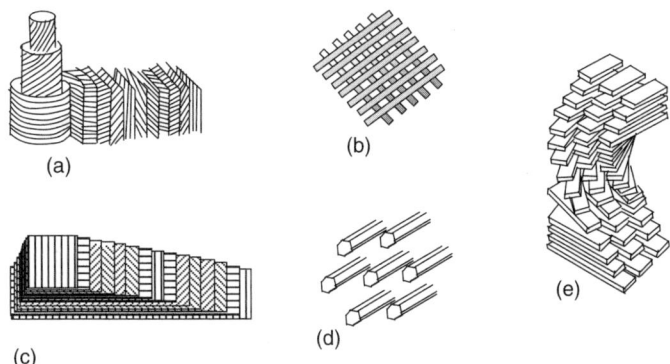

Figure 4.28 A wide range of defect textures are observed naturally in fibrous biopolymers. These include (a) cylindrical helicoidal defects, e.g. bone, (b) orthogonal defects, e.g. basement lamellar in vertebrates, (c) twisted orthogonal defects, e.g. vertebrate cornea, (d) parallel defects, e.g. tendons, and (e) pseudo orthogonal defects, e.g. endocuticle of beetles
[*Reprinted with permission from A.C. Neville, Biology of Fibrous Composites, Copyright (1993) Cambridge University Press*]

collagen), and conversely liquid crystalline phases can induce increased persistence lengths in polymers due to steric constraints. Thus helicity is seen to be intimately involved with the appearance of liquid crystalline phases.

Starch, a storage polysaccharide, and many proteoglycans are naturally occurring examples of *side-chain liquid crystalline polymers* (Figure 4.30). In these materials additional order parameters are required to simultaneously describe the nematic, cholesteric and smectic phases of both the backbone and the side-chains. A wide range of distinct mesophases are conceivable due to all the possible permutations of the values

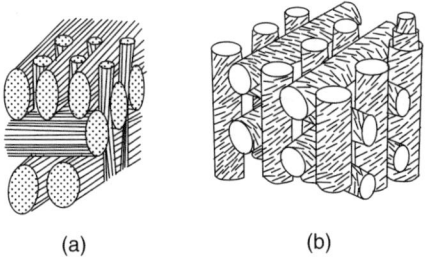

Figure 4.29 (a) The solid cholesteric liquid crystalline blue phase observed in the skin of fish is very similar in morphology to (b) the liquid cholesteric liquid crystalline phase in synthetic small molecule mesogens
[*Ref.: M.M. Giraud, J. Castanet and F.J. Meunier, Tissue and Cell, 1978, 10, 671–686*]

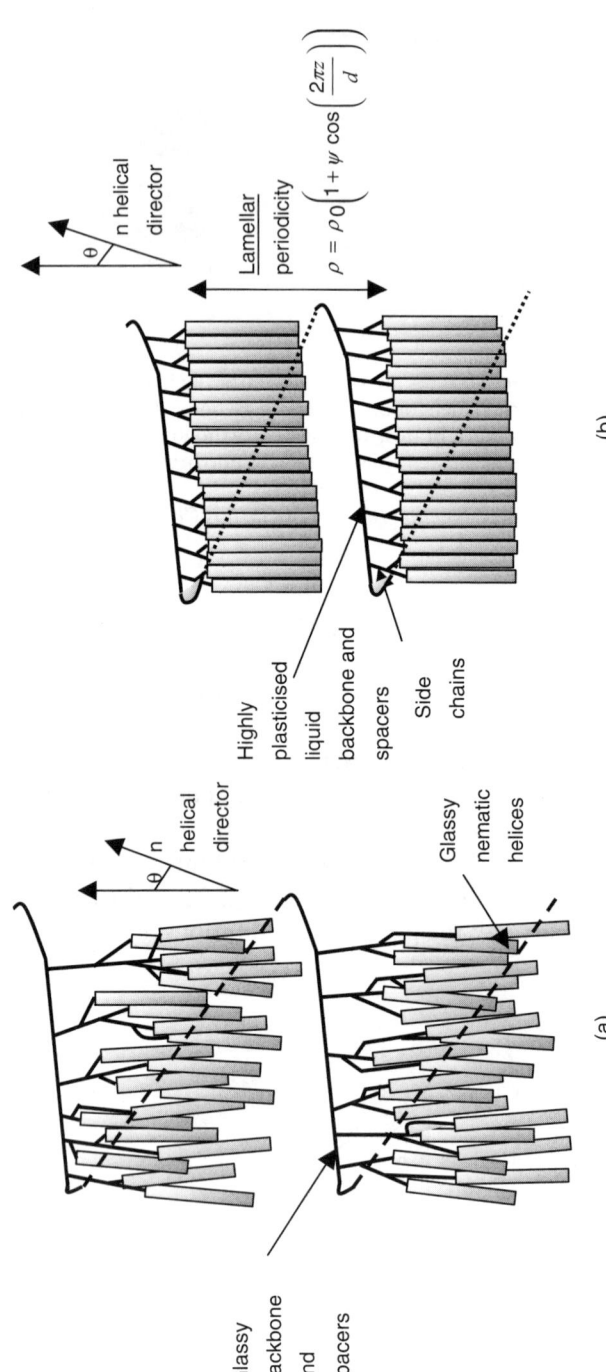

Figure 4.30 (a) Schematic diagram of a side-chain liquid crystalline polymer (amylopectin) showing the process of self-assembly that occurs upon the addition of water (The dry glassy nematic structure (a) and hydrated side-chain liquid-crystalline smectic structure (b) are shown. [*Ref.*: *T.A. Waigh, I. Hopkinson, A.M. Donald et al., Macromolecules, 1997, 30, 3813–3820*])

Figure 4.31 Liquid crystalline elastomers where n is the director for the nematic ordering. The mesogens are cross-linked in a rubbery network

of the order parameters. The intricate inter-relationships of the order parameters for the backbone and side-chains directly relates to the macroscopically observed physical behaviour of these materials.

It is possible for the orientational order of liquid crystals to persist into the solid state without forming a fully crystalline lattice, in so called 'solid liquid crystals'. 'Liquid' in this case describes the highly disordered (amorphous) static distribution of the molecules in their liquid crystalline lattice. Solid nematic elastomer phases are formed by cross-linking mesogens with flexible (rubbery) polymeric chains. Liquid crystalline elastomers models have been proposed for solid biopolymer elasticity (Figure 4.31), e.g. the unusual soft anisotropic elasticity observed in collagen networks and spider silks.

FURTHER READING

H.M. Schey, *Div, Grad, Curl and All That*, W.W.Norton, 2005. A booster course in vector calculus.

P.J. Collings and M. Hird, *Introduction to Liquid Crystals*, Taylor and Francis, 1997. Provides a good introduction to the physics and physical chemistry of small molecule liquid crystals.

A.C. Neville, *Fibrous Composites*, Cambridge University Press, 1993. Provides a large range of examples of biological liquid crystals.

P. Chaikin and T. Lubensky, *Condensed Matter Physics*, Cambridge University Press, 1995. Mathematically advanced treatment of liquid crystals.

M. Kleman and O.D. Lavrentovich, *Soft Matter Physics*, Springer, 2003. Similar in level to Chaikin and Lubensky's treatment.

S.M. Allen and E.L. Thomas, *The Structure of Materials*, John Wiley & Sons Ltd, 1999. An introduction to materials science that includes a good description of defect morphology.

TUTORIAL QUESTIONS

4.1) Hydrated potato starch consists of a smectic ordering of the double helical crystallites (Figures 1.16 and 4.32). If the starch sample is heated to create a chip there are three different order parameters that determine the manner in which it cooks: the helicity (h), the nematic order parameter (P_2) and the smectic order parameter (ψ). What sequence of phase transitions is possible as the chip is cooked (a possible scenario is shown in Figure 4.32)? What physical processes cause coupling between the three order parameters?

4.2) Calculate the nematic order parameter $P_2(\cos\theta)$ for a sample of mucin molecules in the trail of a slug if their orientation follows a top hat function given by:

$$P(\theta) = \frac{2}{\pi} \quad \frac{\pi}{4} < \theta < \frac{3\pi}{4}$$

$$P(\theta) = 0 \quad \text{Otherwise}$$

In a particular trail of slime the mucin molecules have defects of strength two. How many brushes (black lines in the image) would emanate from the defects under a polarising microscope?

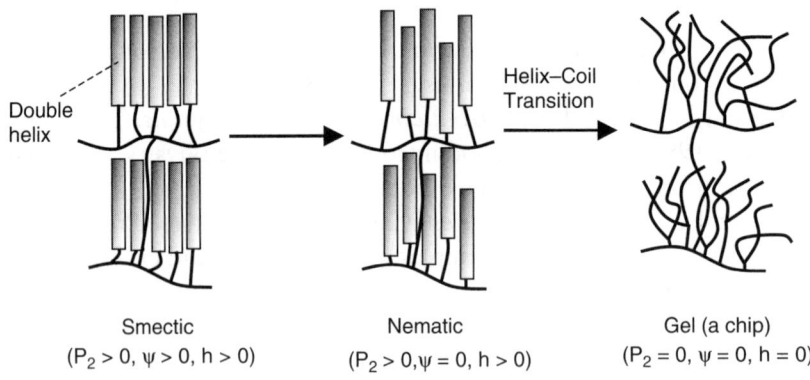

Figure 4.32 A possible sequence of phase transitions that occurs during the break down in structure of a hydrated potato starch granule when heated
[Ref.: T.A. Waigh, M.J. Gidley, B.U. Komanshek and A.M. Donald, *Carbohydrate Research*, 2000, 328, 165–176]

TUTORIAL QUESTIONS

4.3) A cylindrical virus is 200 nm in length and 10 nm in diameter. What is the critical volume fraction of virus particles to observe a nematic liquid crystalline phase according to the Onsager theory?

4.4) The addition of flexible side-chains to a flexible protein backbone induces the protein backbone to become more rigid and can result in the creation of a nematic phase. Can you suggest a reason for the induced rigidity?

5
Motility

Both living and inanimate microscopic objects are subject to *thermal fluctuations* that cause them to jiggle about incessantly when viewed under an optical microscope. Many biological organisms modify these thermal fluctuations to facilitate transport of molecules and to move through their environment. Motility in biological systems is crucially important in a wide range of biological processes including the transcription of DNA, the packaging of DNA in viruses, the propulsion of bacteria as they search for food and the exercise of striated muscle (when a dumb bell is lifted). Initially, an understanding of the undriven process of passive diffusion due to thermal energy is developed, and this is then extended to the analysis of the motions produced by molecular motors (see also Chapter 14).

Owing to the importance of motility in determining biological processes, a series of non-invasive methods have been developed to measure molecular mobility (Chapter 13); these include fluorescence correlation spectroscopy, pulsed laser techniques, dynamic light scattering, neutron/X-ray inelastic scattering, video particle tracking and nuclear magnetic resonance spectroscopy. Due to this impressive range of dynamic experimental techniques, the field of biological motility has been provided with firm foundations. The time scales that can now be probed experimentally in biomolecular liquids span from femtoseconds (10^{-15} s), all the way up to the aging processes of biopolymer glasses, which can be on the order of many years.

Applied Biophysics: A Molecular Approach for Physical Scientists Tom A. Waigh
© 2007 John Wiley & Sons, Ltd

5.1 DIFFUSION

Diffusion is the process by which molecules jiggle around over small distances due to thermal collisions with their neighbours, and equivalently, diffusion can be used to explain how macroscopic concentration gradients in materials evolve with time. Thus a food dye injected into water eventually colours the whole vessel as the dye diffuses throughout the specimen; the jiggling motion at the nanometre level produces a global redistribution of the dye molecules at the macroscale. To obtain a quantitative understanding of the process of diffusion, the phenomenon is first described in a statistical way relevant to short distances. At the macroscopic level an equivalent description is provided by *Fick's law* for the concentration of a diffusing species.

As a first step it is useful to examine the statistical form of translational diffusion in one dimension, which considerably simplifies the analysis. In one dimension the displacement ($x_i(n)$) of a particle as a function of the position of the previous random displacement ($x_i(n-1)$) after n steps is:

$$x_i(n) = x_i(n-1) \pm \varepsilon \qquad (5.1)$$

where ε is the random step size and n is the number of steps. The probability that a particle moves to the right ($+\varepsilon$) is assumed to be equal to that with which it moves to the left ($-\varepsilon$). The average of the displacements (x_i) is zero ($\langle x_i(n) \rangle$), so the square of this quantity needs to be used to create a meaningful measure of the particle's motion:

$$x_i^2(n) = x_i^2(n-1) \pm 2\varepsilon x_i(n-1) + \varepsilon^2 \qquad (5.2)$$

Next the mean square value of the displacement is constructed, the second term in equation (5.2) is seen average to zero, since $\langle x_i(n-1) \rangle = 0$. Therefore the mean square displacement ($\langle x^2(n) \rangle$) is given by:

$$\langle x^2(n) \rangle = \frac{1}{N}\sum_{i=1}^{n} x_i^2(n) = \langle x^2(n-1) \rangle + \varepsilon^2 \qquad (5.3)$$

This expression for the mean square displacement is an iterative equation that relates the mean square displacement at step n ($\langle x_i^2(n) \rangle$) to that of the previous step ($\langle x_i^2(n-1) \rangle$). Equation (5.3) can be iteratively applied all the way down to the first step of the motion ($n = 1$) and it is seen that the mean square displacement scales as the number of time steps:

$$\langle x^2(n) \rangle = n\varepsilon^2 \qquad (5.4)$$

DIFFUSION

where n is proportional to the number of time steps. This linear scaling of the mean square displacement with the time is a basic characteristic of diffusive motion. It can be compared with familiar ballistic motion (e.g. a bullet from a gun) where $\langle x^2 \rangle$ scales as n^2.

The number of time steps over which the particle diffusion is considered is related to the time (t) and the step size (τ). Thus the number of steps is given by $n = t/\tau$ and this expression can be substituted in equation (5.4) to give:

$$\langle x^2(t) \rangle = \left(\frac{\varepsilon^2}{\tau}\right) t \tag{5.5}$$

The *diffusion coefficient* (D) is then defined to quantify the magnitude of the mean square fluctuations during this one dimensional statistical process:

$$D = \frac{\varepsilon^2}{2\tau} \tag{5.6}$$

Particles with large diffusion coefficients fluctuate a considerable amount and vice versa. The factor of a 1/2 is used to tidy up Fick's equation that results from the continuum description of the macroscopic behaviour (equation 5.16). The combination of equations (5.5) and (5.6) gives an expression that relates the diffusion coefficient to the mean square fluctuations of displacement in *one dimension*:

$$\langle x^2(t) \rangle = 2Dt \tag{5.7}$$

Diffusion in one dimension statistically corresponds to the probability distribution of the particle positions broadening with time (Figure 5.1). A sharp point-like distribution of particles at the first time step ($t = 1$) evolves into a broad distribution ($t = 15$) as diffusive motion takes place.

For a small molecule in water at room temperature a typical diffusion coefficient (D) is 10^{-5} cm^2 s^{-1}. The characteristic time (τ_c) for this molecule to diffuse the length of a bacterium (10^{-4} cm) is then $\tau_c \approx x^2/2D = 5 \times 10^{-4}$ seconds.

For *two* or *three dimensions* the extension of the definition of the diffusion coefficient is fairly straightforward:

$$\langle r^2 \rangle = 2NDt \tag{5.8}$$

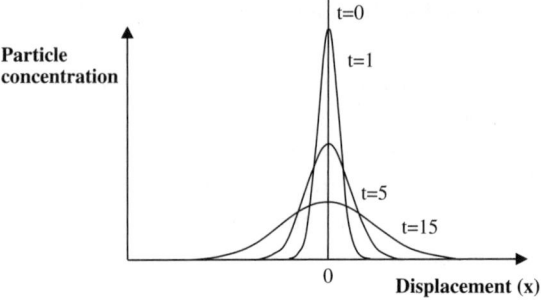

Figure 5.1 The evolution of the Gaussian probability distribution of freely diffusing particles with time (All of the particles start at $x = 0$, when $t = 0$.)

where r is the displacement in N dimensions, t is the time and D is the diffusion coefficient. An example of a two dimensional random walk is shown in Figure 5.2 for a polystyrene sphere moving in both water and a more viscous glycerol solution. Clearly an increase in the viscosity of the

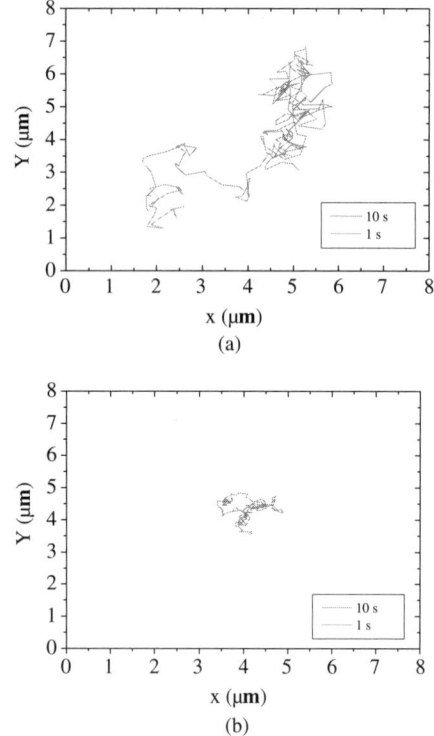

Figure 5.2 Translational diffusion of a 0.5 μm colloidal sphere in (a) water and (b) glycerol for two different periods (1 and 10 seconds)
[*Ref.: PhD A. Papagiannopoulos, University of Leeds, 2005*]

solution decreases the amplitude of the fluctuations of the displacement of the polystyrene spheres. The decrease in the spheres' fluctuations is explained by the increased friction experienced by the particle. The diffusion coefficient is related to the dissipating force using the Einstein relationship (the *fluctuation–dissipation theory*):

$$D = \frac{kT}{f} \qquad (5.9)$$

where kT is the thermal energy, D is the diffusion coefficient and f is the frictional coefficient. The generalisation of this expression to viscoelastic materials is considered in Chapter 13.

For a sphere in a fluid, with the assumption of non-slip boundary conditions, the frictional coefficient (f) can be calculated from the Navier–Stokes equations and is given by the Stoke's relationship:

$$f = 6\pi \eta a \qquad (5.10)$$

where η is the viscosity of the solution and a is the particle radius. Friction coefficients are known (or can be numerically calculated) for a wide range of rigid microscopic objects in solution. The *Stokes–Einstein* equation for a sphere combines equations (5.9) and (5.10):

$$D = \frac{kT}{6\pi \eta a} \qquad (5.11)$$

Thus measuring the fluctuations in a particle's position as a function of time (or equivalently the diffusion coefficient) the size of the particle can be calculated.

It is important to note that there is a difference between *mutual* and *self-diffusion* for the motion of an array of particles. With mutual diffusion the fluctuating rearrangement of particles with respect to their neighbours is considered, whereas with self-diffusion it is the rearrangement of individual particles relative to the laboratory frame of reference that is important. Experimental techniques are often sensitive to one or other of the two types of diffusion. The previous discussion was centred on *translational self-diffusion*. Quasi-elastic scattering techniques often provide information on translational mutual diffusion.

Particles in solution experience fluctuations in their *rotational motion* in much the same way as with translational motion. The particles are constantly being buffeted by the surrounding solvent molecules, which impart angular momentum to them. A similar statistical analysis is possible for their angular motion as that for translational motion. The mean square angular rotation ($\langle\theta^2\rangle$) is related to the time (t) through the *rotational diffusion coefficient* (D_θ):

$$\langle\theta^2\rangle = 2ND_\theta t \tag{5.12}$$

where N is the number of angular degrees of freedom (the Euler angles in three dimensions). The fluctuation dissipation theory can again be used and relates the rotational diffusion coefficient to the thermal energy (kT) and the fricitional coefficient (f_θ) for rotational motion:

$$D_\theta = \frac{kT}{f_\theta} \tag{5.13}$$

For a sphere the frictional coefficient for rotational motion is given by:

$$f_\theta = 8\pi\eta a^3 \tag{5.14}$$

where a is the radius of the sphere and η is the solvent viscosity. Equations (5.12) and (5.13) can be combined to provide an expression for the rotational diffusion coefficient of a sphere:

$$D_\theta = \frac{kT}{8\pi\eta a^3} \tag{5.15}$$

The rotational frictional coefficient is strongly dependent on the particle radius (a) and thus large particles rotate very slowly.

There is a macroscopic description of *translational diffusion* which uses Fick's laws, and is equivalent to the microscopic approach on small distances. *Fick's first equation* relates the flux of particles (J_x) that are diffusing to the gradient of the particle concentration ($\partial c/\partial x$):

$$J_x = -D\frac{\partial c}{\partial x} \tag{5.16}$$

where the particle concentration (c) is in moles per cm^3 and the flux is in moles per cm^2 s. The net flux (at position (x) and time (t)) is proportional to the gradient of the concentration function (at x and t). The constant of proportionality is the negative of the diffusion coefficient ($-D$).

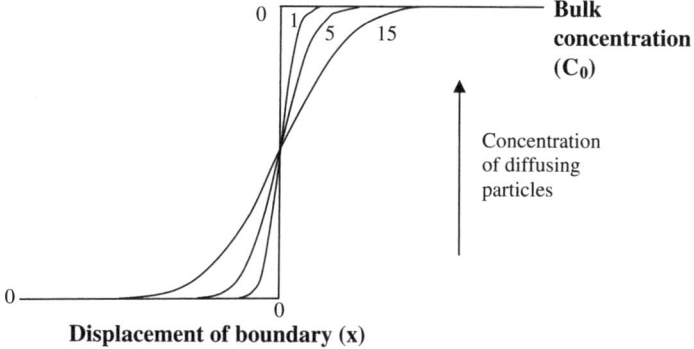

Figure 5.3 The development of the concentration gradient of a step profile as a function of time (at $t = 0, 1, 5$ and 15 seconds)

Fick's second equation is:

$$\frac{\partial c}{\partial t} = D\frac{\partial^2 c}{\partial x^2} \quad (5.17)$$

The time rate of change in concentration (at x and t) is proportional to the curvature of the concentration function (at x and t), and the constant of proportionality is again the diffusion coefficient (D).

A non-uniform distribution of particles redistributes itself in time according to Fick's two laws. This is pictured in Figure 5.3 where a sharp concentration gradient in dye molecules is reduced during time by inter-diffusion of the dye/solvent system.

In three dimensions Fick's two laws can be written in vector notation as:

$$\underline{J} = -D\underline{\nabla}c \quad (5.18)$$

$$\frac{\partial c}{\partial t} = D\nabla^2 c \quad (5.19)$$

For a quick solution to Fick's law partial differential equations in a particular geometry, the most efficient strategy is to look them up in a specialist applied mathematics text book. Solution of such diffusion problems often requires sophisticated mathematical methods. A few results for the diffusive behaviour in some specific geometries are outlined in the following to give a flavour of the basic principles involved.

(a) Diffusion from a **point source**. Consider the injection of some fluorescent dye molecules from a micropipette in a water bath.

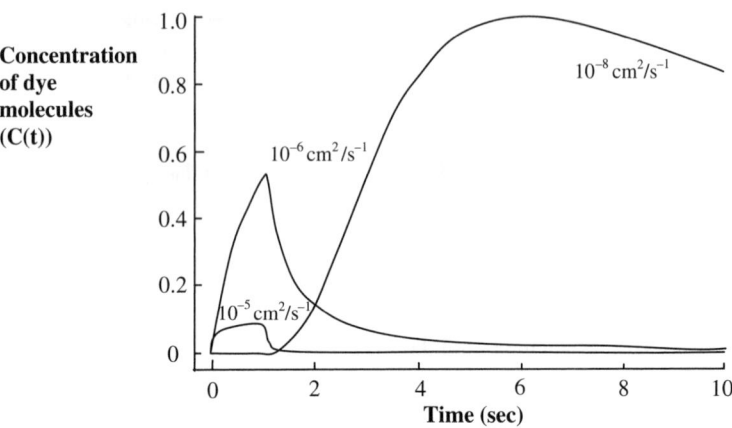

Figure 5.4 Development of the concentration profile with time (t) at a fixed position from a point source as a function of the diffusion coefficient (D) [Reprinted with permission from H.C. Berg, Random Walks in Biology, Copyright (1993) Princeton University Press]

The diffusion of the dye is found to be well described by the Gaussian equation:

$$c(r,t) = \frac{N}{(4\pi Dt)^{3/2}} \exp\left(\frac{-r^2}{4Dt}\right) \qquad (5.20)$$

where $c(r,t)$ is the concentration of the dye molecules as a function of time (t) and position, r is the distance from the point of injection, and N is the total number of dye molecules. The flux of dye molecules can then be calculated from Ficks first law, equation (5.15), using this concentration profile. Experimentally the dye molecules under a microscope will first appear as a bright spot upon injection that spreads rapidly outwards and then fades away as the concentration becomes homogenised at a constant low value (Figure 5.1). How the concentration at a single spatial position from a point source evolves with time is shown in Figure 5.4.

(b) Diffusion to a **spherical adsorber**. It is assumed that every diffusing particle that reaches the surface of a sphere is gobbled up (Figure 5.5(a)). These boundary conditions are slightly artificial (perhaps a good model for a stationary feeding bacterium), but mathematically the concentration at the surface of the sphere ($r = a$) is assumed to be zero and at a long distance from the

DIFFUSION

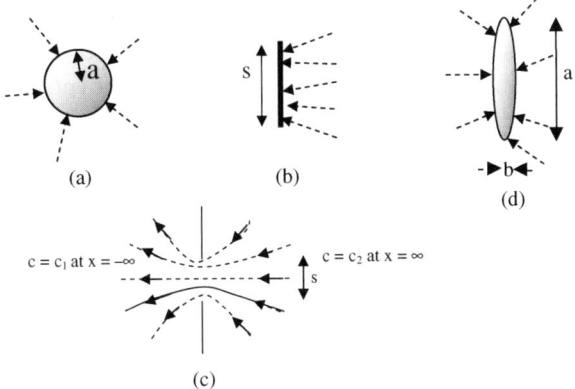

Figure 5.5 Morphologies of different absorbers for diffusing particles ((a) sphere, (b) disk, (c) circular aperture and (d) ellipsoid. Arrows indicate the motion of the diffusing species.)

sphere ($r = \infty$) it is c_0. The solution for the concentration of diffusing particles is found to be:

$$c(r) = c_0\left(1 - \frac{a}{r}\right) \quad (5.21)$$

where a is the radius of the sphere. The flux of diffusing particles can then be calculated from Fick's first law (equation (5.15)):

$$J_r = -D\frac{\partial c}{\partial r} = -Dc_0\frac{a}{r^2} \quad (5.22)$$

The particles are adsorbed by the sphere at a rate equal to the area $4\pi a^2$ multiplied by the inward flux $-J_r(a)$:

$$I = 4\pi Dac_0 \quad (5.23)$$

The adsorption rate (I) is the *diffusion current of particles* per second, and c_0 is the particle concentration per cm^3. Similar results with the current proportional to the size of the particles ($I \sim a$) are found for a wide range of different absorbing geometries and thus the rate of capture from a reservoir of particles is relatively independent of the geometry. This implies a wide range of efficient mechanisms are possible in nature for the absorption of biomolecules, and this is indeed observed experimentally. Such considerations are important for a range of reaction diffusion problems and three more results will be quoted for completeness.

(c) For diffusion to a **disk-like adsorber** the adsorption rate (Figure 5(b)) is:

$$I = 4Dsc_0 \qquad (5.24)$$

where s is the diameter of the aperture.

(d) For diffusion through a **circular aperture** (Figure 5(c)) from a particle concentration of c_1 to c_2 the current is:

$$I_{2,1} = 2Ds(c_2 - c_1) \qquad (5.25)$$

The currents are not proportional to the area of the disk, but to its radius (s).

(e) For diffusion to an **ellipsoidal adsorber** (Figure 5(d)) the concentration at the surface of the ellipsoid is zero and the concentration at large distances of separation ($r = \infty$) is c_0. The length of the ellipsoid is a and its width is b. If the length is much bigger that the width $a \gg b$ the diffusive current is:

$$I = \frac{4\pi Dac_0}{\ln(2a/b)} \qquad (5.26)$$

Again the current is roughly proportional to the length (a).

5.2 LOW REYNOLD'S NUMBER DYNAMICS

In molecular biophysics diffusion predominantly occurs under *low Reynolds number conditions*. This provides some counterintuitive results, since viscous effects dominate the motion and particle inertia is negligible, but the good news is that low Reynold's number conditions greatly simplify the mathematics required to understand the motion of biological molecules.

The *Reynold's number* (R, a dimensionless ratio i.e. it has no units) of a particle moving at a velocity (v) in a fluid is defined by:

$$R = \frac{2vL\rho}{\eta} \qquad (5.27)$$

where L is the size of a particle, ρ is the specific gravity of the fluid and η is the viscosity. The utility of the Reynold's number is found through an analysis of Navier–Stokes 'equations, the equations that predict the general motion of fluids. It is found that when $R < 1$ inertial forces (mdv/dt the product of mass and acceleration) can be neglected at reasonably long time scales ($>$ ms) and furthermore there is no turbulent flow. For a salmon that travels at a velocity (v) of $10^2\,\mathrm{cms^{-1}}$, with a length (L) of 10 cm, specific gravity (ρ) of $1\,\mathrm{gcm^{-3}}$, and through water of viscosity (η) of $10^{-2}\,\mathrm{gcm^{-1}s^{-1}}$, the Reynold's number is 10^5 (large Reynold's number dynamics). However, for a bacterium that travels at $v \approx 10^{-3}\,\mathrm{cms^{-1}}$, where $L = 10^{-4}\,\mathrm{cm}, \rho \approx 1\,\mathrm{gcm^{-3}}$ and $\eta \approx 10^{-2}\,\mathrm{gcm^{-1}\,s^{-1}}$, the Reynold's number is very small $R \approx 10^{-5}$ (small Reynolds number dynamics). Due to the relative importance of the inertial terms the fish and the bacterium have different strategies for swimming. The salmon propels itself by accelerating the water that surrounds it. A bacterium uses viscous shear to propel itself.

A useful example that emphasises the counterintuitive behaviour in low Reynold's number motility is to calculate the length a bacterium can coast in water before it comes to a stop (Figure 5.6). The mathematical analysis is very simple. Without any external forces in a purely viscous material, Newton's second law relates the acceleration (dv/dt) to the frictional force created by the surrounding water:

$$-m\frac{dv}{dt} = 6\pi\eta av \quad (5.28)$$

where m is the mass of the particle, v is the velocity, η is the viscosity of water and a is the particle radius. There is a velocity on both sides of this equation and it can be integrated by parts:

$$\frac{dv}{v} = -\frac{6\pi\eta a}{m}dt \quad (5.29)$$

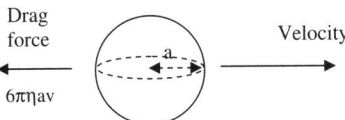

Figure 5.6 The bacterium, which approximates to a spherical colloid (radius a), experiences a drag force ($6\pi\eta av$) from the viscosity of the surrounding water which rapidly decelerates its motion

This equation is solved and it is found that the velocity relaxes to zero with a characteristic time constant (τ):

$$v(t) = v(0)e^{-t/\tau} \quad (5.30)$$

$$\tau = \frac{2a^2 \rho_s}{9\eta} \quad (5.31)$$

where ρ_s is the density of the particles. This result can be integrated once more and provides the distance coasted (d) by the bacterium before it comes to a halt:

$$d = \int_0^\infty v(t)dt = v(0)\tau \quad (5.32)$$

Typical values can be substituted for v, a, ρ_s and η, and the coasting distance of a bacterium is found to be very small, it is in the order of 0.04 Å.

A range of microfluidic experiments depend on the low Reynolds number approximation to interpret the resultant data, e.g. microrheology apparatus and optical/magnetic tweezers at low frequencies (Chapter 13). The full Langevin equation is a useful method for understanding the effects of thermally driven motion in the case of higher Reynold's numbers. Practically it is encountered in situations such as the fluctuations in the position of the atomic force microscope (AFM) tip or the high frequency fluctuations of the bead position with optical tweezers. The equation for a fluctuating particle's motion can be written using Newton's second law as:

$$m\frac{d^2x(t)}{dt^2} + \gamma\frac{dx(t)}{dt} + \kappa x(t) = F(t) \quad (5.33)$$

where $x(t)$ is the particle displacement as a function of time, md^2x/dt^2 is the inertial force that acts on the particle, $\gamma dx/dt$ is the drag force, κx is the elastic force and $F(t)$ is the random force that causes the motion of the particle, e.g. driven by thermal energy.

The *autocorrelation function* of the displacement ($R_x(\tau)$) of the particle displacement with respect to time is very useful practically in a range of spectral applications. The autocorrelation function is defined as:

$$R_x(\tau) = \langle x(t)x(t-\tau)\rangle = \lim_{T \to \infty}\left\{\frac{1}{T}\int_{-T/2}^{T/2} x(t)x(t-\tau)dt\right\} \quad (5.34)$$

where $\langle \rangle$ indicates the process is averaged over time (t) and τ is the delay time. This autocorrelation function satisfies the equation of motion (5.33) with the simplification that the right hand side is zero, since x and F are uncorrelated by definition. Substituting R_x into equation (5.32) gives:

$$m\frac{d^2 R_x(\tau)}{d\tau^2} + \gamma \frac{dR_x(\tau)}{d\tau} + \kappa R_x(\tau) = 0 \qquad (5.35)$$

Such a second order ordinary differential equation can be solved in a standard manner and the resultant autocorrelation function (also its fourier transform, the power spectral density) can be used to interpret experimental data.

5.3 MOTILITY

The absence of inertial forces could at first sight appear to present an insurmountable barrier for a biological organism to propel itself at the micron length scale. Reciprocal motion (e.g. waggling a paddle too and fro) does not lead to motility for micron sized organisms. It is similar to the case of a human trying to do the breast stroke in a swimming pool filled with syrup – they move nowhere. How has evolution overcome this problem?

Flagellated bacteria swim in a manner characteristic of the size and shape of the cells and the number and distribution of the flagella attached to their surface, e.g. an *E. Coli* 10^{-4} cm in diameter and 2×10^{-4} cm in length has six flagellar filaments for propulsion. The flagella are driven by a particularly elegant device for propulsion: the rotatory motor (Chapter 14). The motion of the bacteria is determined by the simultaneous action of the six flagellar filaments. When the flagella turn counter clockwise they form a synchronous bundle that pushes the body steadily forward; the cell 'runs'. When the filaments turn independently the cell moves erratically and the cell 'tumbles'. The cells switch back and forth between the run and tumble modes at random. The distribution of run (or tumble) intervals is exponential and the length of a given interval does not depend on the length of the intervals that precede it. This mechanism enables the bacteria to search for nutrients in its aqueous environment (Figure 5.7). This active swimming motion has completely different statistics to the Brownian

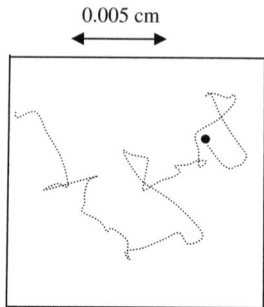

Figure 5.7 Schematic diagram of the trajectory of a Poisson motility process of a bacterial cell in water
[*Reprinted with permission from H.C. Berg and D.A. Brown, Nature, 239, 500–504, Copyright (1972) Macmillan Publishers Ltd*]

process depicted in Figure 5.2 for non-motile colloidal particles; its statistics are Poisson (see box).

Bacterial locomotion is thus achieved by the action of flagellar filaments. The thrust is produced from the component of viscous shear of the helical filament on the surrounding water in the direction of motion (Figure 5.8).

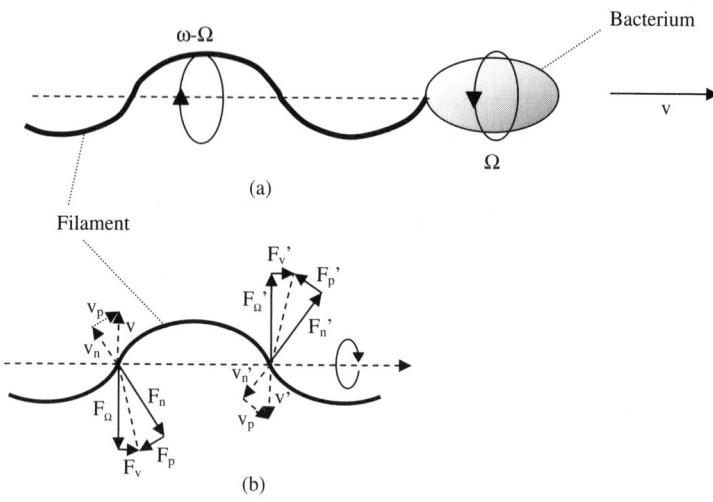

Figure 5.8 Schematic diagram of the process of bacterial locomotion
((a) the bacterium travels at velocity (v), and the forces that propel the filament result from the component (F_v) of the viscous shear
[*Ref.: H.C. Berg, Random Walks in Biology, Princeton, 1993*])

> Consider the probability distribution (P) for a Poisson statistical process. The probability that a particle changes direction once in a time period between t and $t + dt$ is:
>
> $$P(t,\lambda) = \lambda e^{-\lambda t} dt \qquad (5.36)$$
>
> where the probability that there is a change in direction per unit time is λ. The expectation time for the particle to change direction is:
>
> $$\langle t \rangle = \frac{1}{\lambda} \qquad (5.37)$$
>
> The mean squared time interval to change direction is:
>
> $$\langle t^2 \rangle = \frac{2}{\lambda^2} = 2\langle t \rangle^2 \qquad (5.38)$$
>
> And the standard deviation (σ) of the time interval to change direction is equal to the mean:
>
> $$\sigma = (\langle t^2 \rangle - \langle t \rangle^2)^{1/2} = \langle t \rangle \qquad (5.39)$$

It is found for the Poisson distribution of a process such as that which describes bacterial motion, that the apparent diffusion coefficient (D) is given by:

$$D = \frac{v^2 \tau}{3(1 - \alpha)} \qquad (5.40)$$

where α is the mean value of the cosine between successive runs, v is the velocity at which the bacterium propels itself and τ is the mean duration of the straight runs. If the mean angle between successive runs is zero ($\alpha = 0$) the apparent diffusion coefficient is $D = v^2\tau/3$, which is identical to the result for unbiased translational diffusion. Other examples of motility are provided in Chapter 14.

5.4 FIRST PASSAGE PROBLEM

The first passage problem is a basic question in the statistical physics of biological processes. It asks, 'how long does it take for a particle to travel a certain distance?' The diffusion limited rate of the first passage process is thus the reciprocal of the first passage time. For example,

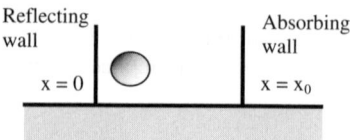

Figure 5.9 Schematic diagram of the first passage time for a freely diffusing particle to travel from $x = 0$ to an absorbing wall at $x = x_0$

this could be the time for a particle released at the origin ($x = 0$) to be absorbed at $x = x_0$ (Figure 5.9). The mean time to capture (t_0) is therefore given by:

$$t_0 = \frac{1}{j(x_0)} = \frac{1}{j(x)} \tag{5.41}$$

where $j(x)$ is the concentration flux as described in Section 5.1. The solution to the reflecting wall and absorbing wall problem is plotted in Figure 5.10. In the absence of an external force the first passage time is the time to diffuse in one dimension (from equation (5.7)):

$$t = \frac{x_0^2}{2D} \tag{5.42}$$

A 500 nm diameter protein diffuses near a fibrin fibre in a blood clot. The first passage time to diffuse the distance between adjacent fibrin monomers (45 nm) is calculated to be 2.33 ms, if an estimated diffusion coefficient from the Stoke's–Einstein equation is used (equation (5.10), $D \approx 4.35 \times 10^{-13}\,\text{m}^2\,\text{s}^{-1}$). A similar calculation can be used to

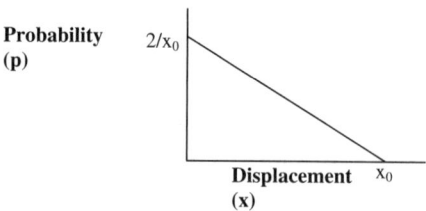

Figure 5.10 Solution of the first passage problem
(The probability is shown as a function of displacement (x) for the situation illustrated in Figure 5.9. A freely diffusing particle in one dimension is placed between a reflecting and an absorbing wall.)

FIRST PASSAGE PROBLEM

understand the mechanism of motion in the lac repressor, e.g. the time to explore the one dimensional length of a DNA chain by a specialist enzyme (Chapter 16).

A more sophisticated problem is how long a molecule that is initially at the origin ($x = 0$) takes to diffuse over an energy barrier that is placed a certain distance away (x_0). For a constant force (F) that acts on the molecule, the potential (U) as a function of distance (x) is given by:

$$U(x) = -Fx \qquad (5.43)$$

The first passage time (t) for the motion is found to be:

$$t = 2\left(\frac{x_0^2}{2D}\right)\left(\frac{kT}{Fx_0}\right)^2 \left\{ e^{-Fx_0/kT} - 1 + \frac{Fx_0}{kT} \right\} \qquad (5.44)$$

This functional form shows that when the diffusion is steeply downhill, the force is large and positive, and the first passage time approaches the distance divided by the average velocity (x_0/v). When the diffusion is uphill the first passage time increases approximately exponentially as the opposing force is increased (Figure 5.11).

For the diffusion of a particle in a parabolic potential well, the first passage time (called the Kramer's time, t_k) now becomes:

$$t_k = \tau \sqrt{\frac{\pi}{4}} \sqrt{\frac{kT}{U_0}} e^{U_0/kT} \qquad (5.45)$$

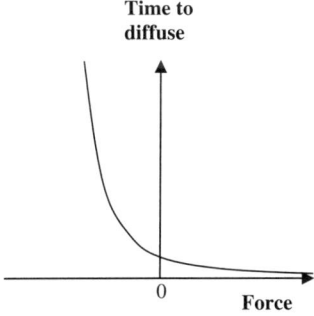

Figure 5.11 First passage time for the diffusion of a particle in one dimension as a function of the applied force, if the particle is attached to an elastic element $U(x) = (\frac{1}{2})\kappa x_0^2$ (a harmonic potential)

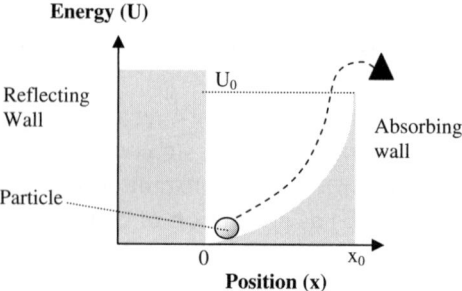

Figure 5.12 Schematic diagram that depicts the energy of a diffusing particle in one dimension out of a parabolic potential well of height U_0 as a function of position (x)

where $\tau = \gamma/\kappa$ is the drag coefficient divided by the spring constant (Figure 5.12) and U_0 is the height of the parabolic potential. The shape of the potential in which a particle is diffusing thus sensitively affects the functional form of the first passage time.

An important example of the first passage problem considers the rate at which a protein changes its conformation. A first possible solution is provided by the *Arrhenius equation* (Figure 5.13), which describes the transition of a protein between two states of free energy, e.g. unfolded and folded states. The probability of the protein being in an activated state is given by a Boltzmann distribution. The rate constant (k_1) is therefore:

$$k_1 = Ae^{-\Delta G_{12}/kT} \tag{5.46}$$

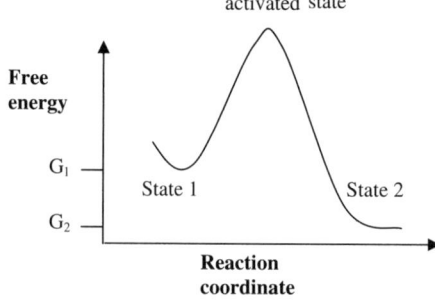

Figure 5.13 The free energy as a function of the reaction co-ordinate for the conformational change of a molecule from a state 1 (free energy G_1) to a state 2 (free energy G_2)

where $\Delta G_{12} = G_1 - G_2$ is the difference in energy between the two transition states (1 and 2). The Arrhenius equation gives no information on the constant prefactor (A) and additional assumptions are required to calculate this coefficient. In the *Eyring rate theory* the reaction constant (A) corresponds to the breakage of a single quantum mechanical vibrational bond and is typically of the order of $kT/h \cong 6 \times 10^{12}\,\mathrm{s}^{-1}$. This approximation applies equally well to the breakage of covalent bonds, but it is not useful for global conformational changes of protein chains.

For global protein conformational changes *Kramers rate theory* is a more realistic calculation of the prefactor A in equation (5.46). This includes the effects of diffusive fluctuations that determine the reaction rate in this case:

$$k_1 = \frac{\varepsilon}{\pi\tau}\sqrt{\frac{\Delta G_{al}}{kT}}e^{-\Delta G_{al}/kT} \quad (5.47)$$

$$\Delta G_{al} = G_a - G_l \quad (5.48)$$

Proteins diffuse into the transition state with a rate equal to the reciprocal of the diffusion time. The efficiency factor (ε) for the transition rate is equal to the probability that the conformational transition is made when the protein is in the transition state. τ is the time over which the protein's shape becomes uncorrelated. According to the Kramer's theory the frequency factor (τ^{-1}) is approximately equal to the inverse of the relaxation time ($\tau^{-1} = \kappa/\gamma$, where κ is the elastic constant and γ is the dissipative constant for the protein).

5.5 RATE THEORIES OF CHEMICAL REACTIONS

The rates of many biochemical processes are determined by the combined diffusion of the reactants. It is assumed that the only interaction which occurs between biomolecules A and B in a mixture is when they collide (Figure 5.14). The flux of matter due to molecule A, assuming B is

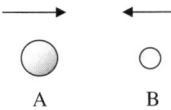

Figure 5.14 Schematic diagram of two biochemical species (A and B) that experience a collisional diffusive reaction

at rest, is given by J_A. Fick's first equation (5.16) can be written for the flux A:

$$J_A = -D_A \frac{\partial c_A}{\partial r} \qquad (5.49)$$

where D_A is the diffusion coefficient of species A and r is the radial position of species A.

There is an excluded region around the two particles equal to the sum of their two radii ($r_0 = r_A + r_B$). Therefore for separation distances less than the combined radii of the two particles ($r < r_0$) the concentration of A is equal to zero ($c_A = 0$). For large separation distances ($r \to \infty$) the concentration of A approaches the bulk concentration ($c_A \to c_A^0$).

The total current of A that flows towards B is the flux multiplied by the area of the surface around B that is considered. dq_A/dt is the current of particles of type A over the surface area of a sphere of radius (r) centred on B:

$$\frac{dq_A}{dt} = -4\pi r^2 J_A = 4\pi r^2 D_A \frac{\partial c_A}{\partial r} \qquad (5.50)$$

where equation (5.49) has been used for the flux of A particles. This expression can be simplified further by adding up all the contributions over space to the current:

$$\frac{dq_A}{dt} \int_{r_0}^{\infty} \frac{dr}{r^2} = 4\pi D_A \int_0^{c_A^0} dc_A \qquad (5.51)$$

where r_0 is the excluded region between the particles. Integration of this variables separable equation gives:

$$\frac{dq_A}{dt} = 4\pi D_A r_0 c_A^0 \qquad (5.52)$$

To calculate the total amount of complex formed per second an additional term for B is required for the total current (dq_A/dt):

$$\frac{dq_{AB}}{dt} = 4\pi r_0 (D_A + D_B) c_A^0 c_B^0 \qquad (5.53)$$

The rate constant (k) for the reaction rate of molecules A and B that is purely due to diffusion is therefore:

$$k = 4\pi f (D_A + D_B) N r_0 \qquad (5.54)$$

where N is Avogadro's number and f is of the order of 1 for hard sphere collisions. This expression will be used in Chapter 16 to understand the interaction of the lac repressor with DNA.

FURTHER READING

J. Howard, *Mechanics of Motor Proteins and the Cytoskeleton*, Sinauer, 2001. Very clear modern account of the motility.
H. Berg, *Random Walks in Biology*, Princeton University Press, 1993. A classic text on biomolecular motion from an expert in the field. Much of the current chapter draws heavily on this clear exposition.

TUTORIAL QUESTIONS

5.1) A mosquito of length 10^{-3} m flies at a speed of 10^{-1} ms^{-1}. What is its Reynolds number, given that the density and dynamical viscosity of air are 1.3 kgm^{-3} and 1.8×10^{-5} Nsm^{-2} respectively?

5.2) The flow of sodium ions in a cell is assumed to satisfy Fick's law and the diffusion equation (the diffusion coefficient is 1.35×10^{-9} m^2 s^{-1}). The flux of sodium is used for signalling inside an organism. How long would it take for the sodium to diffuse the length of a neuron (2.7 mm)? Can you comment on the practicality of such a mechanism of signalling?

5.3) Consider the rotational diffusion of a spherical virus. How does the mean square fluctuation ($\langle \theta^2 \rangle$) of the virus's angle of rotation relate to the rotational diffusion coefficient? Estimate the time for the virus to fluctuate by 90° if the thermal energy (kT) is 4.1×10^{-21} J, the viscosity is 0.001 Pa.s and the virus can be approximated by a sphere of radius 2 μm. How does the time for a point on the circumference of the virus to rotate by diffusion through a distance $2\pi a$ compare with the time to translate by $2\pi a$?

5.4) A Poisson distribution can be used to describe the motion of a bacterium. The apparent diffusion coefficient of the bacterium is 4×10^{-6} cm^2 s^{-1}. If the cell swims at a constant speed of 1×10^{-3} cm s^{-1} and the mean duration of the straight runs is one second, what is the average value of the cosine between consecutive runs?

6
Aggregating Self-Assembly

Biological complexes are often extremely complicated and it was an important advance when many were found to *self-assemble* on a molecular level from their 'raw ingredients'. The molecules arrange themselves spontaneously into aggregates without any outside assistance. If the components, the solvent, the pH and the temperature are correctly chosen, the system will minimise its free energy and organise itself in the correct manner. Such strategies for self-assembly have been invented countless times during biological evolution and appear intimately connected with life itself.

There are a diverse range of examples of self-assembling biological systems. In the construction of the tobacco mosaic virus, RNA attaches itself onto 'pie-shaped' coat proteins to produce a rod-like helical virus which is pathogenic to tobacco plants. Similarly, many globular enzymes can self-assemble from their primary structure into fully functioning chemical factories. This is the extremely complicated Levinthal problem referred to in the discussion of the globule-coil transition in Section 3.3. Actin, tubulin and flagellin can self-assemble to provide a force for cellular locomotion. The gelation of haemoglobin in the interior of red blood cells can disrupt the functioning of the cells and gives them a characteristic sickle shape (Figure 6.1); a first example of a self-assembling disease, sickle cell anaemia. A further medical condition which is currently the subject of intense research is amyloidosis in prion diseases. Self-assembled beta-sheet amyloid plaques are implicated in a large range of diseases including Alzheimer's, bovine spongiform encephalopathy (mad cow disease) and Parkinson's disease (Figure 6.2).

Applied Biophysics: A Molecular Approach for Physical Scientists Tom A. Waigh
© 2007 John Wiley & Sons, Ltd

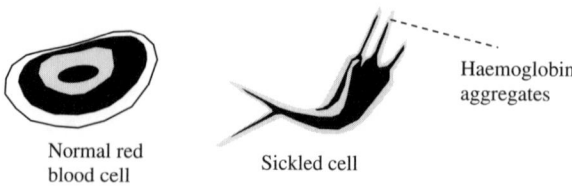

Figure 6.1 Sickle shaped cells are formed when fibrous aggregates of misfolded haemoglobin molecules self-assemble within cells and produce elongated structures on the membrane surfaces. The flow of sickled cells is impaired in the circulation system and this reduces the transport of oxygen in the body (an anaemia)

Cell membranes are found to self-assemble from their raw components. Bilayers are easily created synthetically from a range of lipid molecules and spontaneously arrange themselves into vesicles. Naturally occurring cell membranes follow more complicated schemes of construction (their structures include intramembrane proteins and scaffolding, Figure 6.3); however, the underlying scheme of amphiphilic self-assembly is still thought to hold. Carbohydrates also experience a process of self-assembly; the double helices of starch in plant storage organs are expelled into smectic-layered structures when the carbohydrate is hydrated. An important consideration for the next time the reader cooks a chip.

A distinction is made between examples of *aggregating self-assembly* (e.g. micellisation of lipids) and *non-aggregating self-assembly* (e.g. folding of globular proteins). Aggregating self-assembly has some conceptually sophisticated universal thermodynamic features (e.g. a critical

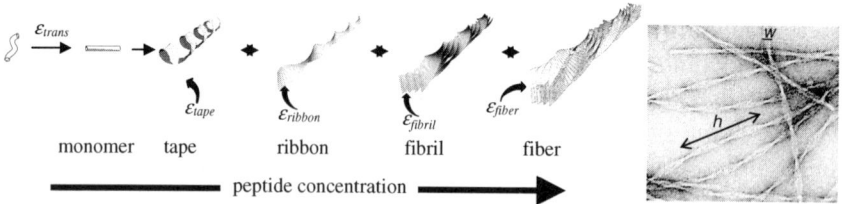

Figure 6.2 Giant self-assembled beta sheet aggregates are thought to be responsible for a range of prion diseases. A hierarchy of structures are found with model peptide systems as a function of peptide concentration. The diameters of the resultant fibres are controlled by the chirality of the peptide monomers. h is the pitch of the self-assembled fibre and w is the width of a fibre
[Reprinted with permission from A. Aggeli, I.A. Nyrkova, M. Bell et al., Proceedings of the National Academy of Sciences, 98, 21, 11857–11862, Copyright (2001) National Academy of Sciences]

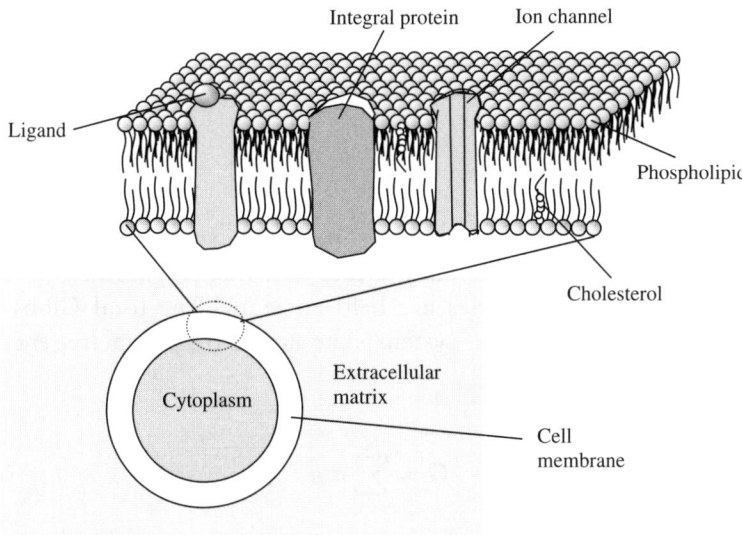

Figure 6.3 Schematic diagram of a cell membrane showing phospholipids, ligands, proteins, cholesterol and ion channels. The cell membrane separates the cytoplasm from the extracellular matrix

micelle concentration) which are considered in detail in this chapter. Non-aggregating self-assembly usually describes the behaviour of a system that moves between some hidden free energy minima, e.g. the subtle molecular origami involved in the folding of globular proteins. Examples of such phase transitions were covered in Chapter 3.

Other more general examples of self-assembly exist in soft-condensed matter physics, such as the morphologies produced in the phase separation of liquids, liquid-crystals, polymers and block copolymers. All of these have analogues in molecular biophysics. Self-organisation is another closely related field of pattern formation in molecular biology and typically is used to describe the results of non-equilibrium thermodynamic processes, e.g. morphogenesis during cell division (how the leopard got its spots). The only examples of driven non-equilibrium self-assembly that will be considered in the current book are the self-assembly of motor proteins described in Chapter 14.

To use thermodynamics to describe processes of aggregating self-assembly, the change in free energy (G) of a system due to the exchange of one of its components needs to be considered. The partial molar Gibbs free energy of a biomolecular system with a number of components is given by the symbol μ (the chemical potential). The chemical potential (μ_i) with respect to one of the species is defined as:

$$\mu_i = \left(\frac{\partial G}{\partial n_i}\right)_{T,P,n_{i\neq j}} \tag{6.1}$$

where n_i is the number of species of type i in the system. The subscripts T, P and $n_{i\neq j}$ on the differential indicate that the temperature, pressure and number of other species are held constant. The total Gibbs free energy (G) of a biomolecular system is the sum of the partial free energies of each of its components:

$$G = \sum_{i=1}^{N} n_i \mu_i \tag{6.2}$$

where μ_i are the potentials that drive chemical reactions or diffusion in which changes in the amounts of chemical substances occur.

Generally the processes of aggregating self-assembly in molecular biophysics have a number of common themes; there exists a critical micellar concentration – a value of the concentration of subunits above which self-assembly occurs (the free monomer concentration is pinned at a single value above this concentration), the entropy change is positive on assembly as the aggregate becomes more ordered (globally the entropy is still maximised due to the increased randomisation of associated solvent molecules), hydrogen bonding and hydrophobicity are often an important driving factor, and the surface free energy is minimised as the self-assembly proceeds.

The general features of self-assembly also depend on the dimensionality of the system. Self-assembly in one dimension produces highly polydisperse polymeric aggregates. In two dimensions self-assembly tends to form an aggregate consisting of a single raft and in three dimensions the aggregate is a single micelle or crystal. Self-assembly is driven by the minimisation of the *surface free energy*. In one dimension the reduction in free energy is independent of polymer length, and thus polydisperse aggregates are formed in the self-assembly of single-stranded fibrous proteins and linear surfactant aggregates. Fusion of two surface rafts in two dimensions reduces the surface area and drives the process of coarsening of raft morphologies, eventually leading to the formation of a single giant raft. Similarly, in three dimensions the process of Ostwald ripening causes a gradual increase in aggregate size as small

6.1 SURFACTANTS

The essential framework of biological membranes in cells is provided by lipid amphiphiles that spontaneously aggregate to form bilayer vesicles (Figure 6.3). The bilayer encapsulates an internal cavity in which the environment for a living cell is maintained (its osmotic pressure, salt concentration and pH). Amphiphilic molecules such as surfactants, lipids, copolymers and proteins can spontaneously associate into a wide variety of structures in aqueous solutions. With naturally occurring lipids the critical micelle concentrations (CMCs) occur at extremely low concentrations and allows stable bilayers to be formed from globally low concentrations of sub-units. Critical micelle concentrations are typically in the range $10^{-2} - 10^{-5}$ M and $10^{-2} - 10^{-9}$ M for single and double chained phospholipids respectively.

A graphical illustration of surfactant self-assembly is shown in Figure 6.4. Surfactants are partitioned between micelles and unassociated sub-units above the CMC (Figure 6.5). The unusual phenomenon above the CMC is that the monomers are pinned at a fixed concentration due to the thermodynamics of the assembly process and this process will be motivated theoretically in the following section.

Normally surfactants are in dynamic equilibrium with their aggregates during the process of micellar assembly. There is a constant interchange between lipids in micelles and those free in solution (Figure 6.4). The

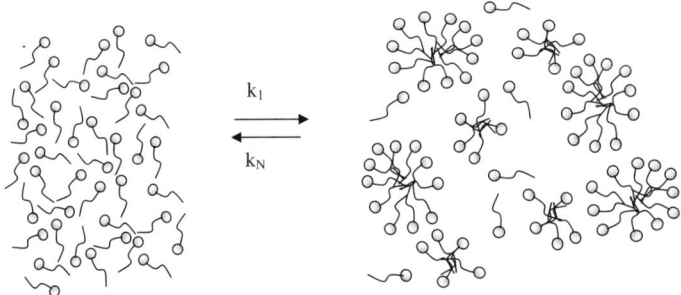

Figure 6.4 Schematic diagram of the self-assembly of surfactants into micellar structures
(Monomers on the left assemble into multimer aggregates (on the right) above the critical micelle concentration. The process is one of dynamic equilibrium.)

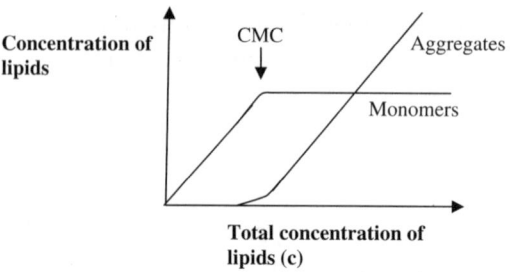

Figure 6.5 Monomer and aggregate concentrations as a function of the total concentration of lipids
(Above the CMC the concentration of lipid monomers is held fixed by the thermodynamics, whereas the concentration of lipids in aggregates increases exponentially.)

morphology of the aggregates is determined by the geometry of the amphiphilic molecules (the head group area and the length of the tails etc), and the hydrophobicity/hydrophilicity of both the head and tail groups.

There is a strong similarity between Figure 6.5, which indicates the CMC for lipid amphiphiles, and that for the assembly of proteins covered in a following section (Section 6.4, Figure 6.13). This indicates that a universal thermodynamic process is in action. In equilibrium, thermodynamics requires that the chemical potential of all identical molecules in the different sized self-assembled aggregates are equal. The chemical potential for monomers, dimers, and trimers can thus be equated:

$$\mu = \mu_1 + kT \log c_1 = \mu_2 + \tfrac{1}{2}kT \log c_2 = \ldots\ldots \qquad (6.3)$$

and more generally for a N-mer aggregate:

$$\mu = \mu_N^0 + \frac{kT}{N} \log\left(\frac{c_N}{N}\right) = const \qquad (6.4)$$

where N is the aggregation number, kT is the thermal energy, c_N is the concentration of the micellar species of aggregation number N and μ_N^0 is the standard part of the chemical potential. The rate of association is $k_1 c_1^N$ and the rate of dissociation is $k_N(c_N/N)$. k_N is the reaction rate for the Nth order association process, and K is the dissociation constant for the equilibrium process. By definition the dissociation constant is given by:

$$K = \frac{k_1}{k_N} = \exp[-N(\mu_N^0 - \mu_1^0)/kT] \qquad (6.5)$$

SURFACTANTS

Solute molecules self-assemble in solution to form clusters of aggregation number (N) per cluster. The chemical equation for the association of monomers (A) into aggregates (B) in solution can be expressed as:

$$A + A + A + \ldots = B \tag{6.6}$$

c_A and c_B are defined to concentrations of A and B in mole fraction units respectively, and c is the total concentration of the solute molecules. A general relationship can be obtained between K, N, c and c_A. For large values of the dissociation constant ($K \gg 1$) and large micellar aggregates ($N \gg 1$) it can be shown that the concentration of monomers c_A can never exceed $(NK)^{-1/N}$. By definition the equilibrium dissociation constant is:

$$K = \frac{c_B}{c_A^N} \tag{6.7}$$

The total concentration of species equals the sum of the concentration of the components:

$$c = c_A + N c_B \tag{6.8}$$

This allows equation (6.7) to be reexpressed as:

$$K = \frac{(c - c_A)}{N c_A^N} = const \tag{6.9}$$

Equation (6.9) can be rearranged to give:

$$c_A = \left[\frac{c - c_A}{NK}\right]^{1/N} \tag{6.10}$$

The maximum possible value of $c - c_A$ is 1, since the calculation is in fractional molar units, e.g. when $c_A \sim 0, c - c_A \sim 1$. Therefore c_A cannot exceed $(NK)^{-1/N}$. This was shown graphically on Figure 6.5. If a large value is taken for the dissociation constant ($K = 10^{80}$) and a reasonable value for the aggregation number ($N = 20$) is chosen the critical concentration (c_A) is 0.86×10^{-4}. Substitution in equation (6.10) gives:

$$c_A = 10^{-4} \left[\frac{(c - c_A)}{20}\right]^{1/20} \tag{6.11}$$

Detailed analysis of this equation shows that for $c \ll 10^{-4}$ $c_A \cong c$, whereas for $c \cong 10^{-4} c_A \cong Nc_B$, and there is an equal partition between micelles and unimers. Thus $(NK)^{-1/N}$ is the critical micelle concentration for this process of self-assembly.

The process of *one dimesional aggregation* can now be considered in detail. αkT is defined to be the monomer–monomer 'bond' energy of the linear aggregate relative to isolated monomers in solution. The total interaction free energy $(N\mu_N)$ of an aggregate of N monomers is therefore (terminal monomers are unbonded):

$$N\mu_N = -(N-1)\alpha kT \qquad (6.12)$$

This can be rearranged as:

$$\mu_N = -\left(1 - \frac{1}{N}\right)\alpha kT \qquad (6.13)$$

and can be written in the equivalent form:

$$\mu_N = \mu_\infty + \frac{\alpha kT}{N} \qquad (6.14)$$

Thus μ_N decreases asymptotically towards μ_∞ the bulk energy of an extremely large aggregate ($N \to \infty$).

In *two dimensional aggregation* the number N of molecules per disc-like aggregate is proportional to the area πR^2. The number of unbonded molecules in the rim of the disc is proportional to the circumference $2\pi R$ and hence $N^{\frac{1}{2}}$. This implies that the free energy of an aggregate is therefore:

$$N\mu_N = -(N - N^{\frac{1}{2}})\alpha kT \qquad (6.15)$$

and the free energy per molecule in an aggregate is:

$$\mu_N = \mu_\infty + \frac{\alpha kT}{N^{\frac{1}{2}}} \qquad (6.16)$$

where $\mu_\infty (= -\alpha kT)$ is again the free energy per particle of an infinitely large aggregate.

For spherical, *three dimensional aggregates*, N is proportional to the volume $((4/3)\pi R^3)$ and the number of unbonded molecules is proportional to the area ($4\pi R^2$), and hence $N^{\frac{2}{3}}$. Therefore the total free energy of an aggregate is:

$$N\mu_N = -(N - N^{\frac{2}{3}})\alpha kT \tag{6.17}$$

This can be rearranged and the free energy per particle is:

$$\mu_N = \mu_\infty + \frac{\alpha kT}{N^{\frac{1}{3}}} \tag{6.18}$$

where $\mu_\infty (= -\alpha kT)$ is the free energy per particle of an infinitely large aggregate. It can be shown that for a spherical micelle the proportionality constant (α) is related to the surface tension (γ) and the size of the aggregate (R):

$$\alpha = \frac{4\pi R^2 \gamma}{kT} \tag{6.19}$$

In general the CMC for the aggregation of surfactants is given by the exponential of the difference in chemical potentials for a monomer and an aggregate:

$$\text{CMC} \approx e^{-(\mu_1 - \mu_N)/kT} \tag{6.20}$$

Therefore, for three dimensional aggregates the CMC is:

$$\text{CMC} \approx e^{-\alpha/N^{\frac{1}{3}}} \tag{6.21}$$

where α is given by equation (6.19).

6.2 VIRUSES

The self-assembled geometrical structure of heptatitis B determined by X-ray crystallography measurements is shown in Figure 6.6. This virus is pathogenic to humans, and the self-assembly (reproduction) of such parasites is of vital importance to medical science. The general process of self-assembly in viruses is thought to originate from an interplay between shorter range hydrophobic and longer range electrostatic forces. However, the details of the mechanism of self-assembly can be very complicated and are specific to the particular variety of virus that is considered.

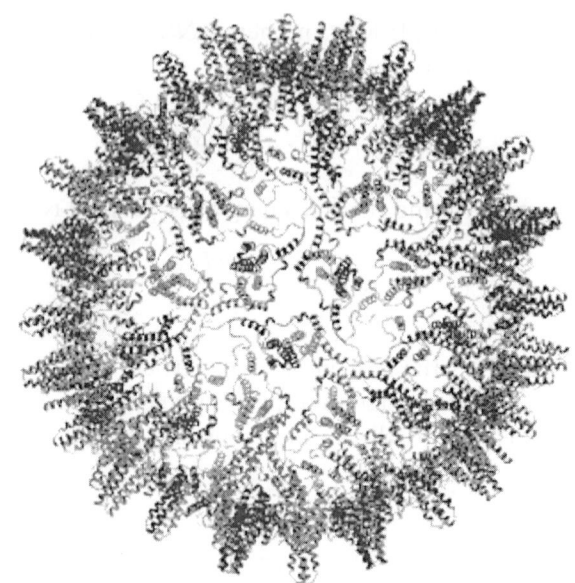

Figure 6.6 Graphical representation of a hepatitis B virus based on X-ray crystallography data

Tobacco mosaic virus (TMV, Figure 6.7) is one of the simplest helical viruses known and consists of a single strand of RNA surrounded by ~2000 identical pie shaped protein subunits. TMV has been a favourite topic of research for physicists, since it is not pathogenic to man and

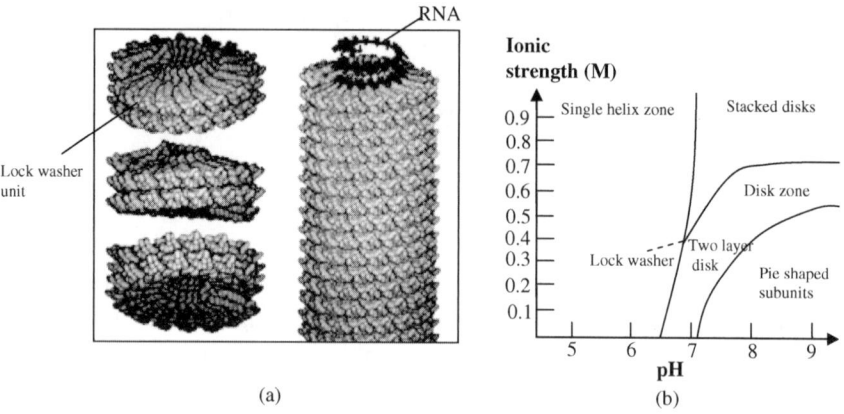

Figure 6.7 (a) Tobacco mosaic virus self-assembles from 'lock washer' units that become attached to a central RNA core. (b) The ionic strength and pH can switch the process of assembly on and off. The self-assembly of a complete TMV virus is favoured at low pH values

presents an ideal monodisperse rod-like system for studying liquid crystalline phases. Assembly of TMV in vitro can occur with and without a chain of RNA. *Without RNA* molecules, the protein monomers of TMV first form double disks of 17 monomer units. The disks contain holes at their centres. If the pH is changed appropriately, it modulates the electrostatic interactions of the disks and they slip with respect to each other and aggregate. The protein disk like sub-aggregate units have a 'lock washer' morphology and slowly stack upon each other to form rods with a high polydispersity in their length. *With RNA* molecules the nucleic acid chain directs the growth of disk aggregation, a monodisperse virus is formed, since the RNA dictates a well defined length for the helical virus.

Many other viruses consist of a nucleic acid core surrounded by a symmetrical shell that is assembled from identical protein molecules (icosohedral viruses such as hepatitis B, Figure 6.6). There are geometrical selection rules for the symmetry of the arrangement of the identical coat proteins. The process of self-assembly in these cases is often much more complicated than with TMV. The self-assembly is sometimes directed by chaperone proteins that guide the process. The complexity of the steps involved in the self-assembly and life cycle of the P4 icosohedral virus is illustrated in Figure 6.8.

The T4 bacteriophage is one of several DNA containing viruses that can infect *E. Coli* bacteria (Figure 6.9) and the process of self-assembly is again slightly more sophisticated than TMV. Separate sections of this virus have been observed to self-assemble from their constituent components. The tail tube forms spontaneously from the core proteins and purified base plates (Figure 6.10). Starting only with purified base plates and core protein monomers the tail tube self-assembles in vitro to a length of ~ 100 nm.

6.3 SELF-ASSEMBLY OF PROTEINS

Two important examples of aggregating protein self-assembly in medical conditions are the aggregation of proteins in amyloid diseases (Alzheimer's, Bovine Spongiform Encephalopathy etc, Figure 6.2) and the aggregation of haemoglobin molecules in sickle cell anaemia (Figure 6.1). There are a series of unusual features found during the aggregation of filamentous proteins in vitro which are indicative of a process of self-assembly as opposed to a conventional chemical reaction. The temperature dependence of the process of polymerisation is very different to that

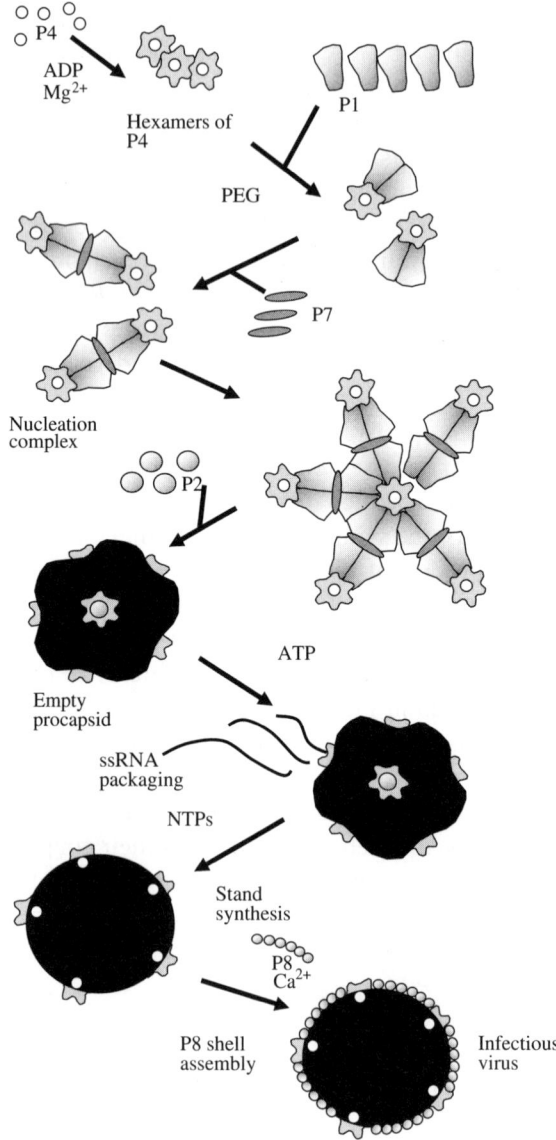

Figure 6.8 Schematic diagram of the sophisticated process of self-assembly found in the P4 virus
[*Ref.: D.H. Bamford, Phil. Trans. R. Soc. Lond. A, 2003, 361, 1187–1203*]

of small inorganic species. At low temperatures there is no polymerisation and at high temperatures polymerisation occurs; normally with synthetic polymerisation reactions the reverse is true. A rise in pressure causes depolymerisation; behaviour opposite to that normally found

SELF-ASSEMBLY OF PROTEINS

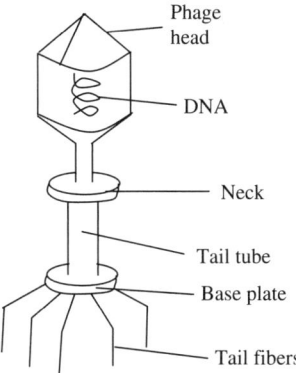

Figure 6.9 Schematic diagram of the self-assembled structure of T4 bacteriophage (The tail tube is found to spontaneously self-assemble in vitro (Figure 6.10).)

with covalent bonds, e.g. polymerisation of polyethylene. Polymerisation of self-assembling proteins only occurs above a critical initial monomer concentration, the CMC of the self-assembly process, as in the case of surfactant self-assembly. The kinetics of protein polymerisation are characterised by a long lag period followed by rapid formation of polymers. This is of particular concern in prion diseases, since the resultant self-assembled amyloid aggregates are implicated in these fatal conditions.

The change in Gibbs free energy (ΔG) during polymerisation is given by the standard thermodynamic equation:

$$\Delta G = \Delta U + P\Delta V - T\Delta S \tag{6.22}$$

where ΔU is the change in free energy, P is the pressure, ΔV is the change in volume, T is the temperature and ΔS is the change in entropy. The

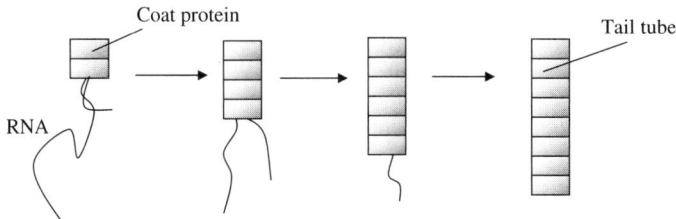

Figure 6.10 Self assembly of the coat proteins of T4 bacteriophage on a single RNA chain template

polymerisation reaction is found to proceed move favourably at high temperatures. From equation (6.22) this signifies that the entropy changes that occur upon polymerisation is positive. This is allowed thermodynamically since the entropy of the associated solvent molecules increases to compensate for the entropy of the protein aggregates which decreases as they self-assemble into ordered structures.

The mechanism of non-aggregating self-assembly that forms the internal structure of globular proteins has been covered in Chapter 3. It is intimately associated with the globule–coil, helix–coil and beta sheet–coil phase transitions. With natural proteins there is the additional complication of frustration during folding which exists due to the large number of closely spaced local minima explored by the protein on its free energy landscape (Levinthal's paradox). This frustration can lead to misfolded proteins if they are improperly chaperoned to their final active states or unfolded by chemical/physical denaturants. The misfolding of a range of proteins is thought to be the nucleation step in a number of amyloid diseases and constitutes a rate limiting step in the development of such conditions.

6.4 POLYMERISATION OF CYTOSKELETAL FILAMENTS (MOTILITY)

The polymerisation of cytoskeletal polymers is an important example of self-assembly, since it can lead to motility with actin polymerisation and has thus been intensively researched. Single stranded polymerisation is found to be an unlikely mechanism for the construction of long fibres due to the surface free energy effects described in the introduction to this chapter (Figure 6.11, 1-D aggregates are short and polydisperse). Actin circumvents this problem by using two interacting (double helical) strands. Similar schemes hold for a range of other helical cytoskeletal filaments, e.g. tubulins.

The addition of a monomer unit to an actin fibre is a process of self-assembly and the equilibrium morphology adopted is a balance between

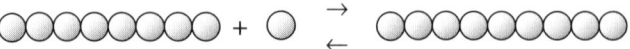

Figure 6.11 Self-assembly of filamentous proteins
(The equilibrium constant is the same for each unit added to the polymer for a single filament that consists of symmetric subunits.)

POLYMERISATION OF CYTOSKELETAL FILAMENTS (MOTILITY) 143

the rate of addition (k_{on}) and dissociation (k_{off}) of a globular protein subunit:

$$A_n + A_1 \underset{k_{off}}{\overset{k_{on}}{\leftrightarrow}} A_{n+1} \qquad (6.23)$$

where A_n is an n-mer aggregate and A_1 is a monomer. There is a dissociation constant (K) attributed to this process of monomer addition on to a n-mer aggregate:

$$\frac{c_n c_1}{c_{n+1}} = K = \frac{k_{off}}{k_{on}} \qquad n \geq 1 \qquad (6.24)$$

where c_1, c_n, and c_{n+1} are the molar concentrations of monomer, n-mer and $n + 1$-mer aggregates respectively. It is assumed that all the individual subunit addition reactions for a length of n-mer have the same dissociation constant (K), dissociation rate constant (k_{off}) and second order association constant (k_{on}). However, the dissociation constants (K) are not held fixed and change with the length of the polymeric aggregate. The equilibrium dissociation constant is given by the standard thermodynamic relationship:

$$K = \exp(\Delta G_0 / kT) \qquad (6.25)$$

The dissociation constant (K) is thus associated with the standard free energy change (ΔG_0) of the reaction. ΔG_0 is the sum of the potential energy (negative) associated with the formation of a monomer–protein bond and an entropy change (positive) associated with the loss of translational and rotational entropy as the subunits are transferred from the standard 1 M concentration in solution to a bound state in a polymeric aggregate (equation (6.22)), at constant volume and pressure, $\Delta V = 0$).

As argued qualitatively in the introduction it is found that *single stranded self-assembled filaments are polydisperse and short*. This result can be deduced more formally by calculating the dependence of the average length of the filamentous aggregates on the total concentration of subunits. The dissociation constant for the reaction is given by equation (6.24). The concentration of different lengths of polymers (c_N) are assumed to follow an exponential distribution given by:

$$c_n = K \exp\left(-\frac{n}{n_0}\right) \qquad (6.26)$$

where K is the equilibrium constant as defined by equation (6.25), n is the length of the polymer and n_0 is defined as:

$$n_0 = -\frac{1}{\ln a_1} \tag{6.27}$$

where a_1 is given by:

$$a_1 = \frac{c_1}{K} \tag{6.28}$$

The exponential distribution can be proven by substitution in equation (6.24), but to calculate the *average number of monomers* in a filament (n_{av}) requires more effort. It is:

$$n_{av} \approx \sqrt{\frac{c_t}{K}} \tag{6.29}$$

where c_t is the total concentration of monomers and K is the equilibrium constant. This result can be derived from the exponential distribution of the n-mer aggregate concentration. The contribution from the monomer lengths is discounted and the definition of an independent probabilistic average gives:

$$n_{av} = \sum_{2}^{\infty} n p_n = \sum_{2}^{\infty} n \frac{a_n}{\sum_{2}^{\infty} a_n} = 1 + \frac{1}{1-a_1} \tag{6.30}$$

where a_n are the statistical weights given by $a_n = c_n/K$, and from equation (6.26) these are given by:

$$a_n = a_1^n = e^{-n/n_0} \tag{6.31}$$

The total number of subunits (a_t) is the algebraic sum:

$$a_t = \sum_{1}^{\infty} n a_n = \frac{a_1}{(1-a_1)^2} \tag{6.32}$$

$$a_1 = 1 + \frac{1}{2a_t} - \sqrt{\frac{1}{a_t} + \frac{1}{4a_t^2}} \tag{6.33}$$

POLYMERISATION OF CYTOSKELETAL FILAMENTS (MOTILITY)

When $a_t \gg 1$, the expression for a_1 simplifies to:

$$a_1 \approx 1 - a_t^{-\frac{1}{2}}$$

From equation (6.27) and using the Taylor expansion $\ln(1+x) = x - x^2/2 + x^3/3 \ldots$ gives:

$$n_0 \approx \sqrt{a_t} \tag{6.34}$$

Finally, using equation (6.30) gives the expression:

$$n_{av} \approx 1 + \sqrt{a_t} \tag{6.35}$$

which is equivalent to equation (6.29).

A similar type of analysis shows that multi-stranded filaments tend to be very long. The problem is slightly more complicated because the geometry of the fibre imposes three separate dissociation constants K, K_1 and K_2 on the process of assembly (Figure 6.12). The average length of multi-stranded filaments is found to be given by:

$$n_{av} \approx \sqrt{\frac{K_1}{K}} \sqrt{\frac{c_t}{K_2}} \tag{6.36}$$

where c_t is the total monomer concentration. This average length is typically much bigger than found with single stranded filaments.

Filaments of actin and microtubules can polymerise and are sufficiently long for their roles in motility because they are multi-stranded. Calculations show that the lengths of the polymeric aggregates are again distributed exponentially, but the average filament length is much greater than predicted by equation (6.36). This is due to the active nature of the

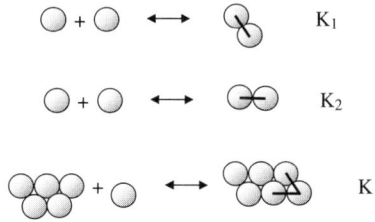

Figure 6.12 Three association constants are required to describe the self-assembly of a two stranded filament formed from symmetric subunits (K_1, K_2 and K_3)

motility process with actin, i.e. ATP (fuel) drives the reaction and the process is not in thermodynamic equilibrium.

The rate of elongation (dn/dt) for filamentous self-assembly is given by the Oosawa equation:

$$\frac{dn}{dt} = k_{on}c_1 - k_{off} \qquad (6.37)$$

The graphical solution is shown in Figure 6.13(a) and the equation is examined in detail in Section 14.1. The partitioning of the monomers between those free in solution and those in the filamentous state that defines the CMC for one dimensional self-assembly of the cytoskeletal fibres is shown in Figure 6.13(b). There is a close resemblance to the equivalent diagram for surfactant self-assembly (Figure 16.5) and this emphasises the general features involved in both processes.

Extensive theoretical development of the assembly of multi-stranded fibres shows that multi-stranded filaments grow and shrink at both their

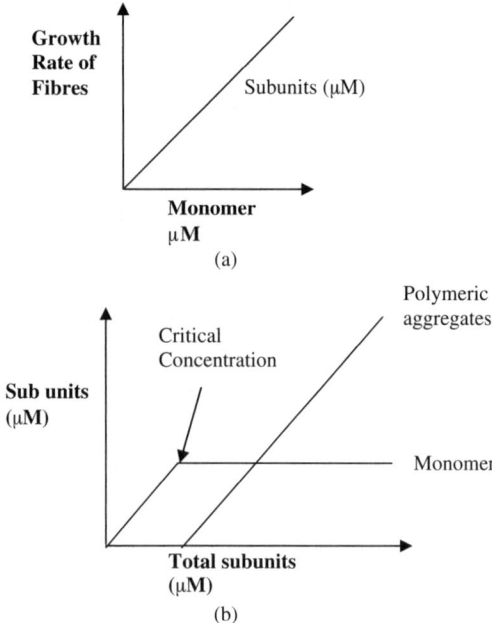

Figure 6.13 The generic self-assembly of filamentous protein aggregates ((a) shows the growth rate of aggregates as a function of the monomer concentration and (b) indicates how the concentration of subunits is partitioned between monomers free in solution and those in fibrous aggregates (note the similarity with Figure 6.5 for surfactant assembly).)

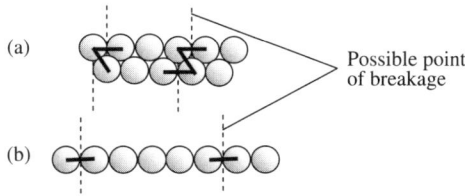

Figure 6.14 Multistanded filaments (a) are stable against breakage whereas single stranded filaments (b) continually break and reform along their length

ends, and there again exists a critical concentration for self-assembly. Multi-stranded filaments are stable to breakage whereas single stranded filaments continually break and reform (Figure 6.14). Other biological examples which follow the same trends in their self-assembly as actin (Figure 6.13) are sickle cell haemoglobin aggregates (double helical fibres) and amyloid aggregates (they often are twisted chiral multi-tape fibres).

The critical concentration (c_c) for self-assembly is given by the minimum in the rate of addition ($dn/dt = 0$) and substitution in the Oosawa equation (6.37) gives:

$$c_c = \frac{k_{\text{off}}}{k_{\text{on}}} = K \tag{6.38}$$

This equation provides a method of determining the dissociation constant experimentally from a plot of the monomer concentration against the total concentration of sub-units.

It is found that the concentration of nuclei are very small during the self-assembly of *multistranded fibres*, the ends of the filaments are blunt due to the extra stability of a snug geometrical fit in these systems, and the mean lengths of the filaments increase very steeply above the critical concentration. Thus slight changes in free monomer concentration give a large change in polymer length for multi-stranded filaments in equilibrium (Figure 6.15).

Many fibrous biopolymers are self-assembled through the intermediate step of protofilaments, in contrast to the mechanism of the direct addition of globular protein sub-units to the end of a fibrous aggregate seen with actin and microtubules. Examples of protofilament assembly are shown in Figure 6.16. Such mechanisms are found to be important in a range of proteins, including the collagens and the intermediate filaments such as the keratins and desmins. Protofilament assembly provides another mechanism for these systems to circumvent the problems with

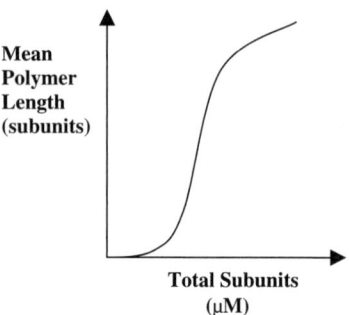

Figure 6.15 The mean length of a self-assembled two stranded filamentous aggregate chains is an S-shaped function of the number of subunits

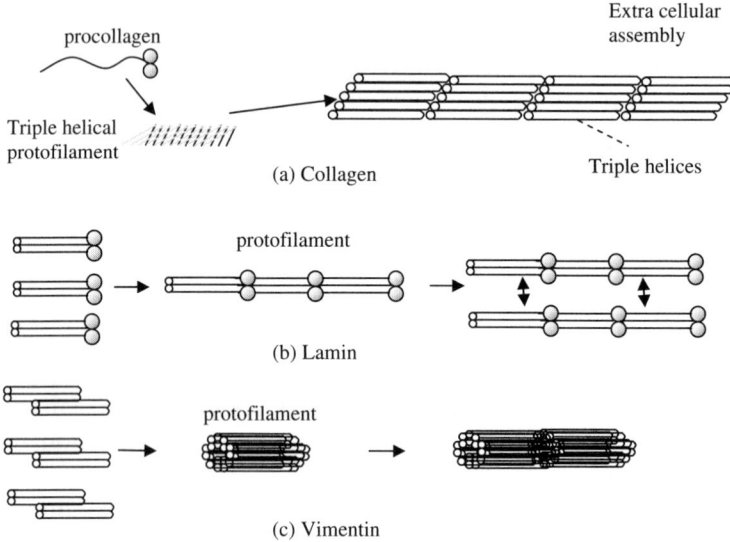

Figure 6.16 Collagen, lamin and vimentin fibres form from extended protofilament units, in contrast to actin and tubulin fibres that self-assemble directly from small spherical subunits
[Ref.: H. Herrmann, U. Aebi, Annu. Rev. Biochem., 2004, 73, 749–789]

one dimensional self-assembly, such as single filament polydispersity and low size, and enables the construction of giant fibrous networks.

FURTHER READING

J. Israelivichi, *Intermolecular and Surface Forces*, Academic Press, 1992. A classic text with a useful account of the self-assembly of lipid molecules.

J. Howard, *Mechanics of Motor Proteins and the Cytoskeleton*, Sinauer, 2001. The self-assembly of motor proteins (actin and microtubules) is well covered in this book.

TUTORIAL QUESTIONS

6.1) What is the critical micelle concentration for spherical lipid micelles if the number of lipid molecules in a micelle is 1000, the surface tension is 20 mJm^{-2} and the micellar radius is 2 nm?

6.2) What is the critical subunit concentration for self-assembly (c_c) of a linear fibrous protein aggregate if the free energy change on the addition of a subunit (ΔG) is –4.6 kT? What is the average filament length if the monomer size is 5 nm and the total monomer concentration is 1 M?

7
Surface Phenomena

The physical phenomena associated with surfaces is a large emerging area of interest in molecular biophysics. A wide range of biophysical problems are currently being examined in relation to surfaces, which could have important industrial applications. Examples include the adhesive hairs on insect legs that allow them to hang upside down (Figure 7.1), the self-cleaning ability of lotus leaves (hydrophobic surface coatings, Figure 7.2), how slugs travel over razor blades (they slide on trails of liquid crystalline polyelectrolytes), how the skin of sharks suppresses short wavelength turbulence (riblet patterned surfaces, Figure 7.3) for efficient locomotion, how molluscs attach themselves to rocks (protein cements that act in a highly hydrated environment) and how dental fillings are glued to hydroxyapatite (the physics of composite materials, their adhesion and fracture).

7.1 SURFACE TENSION

The work done in creating a new surface (w) in a material is proportional to the number of molecules transported to the surface and thus to the area of the new surface:

$$w = \gamma \Delta A \quad (7.1)$$

where γ is a constant of proportionality (the surface tension) and ΔA is the change in surface area. γ has two equivalent definitions as the free energy per unit area (units of [energy]/[length]^2) or as a surface tension (units of [force]/[length]).

Applied Biophysics: A Molecular Approach for Physical Scientists Tom A. Waigh
© 2007 John Wiley & Sons, Ltd

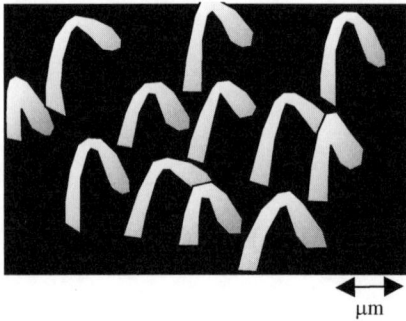

Figure 7.1 Schematic diagram of the hairy appendages found on a range of insects (The hairs encourage adhesion due to Van der Waals attractive forces)

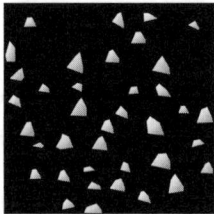

Figure 7.2 Schematic diagram of the textured hydrophobic surfaces that are found on a range of plants and animals
(The wax crystalloids on a plant leaf are shown (μm sized))

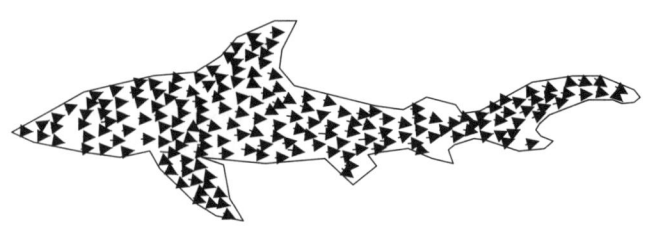

Figure 7.3 Riblets patterns (μm grooves) along the shark's skin are thought to decrease the amount of frictional dissipation due to small wavelength turbulence. The arrows indicate the direction the riblets follow on the skin of the shark

SURFACE TENSION

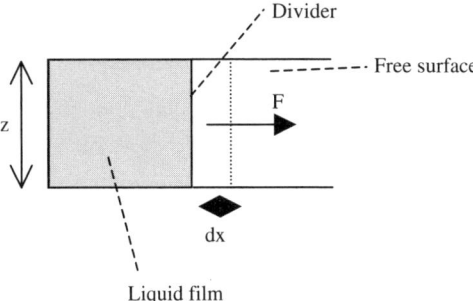

Figure 7.4 The increase in surface area and consequently the surface free energy, is directly related to the amount of force (*F*) placed on the divider (and thus the amount of work it does on the liquid surface as it moves through a distance *dx*). *z* is the length of the divider

To understand the concept of surface free energy in more detail, consider a wire loop that encloses a suspended liquid film (Figure 7.4). The force operates along the entire edge of the film and varies linearly with the length (*x*) of the slide wire. For static equilibrium from Newton's laws the forces on the slide wire must balance:

$$\gamma = \frac{F}{2x} \tag{7.2}$$

The factor of two is included because the force acts on the two sides of the film (front and back).

The work (*dw*) associated with the expansion of the interfacial energy due to a movement of the slide wire by a small distance (*dx*) is therefore given by:

$$dw = Fdx = \gamma 2x dx = \gamma dA \tag{7.3}$$

This work is equal to the change in Gibbs free energy (*dG*) at constant temperature and pressure. The change in Gibbs free energy is given by:

$$dG = \gamma dA \tag{7.4}$$

And thus at constant temperature and pressure an expression for the surface free energy is:

$$\gamma = \left(\frac{\partial G}{\partial A}\right)_{T,P} \tag{7.5}$$

7.2 ADHESION

Adhesion describes the phenomena involved in the cohesion of biological materials. It is useful to estimate the interfacial fracture energy (G) (Figure 7.5) of a simple isotropic material to motivate a quantitative analysis of adhesion. Consider an interface between two surfaces A and B with an interfacial tension, γ_{AB}. If these two surfaces (with free energies γ_A and γ_B respectively in a vacuum) are separated in a thermodynamically reversible manner, the total work of adhesion (W_{AB}) is the difference in surface energies before and after they are separated:

$$W_{AB} = \gamma_A + \gamma_B - \gamma_{AB} \tag{7.6}$$

However, experimentally it is found that fracture energies can be orders of magnitude greater than the value predicted by equation (7.6). The reason for this shortfall is that materials possess the ability to dissipate energy by a number of mechanisms not accounted for in the equation, e.g. plastic deformation at the surface. Indeed, this is the function of the soft filler in many biological composite structures: to maximise the

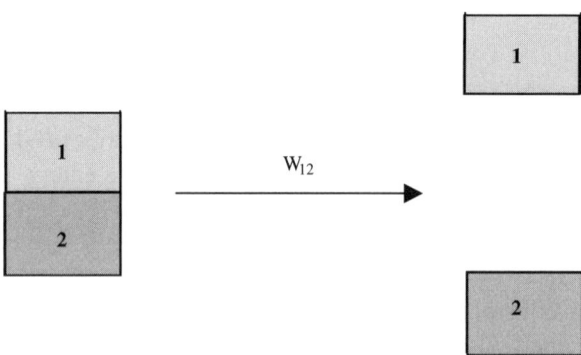

Figure 7.5 The interfacial fracture energy during the separation of two phases (1 and 2) defines the total work of adhesion between the two phases

ADHESION

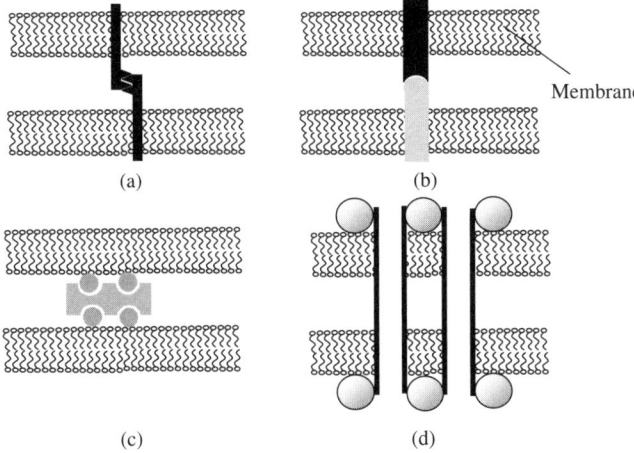

Figure 7.6 A range of organised adhesive links can occur between cellular structures (Non-junctional adhesions are possible such as (a) homophilic, (b) heterophilic and (c) extracellular protein linked to four ligands. Junctional adhesions are also important, such as the (d) gap junction [*Reprinted with permission from G. Forgacs and S.A. Newman, Biological Physics of the Developing Embryo, Copyright (2005) Cambridge University Press*])

dissipation of energy during impacts through a mechanism of plastic deformation (Section 11.2).

In molecular biophysics adhesion is often carefully controlled by biomolecular structures. For example, organised adhesive links form as cells mature and arrange themselves into tissues, a key mechanism during the process of morphogenesis. The adhesive links can act as both junctions and switch yards for cytoskeletal components. To bind membranes, there are homophilic, heterophilic and extracellular proteins that are used to create linkages (Figure 7.6). The proteins used in cell adhesion include the cadherins, integrins and selectins, and their binding energies are typically in the range 5–30 kT. Experiments to study the interaction of cellular adhesive proteins have been made using atomic force microscopy (AFM), optical tweezers and surface force apparatus. Many experiments also consider the adhesion of pure bilayers by measuring the contact angle when two bilayers are pressed together. Micromanipulation techniques have proven particularly successful in this area. An experiment based on the compression of two spherical bilayers is illustrated in Figure 7.7. The contact angle (ϕ) can be related to the force of adhesion between the cells.

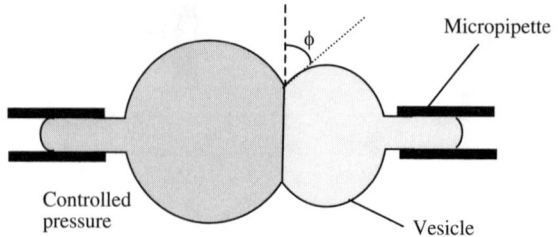

Figure 7.7 Adhesive forces between two vesicles can be measured using micropipettes. The contact angle (ϕ) is related to the pressure exerted on the vesicles

7.3 WETTING

Consider a liquid droplet that is sat (sessile) on a solid surface. Young's equation for the contact angle (θ) on the solid surface (Figure 7.8) is given by:

$$\gamma_{sg} = \gamma_{sl} + \gamma_{lg} \cos \theta \tag{7.7}$$

where γ_{sg}, γ_{sl}, and γ_{lg} are the surface tensions of the solid–gas, solid–liquid and liquid–gas interfaces respectively. Young's equation is deduced from the force balance at the three phase contact line (Newton's law), when the components of the forces are taken in the direction parallel to the surface.

Similarly it is possible to perform a force balance to deduce the structure of liquid–liquid–gas interfaces (Figure 7.9), e.g. water, oil and air. However, all of the components can be deformed in this case and three angles must be introduced to calculate the static conformation:

$$\gamma_{wa} \cos \theta_1 = \gamma_{oa} \cos \theta_2 + \gamma_{wo} \cos \theta_3 \tag{7.8}$$

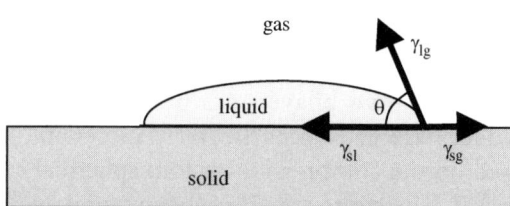

Figure 7.8 A liquid drop sat on a solid surface. In equilibrium the components of the surface tension parallel to the surface of the solid balance to give Young's equation

WETTING

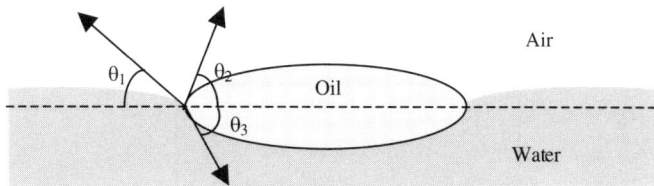

Figure 7.9 An oil (liquid) drop placed on a liquid–air interface requires three contact angles (θ_1, θ_2 and θ_3) to describe the position of the triple line of contact

where the subscripts *w*, *a* and *o* refer to water, air and oil interfaces respectively.

Using Young's equations the subject of *wetting* can be approached. At equilibrium in a phase separated system of a surface immersed in a solution of biomolecules (say oil and water), the surface is expected to be coated by a macroscopic thick layer of the phase of lower surface energy which is called a 'wetting layer', e.g. the oil completely covers the solid surface. The layer is macroscopically thick and the interface between the surface layer and the bulk fluid is identical to that between the two bulk coexisting phases. It is therefore possible to write down an inequality for the surface tension of the two phases and the interfacial tension between the phases. This inequality can be used to understand the spreading of molecules on a surface in terms of a phase transition (a 'wetting' transition). The *wetting transition* has a range of practical applications, including an understanding of biofouling on the membranes used in kidney dialysis machines and how agrochemicals adhere to waxy plant leaves.

To have *perfect wetting* the Young equation (7.7) must not have a solution that corresponds to a finite contact angle (Figure 7.8). Without a finite contact angle, a liquid will completely coat a solid surface with a macroscopically thick layer. When a liquid (l) only *partially wets* a surface (s) a finite contact angle occurs. Writing the surface energies of the two phases as in equation (7.7), the following equation must hold for stable equilibrium of the forces at a point of three phase coexistence (otherwise the triple point will move):

$$\gamma_{sg} > \gamma_{lg} + \gamma_{sl} \qquad (7.9)$$

Young's equation can be rewritten in the form:

$$\gamma_{lg} \cos(\theta_c) + \gamma_{sg} = \gamma_{sl} \qquad (7.10)$$

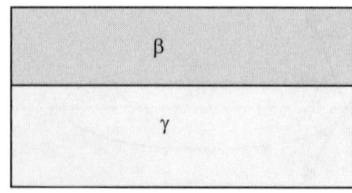

Figure 7.10 Complete wetting of the liquid (β) on the solid surface (γ) is observed with a uniform macroscopically thick film of β

This can be rearranged to form the wetting coefficient (k):

$$k = \frac{\gamma_{sg} - \gamma_{sl}}{\gamma_{lg}} = \cos\theta \qquad (7.11)$$

When the wetting coefficient equals one, the contact angle of the drop is zero and the solid is *completely wetted* as in Figure 7.10. In the intermediate regime the wetting coefficient is between zero and one ($0 < k < 1$) and the contact angle (θ) lies between 0 and $\pi/2$; the solid is *partially wetted* by the liquid (Figure 7.11). Again, when the wetting coefficient is minus one, the contact angle is π and the solid is completely wetted. In the range of wetting coefficients between minus one and zero ($-1 < k < 0$) the solid is *unwetted* by the liquid (experimentally the liquid would form a tight ball with a high contact angle and roll off the surface).

Once the stability of the equilibrium situation has been examined, it is possible to understand the dynamic development of surface morphologies. No equilibrium position of the contact line exists when the spreading coefficient (s) is greater than zero from inequality (7.9):

$$s = \gamma_{sg} - \gamma_{sl} - \gamma_{lg} > 0 \qquad (7.12)$$

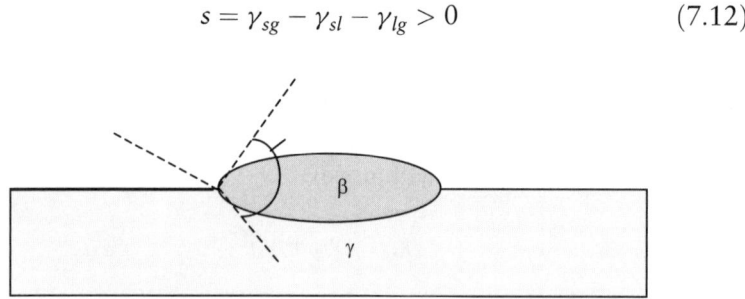

Figure 7.11 Partial wetting of the liquid (β) on the solid surface (γ) causes the solid surface (γ) to be inhomogeneously wet

WETTING

Thus the angle of contact will change dynamically when the spreading coefficient is greater than zero, a measure of the driving force for this non-equilibrium phenomenon. The spreading coefficient (s) is related to the wetting coefficient (k):

$$k = 1 + \frac{s}{\gamma_{lv}} = \cos\theta \qquad (7.13)$$

If the spreading coefficient is greater than zero ($s > 0$) then the wetting coefficient is greater than one ($k > +1$). The larger the positive value of the spreading coefficient (s) the better the spreading over the solid surface. The transformation from a partially to a totally wet condition has the form of a phase transition (Section 3.1). This is normally a first order process, but can become continuous with the correct choice of interaction potential and state variable.

Consider the wetting of a protein on a water/air interface. Initially the spreading coefficient is found to be positive ($s_{initial} = 13\,\text{mJm}^{-2}$) and at equilibrium it is negative ($s_{final} = -2\,\text{mJm}^{-2}$). This protein is therefore strongly surface active and leads to an important reduction of the surface tension. Trace impurities on a surface reduce its free energy and can cause a dramatic change in the spreading behaviour, e.g. surface active proteins will spread on a clean water surface but not on a contaminated one. The process of wetting can also be considered in terms of the form of the energy of the protein/surface interaction (P) on the thickness of the film (t) (Figure 7.12). A positive curvature for the functional form of the energy on the film's thickness ($P(t)$) gives unstable films, whereas a negative curvature gives a stable film.

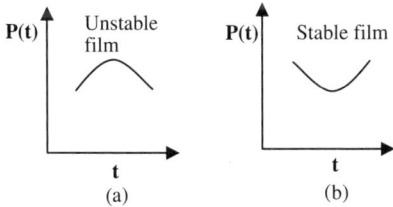

Figure 7.12 The stability of surface absorbed films depends on the form of the surface potential $P(t)$ on the thickness of the film (t). (a) A positive curvature of $P(t)$ leads to an unstable film, whereas (b) a negative curvature provides film stability

7.4 CAPILLARITY

The *Young–Laplace* equation can be used to relate the pressure difference across a surface to its curvature. For an isolated particle the surface tension is balanced by the stresses within the particle.

Consider a *spherical gas bubble* suspended in a liquid as shown in Figure 7.13. The bubble is expanded infinitesimally by a radial distance dR and the change in surface area (dA) is given by:

$$dA = 4\pi\{(R + dR)^2 - R^2\} = 8\pi R dR \quad (7.14)$$

The corresponding change in volume (dV) is:

$$dV = \frac{4\pi}{3}\{(R + dR)^3 - R^3\} = 4\pi R^2 dR \quad (7.15)$$

The free energy change by the pressure/volume variation is $dw = \Delta p dV$, simply the force multiplied by the distance moved. The corresponding free energy (dG) due to the surface tension from equation (7.4) is:

$$dG = \gamma 8\pi R dR \quad (7.16)$$

where the geometric relationship for the change in surface area has been substituted. The work done to increase the surface area is balanced by the pressure volume work to give:

$$\gamma 8\pi R dR = \Delta p 4\pi R^2 dR \quad (7.17)$$

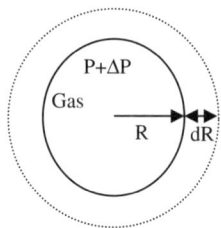

Figure 7.13 The expansion of a spherical gas bubble in a liquid is controlled by its surface tension. A change in pressure (ΔP) causes the bubble to increase its radius (dR)

CAPILLARITY

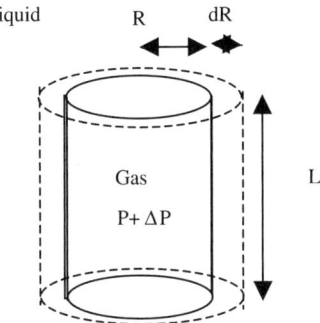

Figure 7.14 The expansion of a cylindrical gas bubble is controlled by its surface tension. A change in pressure (ΔP) causes the radius of the bubble to increase (dR)

Therefore the pressure difference across the bubble is given by:

$$\Delta p = \gamma \frac{2}{R} \tag{7.18}$$

For *cylindrical bubbles* the physical behaviour is very similar. The bubble has the morphology of a cylinder of length L (Figure 7.14). The area change (dA) for a small variation in the radius is given by:

$$dA = 2\pi L(R + dr - R) = 2\pi L dR \tag{7.19}$$

And the volume change (dV) is:

$$dV = \pi L\{(R + dR)^2 - R^2\} = 2\pi L R dR \tag{7.20}$$

A balance between the surface and pressure volume work now gives the pressure drop to be:

$$\Delta p = \frac{\gamma}{R} \tag{7.21}$$

This expression is almost identical to that of a sphere (equation (7.18)) with the inclusion of an extra factor of two.

For interfaces of *arbitrary shape* a general expression is found that relates the pressure drop across the surface to its curvature:

$$\Delta p = \gamma \left(\frac{1}{R_1} + \frac{1}{R_2} \right) = \gamma 2H \qquad (7.22)$$

where R_1 and R_2 are the radii of curvature and H is the mean curvature given by:

$$H = \frac{1}{2} \left(\frac{1}{R_1} + \frac{1}{R_2} \right) \qquad (7.23)$$

Equations (7.21) and (7.18) for spherical and cylindrical bubbles are both then special cases of equation (7.22).

For simple liquids such as cyclohexane the macroscopic approach to surface tension has been shown to be accurate down to distances of seven times the molecular diameter with surface force apparatus experiments. This is exceedingly good agreement for such a simple continuum theory. The good agreement with experiment lends confidence for the utility of equation (7.22) to gauge the strength of capillary forces on the nanoscale.

Surface tension can govern the rise of a liquid in a capillary tube. The capillary forces that drive this motion are important in a number of biological processes, such as the rise of sap in plants, mucin in the trachea of the lungs and urine in the kidneys. However, although it is a classic example, it needs to be stressed that the main driving force for the motion of sap in plants is a reduction in pressure due to the evaporation through the leaves, i.e. a liquid–gas phase transition. Capillary forces alone are unable to drive a fluid up the hundred metres of trunk in a giant redwood tree.

A simple relation exists between the surface tension (γ), the height of capillary rise (h), the capillary radius (r) and the contact angle (θ) on the surface of a capillary. This expression can be used to measure the surface tension of simple liquids (Figure 7.15). To perform these experiments all that is required is a capillary tube with a well defined clean surface. From the capillary geometry the radius of curvature of the meniscus (R) is related to the contact angle and the radius of the capillary tube:

$$R \cos \theta = r \qquad (7.24)$$

CAPILLARITY

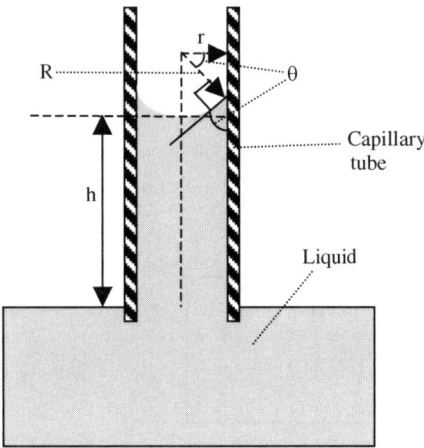

Figure 7.15 The geometry used to describe the phenomena of capillary rise in cylindrical tubes
(h is the height of capillary rise, R is the radius of curvature of the meniscus, r is the radius of the tube and θ is the contact angle of the liquid on the wall of the capillary tube.)

There is a Laplace pressure (Δp) across the interface of the cylindrical tube due to the curvature of the liquid's surface and it is given by equation (7.18):

$$\Delta p = -\frac{2\gamma}{R} = -\frac{2\gamma \cos\theta}{r} \qquad (7.25)$$

In equilibrium the Laplace pressure is balanced by the hydrostatic pressure due to the height of the fluid (h) above the height of the reservoir. This hydrostatic pressure is linearly related to the height of the fluid:

$$\Delta p = \rho g h \qquad (7.26)$$

where ρ is the density of the fluid and g is the acceleration due to gravity. Equations (7.25) and (7.26) can be combined to provide an expression for the surface tension:

$$\gamma = \frac{\rho g h r}{2 \cos\theta} \qquad (7.27)$$

The equation is particularly simple to apply when the fluid wets the surface of the capillary, since $\cos\theta = 1$.

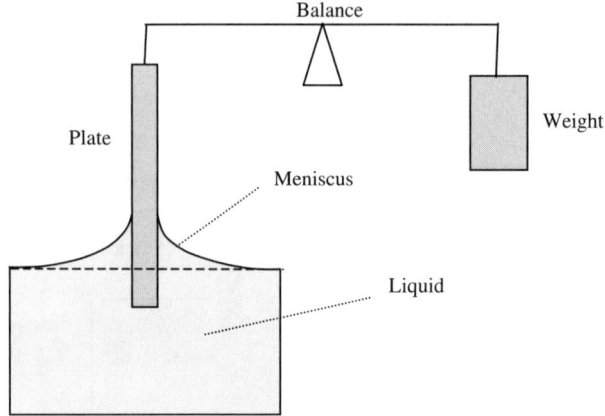

Figure 7.16 Schematic diagram of a Wilhemy plate used to measure the surface tension of a liquid. The weight required to balance the surface tension is measured

7.5 EXPERIMENTAL TECHNIQUES

The *Wilhemy plate method* is a simple robust technique for the measurement of the surface tension of liquid films (Figure 7.16). The apparatus uses the balance of the torques on a pivoted beam to measure the force due to a liquid surface. The apparatus is balanced before it is brought into contact with the liquid and the increase in weight is measured as a result of the entrained meniscus. The weight (w) is therefore given by:

$$w = \gamma P \cos\theta \tag{7.28}$$

where P is the perimeter of the plate and γ is the surface free energy. A range of methods that can be used for the measurement of surface tensions is listed in Table 7.1.

Table 7.1 Methods for measuring the surface tension of liquids

Method	Principle
Capillary height	Capillary rise
Wilhemy plate	Capillary force on a plate
Drop profile	Digital analysis of drop geometry
DuNouy ring	Capillary forces on a wire ring
Spinning drop tensiometer	Digital analysis of drop geometry
Oscillatory bubble	Capillary forces on an air bubble

7.6 FRICTION

The coefficient of friction (μ) between two solids is defined as the ratio of the frictional force (F) to the normal force to the surface (W, $\mu = F/W$). *Amontons' law* states that the coefficient of friction is independent of the apparent area of contact (a surprising counterintuitive result). Amontons' law is an empirical result that has been shown to hold for a wide variety of materials. A corollary is that the coefficient of friction is independent of load. Furthermore, if two objects of equal weight ($W_1 = W_2$) are made of the same material then their frictional forces are equal ($F_1 = F_2$) (Figure 7.17). To illustrate the bizarre nature of this law consider a rectangular block of wood that is initially balanced on an end with a small surface area (Figure 7.18). The block is then toppled onto its side, which results in a large increase in the area of contact. However, this *does not change* the horizontal frictional force that opposes motion, since the normal reaction to the block's weight is unchanged.

A qualitative microscopic explanation can be made for Amonton's law. As the two surfaces are brought together the pressure is initially

Figure 7.17 The frictional forces on two objects made from the same material, placed on the same surface, are equal if they have the same weights, i.e. if $W_1 = W_2$ then $F_1 = F_2$

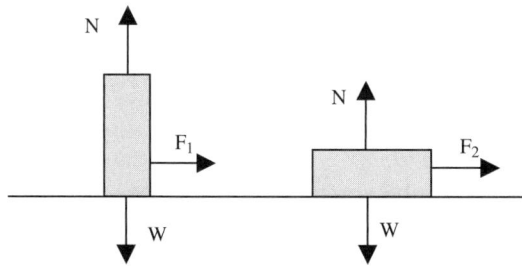

Figure 7.18 Topple a solid block on its side and Amonton's law gives the same frictional force opposing the motion, $F_1 = F_2$

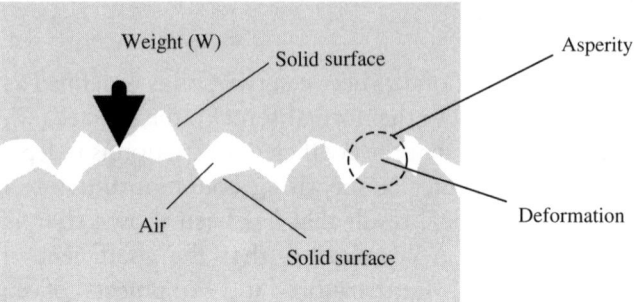

Figure 7.19 Schematic diagram of the microscopic contacts (asperities) between two solid surfaces that give rise to the frictional force that opposes lateral motion

extremely large at the first few points of contact that develop, and plastic deformation immediately occurs to allow more and more contacts to form (Figure 7.19). This plastic flow continues until the total area of contact is such that the local pressure has fallen to a characteristic yield pressure (P_m) of the softer material. The actual contact area (A) is determined by the characteristic yield pressure of the material:

$$A = \frac{W}{P_m} \quad (7.29)$$

where W is the weight of the upper solid object. By definition the force to shear the junctions at the point of contact is:

$$F = A s_m \quad (7.30)$$

where s_m is the shear strength per unit area. As an approximation the contact area (A) can be eliminated, giving the result:

$$F = W \frac{s_m}{P_m} \quad (7.31)$$

$$\text{Or} \quad \mu = \frac{s_m}{P_m} = const \quad (7.32)$$

Amonton's law is a useful starting point for the study of friction. However, many non-Amonton's law materials have evolved in nature to provide carefully optimised frictional properties.

Dynamic frictional effects are also an important but extremely complicated area. From detailed studies of lubrication in automotive

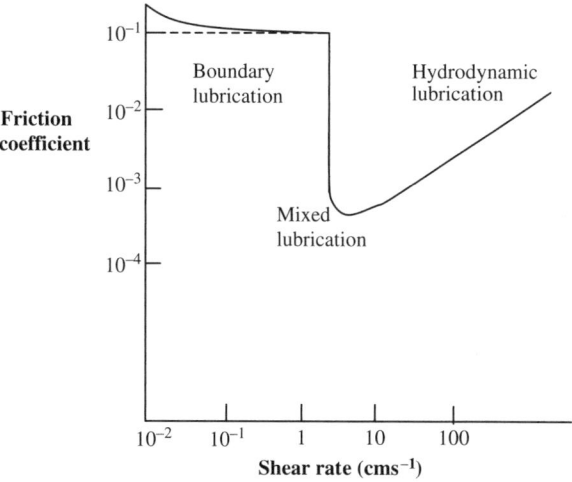

Figure 7.20 Schematic diagram of a Stribeck curve that illustrates the regions of hydrodynamic, mixed and boundary lubrication for liquid films held between two solid surfaces. The frictional coefficient is shown as a function of the relative shear rate of the two identical solid surfaces

applications (a billion pound industry that has invested considerably into research in the field) the form of the frictional force as a function of shear rate is well known (Figure 7.20). These Stribeck curves have two important regimes: *hydrodynamic lubrication* at high shear rates, in which the details of the lubricant are not critically important, and *boundary lubrication* at low shear rates, which is very sensitive to the molecular properties of the lubricant. Engine oils are specifically designed to modify the region of low shear rate lubrication and it is believed that proteoglycans and glycoproteins have evolved naturally to reduce frictional effects in this regime in a wide variety of biological processes, e.g. ocular mucins in the eye.

An important example of lubrication is the ultra low friction exhibited by cartilage on cartilage contacts in synovial joints. Cartilage has an extremely low friction coefficient, lower than Teflon on Teflon, and there are large deviations from Amonton's law for cartilage surfaces (Section 15.1). Anisotropic frictional behaviour has evolved in the textured skin of snakes. This provides low frictional resistance to forward motion, but allows the skin of the snake to grip the surface upon muscular retraction to produce the explosive reactive force needed to strike the snake's prey. The frictional losses of fibres (actin and myosin) in striated muscle are minimised as they slide past one another (Section 14.2). Non-Amonton's law friction occurs again in this system and frictional losses are reduced by a combination of the electrostatic stabilisation of the fibrillar array and hydrodynamic lubrication of fluids held within the sacromere.

7.7 OTHER SURFACE PHENOMENA

When two surfaces approach one another surface energy effects can cause a liquid to condense on a surface prior to its saturation in the bulk phase. This effect can cause a large change in the adhesive energies and is termed *capillary condensation*.

Ostwald ripening is the process by which surface free energies govern the growth of colloidal particles. Small crystals are subsumed by larger ones during their growth in order to globally minimise the surface free energy. Such behaviour is important with the production of ice cream (large crystals are considered unpalatable), and in the creation of large defect free protein crystals for crystallographic structural determination (Section 3.5).

During the *nucleation* of a new phase of material the surface free energies oppose the nucleation of a new droplet or microcrystal. Thus the surface forces are the critical factor determining the dynamics of phase separation (Section 3.5).

Gradients in surface tension can drive diffusion in mixed liquids and this process is called the *Marangoni effect*. The effect can be easily observed in the production of tear drops of brandy if a thin meniscus of this alcohol/water mixture is deposited on the side of a glass and the alcohol begins to evaporate.

There are a range of surface effects relating to *liquid crystallinity*. Phase transitions can be induced by the interaction of liquid crystalline order parameters with a surface, e.g. smectic terraces often form at solid/nematic interfaces from a bulk nematic phase.

FURTHER READING

M. Scherge, S. Gorb, *Biological Micro and Nano Tribology*, Springer, 2001. A fascinating discussion of the biological implications of surface forces.

J. Israelachvili, *Intermolecular and Surface Forces*, Academic Press, 1994. A classic readable text.

P.G. De Gennes, D. Quere and F. Brochard-Wyart, *Capillarity and Wetting Phenomena*, Springer, 2004. Clear insightful theoretical account of physical phenomena at surfaces.

A.W. Adamson and A.P. Gast *Physical Chemistry of Surfaces*, John Wiley & Sons Ltd, 1997. Extensive coverage of the physical processes at surfaces.

TUTORIAL QUESTION

7.1) A water droplet sits on a lotus leaf. If the surface free energy of the leaf/air, leaf/water and air/water interfaces are $18\,\text{mJm}^{-2}$, $73.2\,\text{mJm}^{-2}$ and $72\,\text{mJm}^{-2}$ respectively what is the equilibrium contact angle of the drop? What is the wetting coefficient for the system? Is the surface unwetted, partially wetted or completely wetted by the droplet?

8
Biomacromolecules

Macromolecules (polymers) are long chain molecules built of repeating sub-units (monomers). All proteins, nucleic acids and many carbohydrates are polymeric, and polymers are therefore given a thorough discussion in this chapter. Synthetic polymers have been extensively investigated for their diverse range of industrial applications over many years. This provides the field of biomacromolecules with a rich variety of quantitative models that can be readily transported from their original synthetic origins into the analysis of biological problems.

8.1 FLEXIBILITY OF MACROMOLECULES

The *persistence length* of a polymer is a quantity commonly used to measure its flexibility. There are three standard classifications for the *persistence length* of isolated polymer chains: *flexible*, *semi-flexible*, and *rod-like*. Each of these different classes of polymers requires a separate theoretical model to develop an understanding of their structure and dynamics in solution. *Rigid* polymeric rods form liquid–crystalline phases and have no internal dynamic modes. At the opposite extreme *flexible* polymers are adequately described by blob models, and the chain conformations are dominated by thermal fluctuations inside the blobs at small length scale and the solvent quality (chain–solvent interaction) at large length scales. The internal dynamic modes of flexible polymers are well described in terms of the Zimm model. The intermediate *semi-flexible* class of polymers has been subject to a series of recent developments. The role of both transverse and longitudinal fluctuations of the filaments is highlighted in these developments, since these fluctuations determine the unusual structures and hydrodynamic modes of the chains. It is possible for a long polymeric chain to exhibit dynamics from all three regimes as a function of increasing length scale, e.g. *rod-like* ($\sim$ Å), *semi-flexible* (nms) and *flexible* (100 nms).

The *persistence length* of a macromolecule can be measured using a host of techniques, such as dynamic light scattering, electron microscopy, optical microscopy (optical tweezers), small-angle X-ray scattering, static light scattering, small-angle neutron scattering, atomic force microscopy and fluorescence microscopy. The effect of the persistence length on the conformation of giant protein molecules (titin) in small-angle neutron scattering experiments is shown in Figure 8.1. The cross-over of the dependence of the scattered intensity on the momentum transfer (q) from q^{-1} to q^{-2} (Section 13.1) corresponds to the change in chain conformation of rigid rod over small lengths scales (< 10 nm) to flexible Gaussian conformations over larger length scales (> 10 nm).

The geometrical construction of the persistence length of a macromolecule is shown in Figure 8.2. The persistence length (l_p) is the decay length of the cosines between the tangent vectors ($t(s)$) along the monomers of the chain. Mathematically it is found that the correlation decays exponentially:

$$\langle t(0).t(s) \rangle \sim e^{-s/l_p} \tag{8.1}$$

where s is the distance along the contour of the polymer.

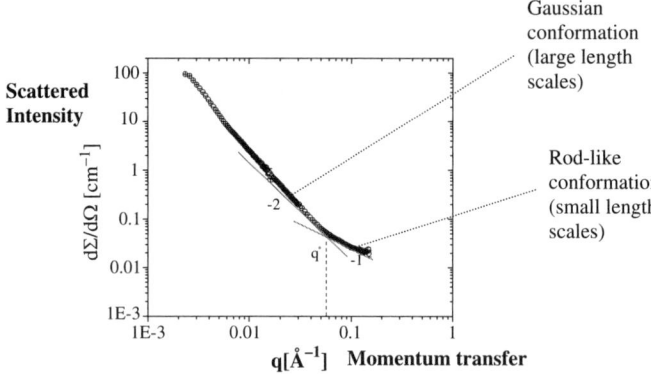

Figure 8.1 Small angle neutron scattering experiment to study the elasticity of titin (The scattered intensity is shown as a function of the momentum transfer (q). $q*$ corresponds to the persistence length of the chains [*Reprinted with permission from E. Di Cola, T.A. Waigh, J. Trinick et al., Biophysical Journal, 88, 4095–4106, Copyright (2005) Biophysical Society*])

FLEXIBILITY OF MACROMOLECULES

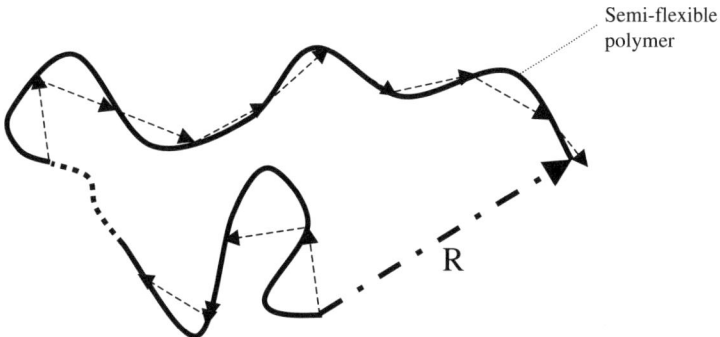

Figure 8.2 Schematic diagram of the conformation of a semi-flexible polymeric chain (R is the end-to-end distance. The dashed arrows indicate the direction cosines along the chain.)

For a rigid rod the persistence length is infinite ($l_p \to \infty$), a flexible chain has a small persistence length ($l_p \to 0$), and with a real semi-flexible chain the persistence length takes intermediate values between the two extremes. The global conformations of polymer chains can be classified by comparison of the persistence length (l_p) with the contour length (L). Thus:

$$l_p \ll L \quad \text{Flexible chain}$$
$$l_p \sim L \quad \text{Semi-flexible chain}.$$
$$l_p \gg L \quad \text{Rigid chain}$$

It is useful to connect the persistence length with the global size of the macromolecule. The average end-to-end distance ($\langle R \rangle$) of a semi-flexible chain is the sum of the cosine components along the chain:

$$\langle R \rangle = a \sum_{k=0}^{N} x^k = \frac{a(1 - x^N)}{1 - x} \qquad (8.2)$$

$$x = \cos \theta \qquad (8.3)$$

where a is the monomer length and N segments make an angle θ with adjacent segments. The persistence length (l_p) can be defined as the limiting value of $\langle R \rangle$ as the chain becomes infinitely long ($N \to \infty$):

$$l_p = \frac{a}{1 - x} \qquad (8.4)$$

Since θ is small for semi-flexible rods, the small angle expansion can be used for the cosine, only the first two terms are kept:

$$\cos\theta \approx 1 - \frac{\theta^2}{2} \tag{8.5}$$

Substitution of equation (8.5) in (8.4) gives:

$$l_p = \frac{2a}{\theta^2} \tag{8.6}$$

It is possible to write x^N in the form of an exponential if the following mathematical identity is noted:

$$x^N = \left(1 - \frac{\theta^2}{2}\right)^N \approx e^{-N\theta^2/2} \tag{8.7}$$

The contour length is equal to the number of segments (N) multiplied by the segment length (a):

$$L = Na \tag{8.8}$$

Using equations (8.6) and (8.8) allows equation (8.7) to be reexpressed as:

$$x^N = e^{-L/l_p} \tag{8.8}$$

From equations (8.2) and (8.3) it follows that:

$$\langle R \rangle = l_p(1 - e^{-L/l_p}) \tag{8.9}$$

As the length of the chain becomes very large ($L \to \infty$) the persistence length becomes equal to the end-to-end distance ($\langle R \rangle \to l_P$). Furthermore, if the persistence length is very large ($L \ll L_P$) the end-to-end distance becomes equal to the contour length ($\langle R \rangle \cong L$ for a rigid rod).

The radius of gyration is another useful quantity for sizing chains, since the average chain size is zero ($\langle R \rangle = 0$) for completely flexible chains. The radius of gyration is the expected value of R^2 ($\langle R^2 \rangle$), and can be calculated using integral calculus:

$$d\langle R^2 \rangle = d\langle R.R \rangle = 2\langle R.dR \rangle = 2\langle R \rangle dL \tag{8.10}$$

FLEXIBILITY OF MACROMOLECULES

Therefore, integration of all the infinitesimal components of $\langle R^2 \rangle$ combined with equation (8.9) gives:

$$\langle R^2 \rangle = 2l_p \int_0^L [1 - \exp(-L/l_p)]dL \qquad (8.11)$$

The integral can be evaluated and gives the expression:

$$\langle R^2 \rangle = 2l_p(L - l_p(1 - e^{-L/l_p})) \qquad (8.12)$$

In the limit of very long chains the radius of gyration of a semi-flexible chain is given by:

$$\langle R^2 \rangle = 2Ll_p \qquad (8.13)$$

where $2l_p$ is the length of the rigid subunits in a equivalent flexible chain. Equation (8.13) provides a useful method for estimating the size of biopolymers in solution. For example, with a B-type variety of DNA the persistence length (l_p) is 450 Å, and the contour length is 800 Å. The radius of gyration of the chain can therefore be calculated to be 600 Å. The radius of gyration is fairly compact and highlights the significant impact of the flexibility on the conformation of the DNA chain.

Using the *Kratky–Porod model* it is possible to connect the persistence length of a polymeric material with its bending rigidity (an intrinsic property of the material). The elementary change in free energy (dG) of a semi-flexible chain for small values of the curvature (ds) is:

$$\frac{dG}{ds} = \frac{dG}{d\theta}\frac{d\theta}{ds} + \frac{1}{2}\left(\frac{d^2G}{d\theta^2}\right)\left(\frac{d\theta}{ds}\right)^2 \qquad (8.14)$$

In the absence of a permanent bending moment for the chain:

$$\frac{dG}{d\theta} = 0 \qquad (8.15)$$

And therefore:

$$\frac{dG}{ds} = \frac{\kappa}{2}\left(\frac{d\theta}{ds}\right)^2 \qquad (8.16)$$

where the bending rigidity (κ) is defined as:

$$\kappa = \frac{d^2 G}{d\theta^2} \tag{8.17}$$

The total energy (ΔG) to bend a finite chain is therefore the integral of equation (8.16) over the length of the chain (L):

$$\Delta G = \frac{\kappa}{2} \int_0^L \left(\frac{d\theta}{ds}\right)^2 ds \tag{8.18}$$

For small displacements of the chain the angle of deviation is proportional to the size of a step along the contour length ($\theta = ks$ when k is a constant) and substitution in equation (8.18) gives:

$$\Delta G = \kappa k^2 \int_0^L \frac{ds}{2} \tag{8.19}$$

$$k = \frac{d\theta}{ds} \tag{8.20}$$

If θ_L is the total angle between the ends of the chains it can be found by integraton of equation (8.20):

$$\theta_L = \int_0^L k \, ds = kL \tag{8.21}$$

where L is the total contour length. And therefore the bending energy of the chain is given by:

$$\Delta G = \frac{\kappa \theta_L^2}{2L} \tag{8.22}$$

The bending coefficient (κ) is equal to twice the energy required to bend a unit length of polymer chain through one radian. The average mean square value of the angle of deviation is given by a Boltzmann average if the system is assumed to be in thermodynamic equilibrium:

$$\langle \theta_L^2 \rangle = \frac{\int_0^\pi e^{-\Delta G/kT} \theta_L^2 \, d\theta_L}{\int_0^\pi e^{-\Delta G/kT} \, d\theta_L} = \frac{LkT}{\kappa} \tag{8.23}$$

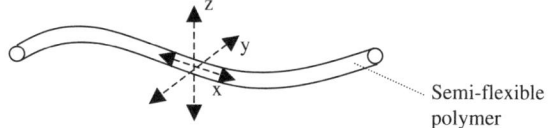

Figure 8.3 Two modes of transverse fluctuations (z and y) are possible with semi-flexible polymers

There are two transverse modes for bending the semi-flexible chain (Figure 8.3), so an additional factor of 2 is included in equation (8.23) to account for them both:

$$\langle \theta_L^2 \rangle = \frac{2LkT}{\kappa} \qquad (8.24)$$

It is then possible by comparison with equation (8.6) to relate the persistence length to the bending rigidity and this gives a final expression:

$$l_P = \frac{\kappa}{kT} \qquad (8.25)$$

This is a successful theory for the conformation of semi-flexible chains that only requires a single elasticity constant (κ) and remains a good first approximation for the conformational statistics of a wide variety of biopolymers. The theory predicts the size of the chains and the effect of temperature on the persistence length and modulus. More modern theories often include chirality (spontaneous torsional twisting of the chains) and even twist-bend coupling, but they continue to be based on the Kratky–Porod model.

8.2 GOOD/BAD SOLVENTS AND THE SIZE OF POLYMERS

Excluded volume is an important parameter which determines the configuration of flexible polymer chains. Steric interactions cause an expansion of a chain in solution when compared with a phantom equivalent (a random walk of monomers with no excluded volume). The quality of a solvent for a polymer also affects its degree of expansion. There are typically three regimes of solvent quality that are defined: *good* (the conformation of a polymer chain expands as it tries to increase the number of contacts with the solvent), *bad* (the chain forms a compact globular conformation as it decreases the number of contacts with the solvent) and *theta* (the excluded volume is balanced by the intermonomer attractive potential) conditions. Varying the solvent quality from good to bad can induce phase transitions as described in Chapter 3, such as the globule–coil transition and liquid–liquid phase separation. In most cases chirality has a secondary impact on the size of flexible polymer chains.

The *end-to-end distance* ($\langle R^2 \rangle^{\frac{1}{2}} = R_{ee}$) for flexible polymer chains is a useful measure of the expansion of a chain. With ideal flexible chains, that neglect excluded volume interactions, accurate models for chain statistics have been known to synthetic polymer physicists for over fifty years. Flexible chains have random walk statistics in three dimensions which lead to a characteristic scaling exponent ($\frac{1}{2}$) on the number of monomers (identical to that for diffusion in Section 5.1):

$$R_{ee} \sim aN^{\frac{1}{2}} \quad \text{flexible} \quad (8.26)$$

where a is the monomer length and N is the number of monomers in a chain. For a rigid rod the interpretation of the chain size is also straightforward, its length is simply the number (N) of monomers multiplied by their size (a):

$$R_{ee} \sim aN \quad \text{rigid rod} \quad (8.27)$$

For self-avoiding walks (with excluded volume and good solvent statistics) the calculation for the end-to-end distance is more complicated. The problem requires the renormalisation group technique, a sophisticated mathematical method from the theory of phase transitions. An intermediate result between that of flexible and rigid rod statistics is obtained for the size of the chains in a good solvent:

$$R_{ee} \sim aN^{\frac{3}{5}} \quad \text{good solvent} \quad (8.28)$$

At the theta point the attractive interchain forces induced by the solvent and the repulsive forces due to the excluded volume of the chains balance. Phantom gaussian statistics occur (as in equation 8.27), but this time in a real experimentally realisable system:

$$R_{ee} \sim aN^{\frac{1}{2}} \quad \text{theta solvent} \quad (8.29)$$

GOOD/BAD SOLVENTS AND THE SIZE OF POLYMERS

For a globule (a chain in a bad solvent) the chain forms a compact hard sphere whose radius is equivalent to a sphere constructed from N smaller spheres of volume $4/(3\pi a^3)$:

$$R_{ee} \sim aN^{\frac{1}{3}} \quad \text{bad solvent} \quad (8.30)$$

The range of scaling laws of the end to end distance that occur with polymeric chains is summarised in Figure 8.4.

A better understanding of the factors that affect chain conformation as a function of solvent quality can be developed using the Flory calculation for the radius of a chain. This is not a completely accurate calculation, but is a useful starting point to understand the interplay between entropic forces and the monomer–monomer interaction that determines the conformation of a chain. In this model, swelling of a polymer chain is due to a balance between the repulsion of the segments inside the coil (binary collisions) and the elastic forces that arise from the monomer entropy. The internal energy (U) due to monomer–monomer collisions in the chain is given by:

$$U(\alpha) \sim \frac{kTBN^{\frac{1}{2}}}{a^3\alpha^3} \quad (8.31)$$

where α is the expansion coefficient ($\alpha = R/R_0$, radius of the chain/initial radius), N is the number of monomers, a is the monomer length and B is

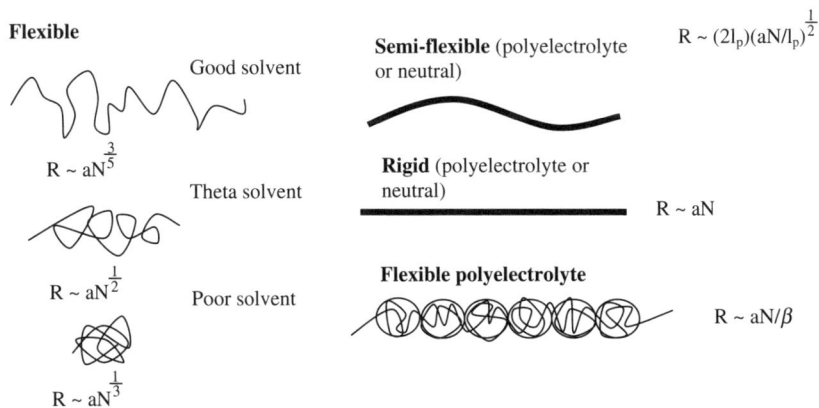

Figure 8.4 The typical range of polymer chain conformations as a function of solvent quality, charge and backbone rigidity. The end-to-end distance (R) is shown for each class of conformation and β is a solvent quality dependent parameter.

the second virial coefficient. The entropy of a chain is related to the expansion coefficient and is given by:

$$S(\alpha) = S_1 - \tfrac{3}{2}k\alpha^2 \qquad (8.32)$$

where k is Boltzmann's constant and S_1 is a constant. The total free energy (F) of the chain can then be calculated as a function of the expansion coefficient by the addition of the two terms, equations (8.31) and (8.32):

$$F(\alpha) = U(\alpha) - TS(\alpha) = \text{const} + kT\frac{BN^{\tfrac{1}{2}}}{l^3\alpha^3} + \tfrac{3}{2}kT\alpha^2 \qquad (8.33)$$

This can be solved graphically, as shown in Figure 8.5. The minimum on the figure corresponds to the equilibrium conformation of the polymer chain. However, the renormalisation group technique is required for exact quantitative results on the size of a polymer chain, since the excluded volume effect has not been accurately incorporated into the Flory model.

Blob models are very useful for determining the physical properties of flexible polymers. Also, they are a useful vehicle for learning scaling ideas, which often present the first best guess for a biophysical model, before a complete quantitative solution can be developed. A blob is a small section of a flexible polymeric chain that has some well defined statistical properties

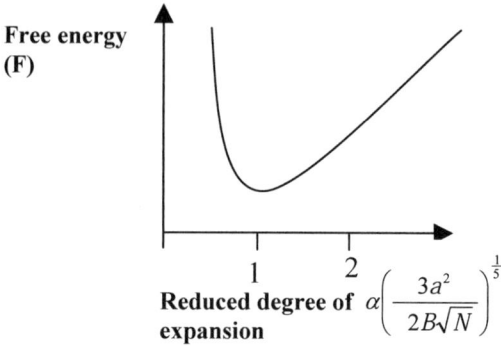

Figure 8.5 Free energy of a single Flory chain as a function of the degree of expansion (α in reduced units)
(The single minimum corresponds to the equilibrium conformation of the chain
[*Reprinted with permission from A.Y. Grosberg and A.R. Khokhlov, Giant Molecules, Copyright (1997) Elsevier*])

GOOD/BAD SOLVENTS AND THE SIZE OF POLYMERS

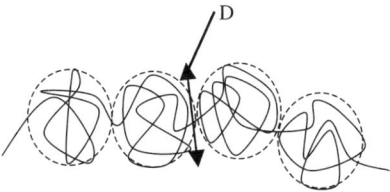

Figure 8.6 A blob (diameter D) is a small section of a flexible polymer chain that has a well defined statistical property

(Figure 8.6). There are three varieties of blob that are typically used in polymer physics: *thermal blobs, electrostatic blobs* and *tension blobs*.

A *thermal blob* (the simplest that will be encountered) is defined as a region of polymer chain in which the chain conformation is unperturbed from its thermal statistics. For example, consider a polymer chain in a capillary of diameter D (Figure 8.7). The chain is the same over a range of lengths, so the size of a small chain segment (R_s) scales in the same manner as the whole chain:

$$R_s \sim ag^v \quad (8.34)$$

where g is the number of monomers in a blob, a is the Kuhn segment length and v is an exponent, which depends on the quality of the solvent (compare with equation (8.28)). When a polymer chain is confined to a cylindrical pore the size of the chain segment (R_s) must be equal to the diameter of the pore (D) and from equation (8.34):

$$ag^v \sim D \quad (8.35)$$

This equation can be rearranged to give the number of monomers in a thermal blob:

$$g \sim \left(\frac{D}{a}\right)^{1/v} \quad (8.36)$$

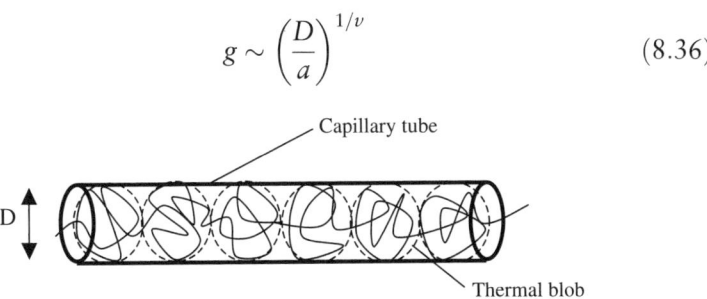

Figure 8.7 A flexible polymer chain inserted into a capillary, e.g. a titin molecule inserted in an actin cage in striated muscle. The blob size (D) is equal to the size of the capillary

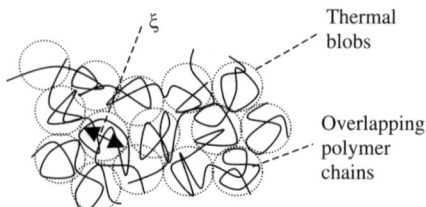

Figure 8.8 Schematic diagram of a close packed array of polymeric blobs in a semi-dilute solution with an average correlation length ξ

Thus with very little mathematical effort the size of the chain in the capillary (R_c) can be calculated, the chain is decomposed into a string of equally sized blobs, i.e. $R_c = (N/g)D$.

The idea of thermal blobs can be extended into *semi-dilute solutions* (Figure 8.8); calculations can thus be performed when chains overlap. A semi-dilute solution of flexible polymer chains can be pictured as a close packed array of blobs. The number of contacts that a blob has with other chain sections is approximately one and the free energy of the solution has by definition thermal energy (kT) per blob. Through a dimensional scaling argument it is possible to show that the semi-dilute correlation length (ξ), the mesh size of the polymer solution, scales with concentration (c) as:

$$\xi = ac^\nu \qquad (8.37)$$

The exponent (ν) again depends on the quality of the solvent for the polymer chains. The exponent is a $\frac{1}{2}$ for the extreme limit of completely extended chains (found with polyelectrolytes and liquid crystalline polymers) and $\frac{3}{4}$ for flexible chains in a good solvent. The predictions are in good agreement with scattering and atomic force microscopy (AFM) experiments for a wide range of polymeric systems.

The concept of a *charged blob* is very useful for calculating chain statistics with weakly charged flexible polymeric chains (Figure 8.9). The

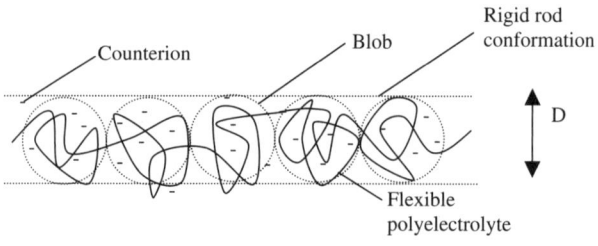

Figure 8.9 Schematic diagram showing the rigid bayonet conformation of a flexible polyelectrolyte whose subsections are arranged in a series of electrostatic blobs (size D)

ELASTICITY

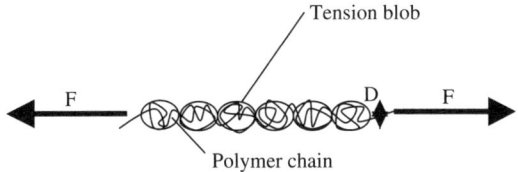

Figure 8.10 Tensions blobs (size D) are determined by the size of the external force (F) and the quality of the solvent

electrostatic repulsion of two neighbouring blobs in the chain is of the order of the thermal energy (kT). Consider the electrostatic energy of repulsion between charged monomers along a chain. The dominant Coulombic term is equal to $g^2e^2/\varepsilon D$, where g is the number of monomers in a blob, e is the charge, ε is the dielectric constant and D is the electrostatic blob size. The electrostatic energy can be equated to the thermal energy (kT) and gives:

$$\frac{g^2e^2}{\varepsilon D} \sim kT \tag{8.38}$$

This equation can be rearranged and the size of an electrostatic blob is therefore:

$$D \sim \frac{g^2e^2}{\varepsilon kT} \tag{8.39}$$

Tension blobs are defined for a chain with an applied force (consider a single molecular elastic spring). The applied force reduces the size of the lateral entropic fluctuations of the flexible chain and consequently the blob size (Figure 8.10).

8.3 ELASTICITY

An analysis of the *elasticity of rubbery networks* is useful to understand both the static and dynamic properties of biopolymers (Figure 8.11). Ideal rubbers (elastomers) consist of flexible chains connected together at a series of junction sites to form a solid network. The elasticity of such networks is crucial to the functioning of a range of biological materials, e.g. resilin in the hinges of dragonfly wings and abductin in the hinges of clams. The behaviour of cross-linked flexible polymer chains is relatively simple and

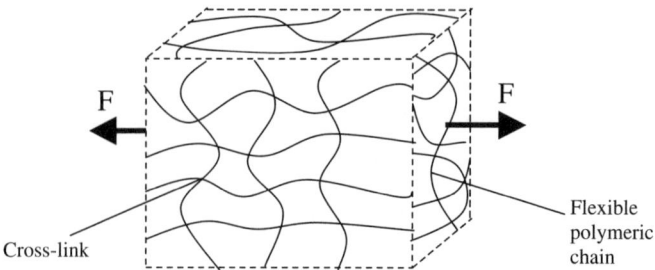

Figure 8.11 Schematic diagram of a rubber network under a tension (F) (The extending force is resisted by the entropy of the flexible chains of the polymeric network.)

will be covered here, whereas semi-flexible polymer chains exhibit more subtle phenomena related to the anisotropy of the network (Section 12.3).

To calculate the elasticity of a network of completely flexible chains the Flory approach will be followed. An expression for the change in entropy of the polymer network when it is stretched is required. The conformation entropy of a single polymer chain has been introduced previously (equation (8.32)) and the chain expansion factor is $\alpha = R/R_0$. The entropy of a flexible Gaussian chain (unperturbed radius $R_0 = N^{\frac{1}{2}}b$) is therefore:

$$S(\overline{R}) = -k\left(\frac{3\overline{R^2}}{2Nb^2}\right) + \text{const} \qquad (8.40)$$

where b is the monomer length, N is the number of monomers in a chain and k is Boltzmann's constant. The change in entropy ($\Delta S(R)$) upon extension of a single chain in three dimensions is the difference of two entropies before and after it is stretched:

$$\Delta S(\overline{R}) = S(R) - S(R_0) \qquad (8.41)$$

In three dimensions this entropy difference can be calculated as:

$$\Delta S(\overline{R}) = -\frac{3k}{2Nb^2}\left[(R_x^2 - R_{0x}^2) + (R_y^2 - R_{0y}^2) + (R_z^2 - R_{0z}^2)\right] \qquad (8.42)$$

It is useful to define the draw ratios ($\lambda_x = R_x/R_{0x}$ etc), the ratio of the radius of the chain (R_x) to the unperturbed radius (R_0), to give

ELASTICITY

another equation equivalent to equation (8.42) for the entropy difference:

$$\Delta S(\overline{R}) = -\frac{3k}{2Nb^2}\left[(\lambda_x^2 - 1)R_{0x}^2 + (\lambda_y^2 - 1)R_{0y}^2 + (\lambda_z^2 - 1)R_{0z}^2\right] \quad (8.43)$$

The elasticity of a polymer network is mainly entropic in nature and the agreement of the prediction of this type of purely entropic model with experiment is very good with simple synthetic polymers, e.g. the contraction of a synthetic rubber band when heated. The change in entropy can be summed over all the subchains to give the total entropy when the network is stretched:

$$\Delta S = -\frac{3kvV}{2Nb^2}\left[(\lambda_x^2 - 1)\langle R_{0x}^2\rangle + (\lambda_y^2 - 1)\langle R_{0y}^2\rangle + (\lambda_z^2 - 1)\langle R_{0z}^2\rangle\right] \quad (8.44)$$

where V is the volume of the sample and v is the concentration of subchains per unit volume. For a rubber network of ideal Gaussian chains the radius of gyration ($\langle R_0^2\rangle^{\frac{1}{2}}$) is proportional to the square of the number of monomers (equation (8.26)) and therefore:

$$\langle R_0^2\rangle = \langle R_{0x}^2\rangle + \langle R_{0y}^2\rangle + \langle R_{0z}^2\rangle = Nb^2 \quad (8.45)$$

where b is the size of a monomer in the chain. For isotropic Gaussian statistics of the chains, the size of the unperturbed chains in each of the draw directions are equal and therefore:

$$\langle R_{0x}^2\rangle = \langle R_{0y}^2\rangle = \langle R_{0z}^2\rangle = \frac{Nb^2}{3} \quad (8.46)$$

The chain entropy from equation (8.43) is:

$$\Delta S = -\frac{kvV(\lambda_x^2 + \lambda_y^2 + \lambda_z^2 - 3)}{2} \quad (8.47)$$

The condition of uniaxial deformation along the x axis for an incompressible material (volume conservation) introduces an additional condition on the draw ratios:

$$\lambda_y = \lambda_z = \lambda_x^{-\frac{1}{2}} = \lambda \quad (8.48)$$

Therefore the change of entropy can be simplified:

$$\Delta S = -k\nu V \frac{(\lambda^2 + 2)}{2(\lambda - 3)} \qquad (8.49)$$

The free energy (f) is proportional to the rate of change of entropy with extension ($\partial S/\partial \lambda$) and therefore:

$$f = -T\frac{\Delta S}{\Delta a_x} = -\frac{T}{a_{0x}}\frac{\partial S}{\partial \lambda} \qquad (8.50)$$

The stress (σ) experienced by the elastic network is the force per unit area:

$$\sigma = \frac{f}{a_{0x}a_{0y}} = -\frac{T\partial S/\partial \lambda}{a_{0x}a_{0y}a_{0z}} = -\frac{T\partial S/\partial \lambda}{V} \qquad (8.51)$$

where a_{0x}, a_{0y} and a_{0z}, are the length, breadth and height of the unstretched sample respectively. A final expression for the stress of an isotropic elastic network of flexible polymer chains in terms of the draw ratio is:

$$\sigma = kT\nu\left(\lambda - \frac{1}{\lambda^2}\right) \qquad (8.52)$$

There is no dependence on the number of monomers (N) or on the size of the monomers (b) in this expression for the stress. The only parameters that are important for the determination of the response of the rubbery network to a stress are the extension ratio (λ) and the density of cross-links (ν). For small amounts of extension the draw ratio is approximately one ($\lambda \cong 1$) and therefore:

$$\lambda - \frac{1}{\lambda^2} \approx 3(\lambda - 1) \qquad (8.53)$$

By definition the Young's modulus (E) is given by the ratio of the stress (σ) to the strain ($\Delta l/l$) (Section 13.1) and therefore:

$$\sigma = E\frac{\Delta l}{l} \qquad (8.54)$$

The draw ratio is simply related to the strain:

$$\lambda - 1 = \frac{\Delta l}{l} \tag{8.55}$$

Thus it is found that the Young's modulus is directly proportional to the number of cross links:

$$E = 3kT\nu \tag{8.56}$$

where ν is the density of cross-links and kT is the thermal energy. This expression for the modulus is compact, simple and very useful for estimating the elasticity of polymer networks. It can also be used to calculate the plateau of the shear modulus in viscoelastic networks of entangled polymer chains in which the cross-links are topological in nature (no interchain chemical bonds are involved). A complication for a detailed description of most biological systems is that polymeric networks often include rigid rod and semi-flexible sections (nematic elastomers). The materials act as composites of their molecular components with unique strain hardening mechanisms that need to be considered separately (Chapter 12).

8.4 DAMPED MOTION OF SOFT MOLECULES

A first step in understanding the dynamics of polymers can be developed through an inspection of the decay of the forced vibrations of a rigid rod. This could be a fibre rigidly attached at one end (Figure 8.12), e.g. a hair in the cochlea of the ear. The equation of motion for the rod's damped spring-like motion (Section 5.2) is given by:

$$m\frac{d^2x}{dt^2} + \gamma\frac{dx}{dt} + \kappa x = F \tag{8.57}$$

where x is the displacement of the end of the rod, γ is the viscous coefficient, κ is the elastic modulus and F is the external force. Through analysis of equation (8.57) it is found there are two distinctly different forms of motion that occur for the fibre. When the damping is small:

$$\gamma^2 < 4m\kappa \tag{8.58}$$

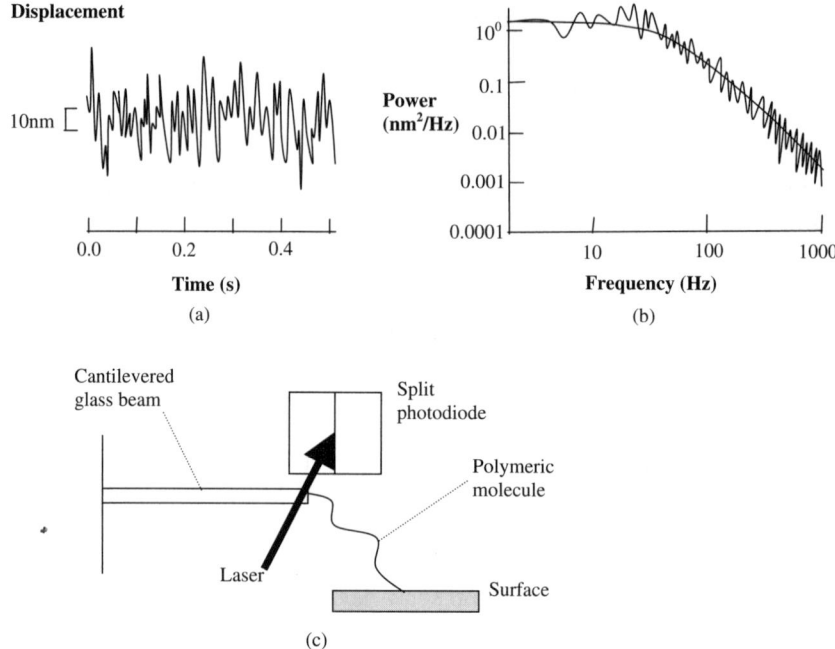

Figure 8.12 The fluctuations in the displacement of a cantilevered glass fibre as a function of time (a) the resultant power spectrum as a function of frequency (b) and the experimental arrangement for the measurement of the displacement fluctuations of the glass fibre (c)
[Ref.: E. Meyhofer and J. Howard, Proc. Nat. Acad. Sci. USA, 1995, 92, 574–578]

The motion is oscillatory and *underdamped* (Figure 8.13(b)). However, for large damping:

$$\gamma^2 > 4m\kappa \qquad (8.59)$$

The motion is *overdamped* (Figure 8.13(c)). There are two time constants associated with the damped oscillatory motion. A fast time constant occurs for the mass to accelerate to the maximum velocity of F/γ, and a slow time constant is introduced for the relaxation time of the spring and dashpot, i.e. the elastic and dissipative components of the rigid rods motion.

Global motions of small comparatively soft objects such as proteins, polysaccharides and nucleic acids in aqueous solutions are overdamped. This can be shown using a crude mechanical model of a globular protein. The protein is pictured as a homogeneous isotropic cube of material with

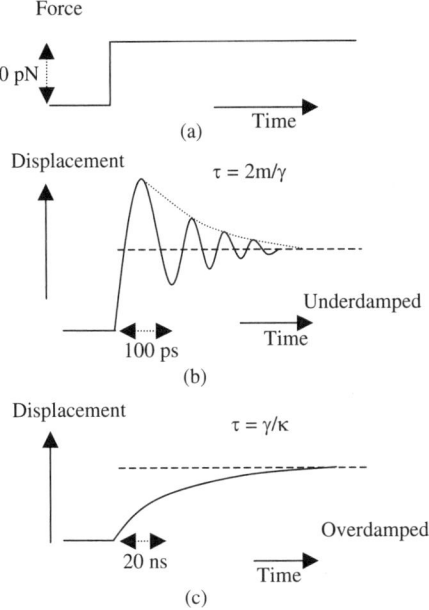

Figure 8.13 (a) A mass (m) attached to an elastic spring (elastic constant κ) embedded in a liquid with viscous dissipation (γ) is subjected to a force of 10 pN at time $t = 0$. The resultant motion of the mass can be (b) underdamped or (c) overdamped. The time constant for the decay of the underdamped motion is $2\,m/\gamma$ and for the overdamped case it is γ/k

a given side length (L), density (ρ), Young's modulus (E) and viscosity (η). The mass of the cube is:

$$m = \rho V = \rho L^3 \qquad (8.60)$$

where ρ is the density and V is the volume. The stiffness (κ) of the cube is related in the standard way to the length (L) and the Young's modulus (E):

$$\kappa = EL \qquad (8.61)$$

The drag coefficient on the cube (γ) is given by Stoke's law as:

$$\gamma = 3\pi \eta L \qquad (8.62)$$

From the previous analysis, it is known that the motion of the cube is overdamped if the ratio $4m\kappa/\gamma^2$ is less than one, and this can now be related to the material properties of the protein:

$$\frac{4m\kappa}{\gamma^2} = \frac{4\rho L^3 E L}{(3\pi \eta L)^2} = \left(\frac{2}{3\pi}\right)^2 \frac{\rho E}{\eta^2} L^2 \qquad (8.63)$$

How the characteristic ratio scales with the dimension L is particularly important. The smaller the physical dimension of the protein the higher the tendency for its motion to be overdamped. The motions of protein domains of diameter less than the characteristic length L_c can be analysed. Using equation (8.63) this characteristic length can be calculated, $\eta \sim 1$ mPa.s, $\rho \sim 10^3$ kg/m^3, $E \sim 1$ GPa:

$$L_c \approx \frac{3\pi}{2} \left(\frac{\eta^2}{\rho E}\right)^{\frac{1}{2}} \approx 5 \text{ nm} \qquad (8.64)$$

Small motions of globular proteins such as the ribosome (1 nm) are thus overdamped. Internal friction of the proteins will tend to accentuate the degree of damping over and above the previous analysis. Elongated proteins are more highly damped than globular proteins because the increase in aspect ratio causes the damping to increase (increased L) and the stiffness to decrease (E). Thus the motion of the cytoskeleton is overdamped. Similarly the motions of most biological polymers are overdamped.

A useful calculation is to solve the expression for an oscillatory damped fibre using a power spectrum of its displacement. Consider the motion of a glass fibre undergoing stochastic fluctuations, e.g. the glass fibre is attached to a DNA chain which is undergoing a helix–coil transition (Figure 8.12). The autocorrelation function of the displacement of the end of the fibre $(R(\tau))$ is introduced to help solve the equation of motion and it is the time dependent quantity often measured electronically in experiments (equation (5.34)). The autocorrelation function is defined as:

$$R(\tau) = \langle x(t)x(t-\tau) \rangle = \lim_{T \to \infty} \left\{ \frac{1}{T} \int_{-T/2}^{T/2} x(t)x(t-\tau)dt \right\} \qquad (8.65)$$

where T is the time over which the motion of the rod is sampled. Using the autocorrelation function the Langevin equation (8.57) can be

DYNAMICS OF POLYMER CHAINS

accurately solved and it is found that the fibre fluctuates around its average position with a characteristic spectra:

$$R(\tau) = \frac{kT}{\kappa} e^{-|\tau|/\tau_0} \qquad (8.66)$$

and the characteristic time constant is given by:

$$\tau_0 = \frac{\gamma}{\kappa} \qquad (8.67)$$

Typically a power spectrum $(G(f))$ is experimentally used to analyse the fluctuations in displacement of a probe, and it is defined as the Fourier transform of the autocorrelation function $((R(\tau))$. The power spectrum of the position of the glass fibre as a function of frequency (f) is found to be:

$$G(f) = \frac{4kT\gamma}{\kappa^2} \frac{1}{1 + (2\pi f \tau_0)^2} \qquad (8.68)$$

This theoretical power spectrum can then be compared with that calculated using a numerical fourier transform of the raw experimental data. The mean square displacement of the fibre end at long times is given by $\langle x^2 \rangle = kT/\kappa$ and the correlation time (τ_0) can be related to the cut off frequency of the power spectrum, $f_0 = \frac{1}{2}\pi\tau_0$, (Section 5.3). The elastic (κ) and dissipative (γ) constants for a biological molecule can thus be directly calculated from the experimental data.

8.5 DYNAMICS OF POLYMER CHAINS

The Stokes–Einstein equation for the diffusion coefficient (D) for the motion of a spherical test particle in solution (Section 5.1) can be used if all the surrounding molecules (radius a) are treated as a continuum of constant viscosity (η_0):

$$D = \frac{kT}{6\pi\eta_0 a} \qquad (8.69)$$

where kT is the thermal energy. The structural relaxation time (τ) of a liquid that consists of these spherical particles is defined as the time for the particles to diffuse their molecular size (a):

$$\tau = \frac{a^2}{D} = \frac{6\pi\eta_0 a^3}{kT} \qquad (8.70)$$

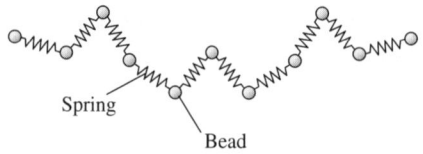

Figure 8.14 Schematic diagram of a Rouse chain, a bead spring model for the dynamics of flexible chains with no hydrodynamic interactions

However, polymer chains can have internal relaxation modes due to their flexibility, and they can have topological entanglements with their neighbours. These two effects significantly complicate the relaxation spectrum at short times with polymeric solutions compared with the case of simple rigid spherical colloids.

The *Rouse model* describes the spectrum of relaxation times of a flexible phantom chain in an immobile solvent (Figure 8.14). It is the simplest successful theory of polymer chain dynamics, formulated for a chain with a Gaussian conformation. The Rouse model assumes the chain statistics are ideal and the solvent is immobile. The mathematical description of the Rouse model is based on coupled equations of motion of the links (beads connected with idealised elastic springs), which include the effects of the random thermalised forces that act upon them.

The Fourier transform of the basic equations of the Rouse model shows that the motion of a polymer chain can be represented as a superposition of independent Rouse modes, much like the harmonic spectrum used to describe the motion of a plucked guitar string. In the Rouse model, the maximum relaxation time of the polymer coil and the diffusion coefficient of the coil as a whole vary with the number of chain links (N) as N^{-2} and N^{-1} respectively. The slowest intramolecular relaxation and the diffusive motion of the coil as a whole conform to the first and fundamental Rouse modes respectively. The root mean square displacement of a link on a Rouse chain varies as $\sim t^{\frac{1}{4}}$ over time intervals (t) less than the maximum relaxation time of the chain (τ_1), and only for times greater than the maximum relaxation time ($t > \tau_1$) does it become proportional to $t^{\frac{1}{2}}$ as in the case of ordinary diffusion of a Brownian particle (Section 5.1). This behaviour has been well demonstrated experimentally with fluorescent microscopy and incoherent quasi-elastic neutron scattering experiments.

The Rouse model is found to be useful for dense polymeric melts in which the hydrodynamic interactions are screened. Similarly, in semi-dilute solutions Rouse dynamics describe the motion of the polymer chains beyond the mesh size of the network, i.e. the long length-scale dynamics (screened hydrodynamics) are well described.

The Rouse model yields results that differ from experimental observation for chains in dilute solution; one problem is the neglect of the hydrodynamic interaction caused by solvent entrainment between different sections of the polymer chain. The Zimm model includes a coarse grained hydrodynamic interaction and is often in reasonable agreement with experimental data for isolated flexible chains in solution. The root mean square displacement of a link of a Zimm chain varies with time as $\sim t^{\frac{1}{3}}$ and only for times greater than the longest relaxation time $(t > \tau_1)$ does it become proportional to $t^{\frac{1}{2}}$ as expected for diffusion (equation (5.8)).

The Rouse and Zimm models are useful for very flexible chains. For *semi-flexible* chains the hydrodynamic beam model is required to explain their dynamics. Such a model can be used to describe relatively rigid polymers such as actin filaments, cochlea, keratin filaments, and microtubules (Figure 8.15). The drag force per unit length $(f_\perp(x))$ on the transverse fluctuations of a semi-flexible filament is given by:

$$f_\perp(x) = -c_\perp v_\perp(x) = -c_\perp \frac{\partial y(x)}{\partial t} \quad (8.71)$$

where $c_\perp$ is the mass density of the filament, $v_\perp$ is the transverse velocity and y is the transverse displacement. The drag force can be balanced with the elastic restoring force of the filament and leads to the *hydrodynamic beam equation*:

$$\frac{\partial^4 y}{\partial x^4} = -\frac{c_\perp}{EI} \frac{\partial y}{\partial t} \quad (8.72)$$

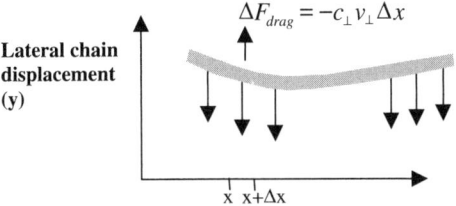

Figure 8.15 The lateral drag force on a semi-flexible fibre is a function of the perpendicular velocity $(v_\perp)$, size of the element (x), and the mass density $(c_\perp)$

where E is Young's modulus and I is the moment of inertia of the fibre. The relaxation time of a semi-flexible chain can then be calculated as the time for a bent rod to relax back to it straight conformation. The amplitude of the bend is found to decrease exponentially with time, with a time constant (τ_n) that depends on the mode number (n) (Figure 8.16):

$$\tau_n \approx \frac{c_\perp}{EI}\left(\frac{L}{\pi(n+1/2)}\right)^4 \tag{8.73}$$

$n = 1, 2, 3, \ldots$, and L is the length of the rod. Such predictions for the dynamic modes of semi-flexible chains are in reasonable agreement with experiment.

The dynamics of both flexible and semi-flexible polymers in *concentrated solutions* have a simple explanation in terms of the one dimensional diffusion of each chain in a tube created by its neighbours. This type of model is called 'reptation' and the name refers to the snake-like motion of the chains in their tubes (Figure 8.17). The utility of the reptation model can be shown with a scaling calculation of the dynamics of a flexible polymer in an entangled solution of other similar chains. Consider the *longest relaxation time* of the concentrated polymer solution. Initially (time, $t = 0$) a constant elongating stress (σ) is applied and the resultant relative deformation ($\Delta l/l$) is measured. If the stress (σ) is small the compliance ($J(t)$ = strain/stress) by definition is given by:

$$\frac{\Delta l(t)}{l} = \sigma J(t) \tag{8.74}$$

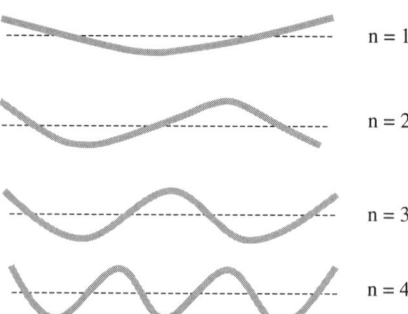

Figure 8.16 The hydrodynamics modes ($n = 1, 2, 3, 4$) of a semi-flexible rod in solution. The free ends act as anti-nodes for the chain motion

DYNAMICS OF POLYMER CHAINS

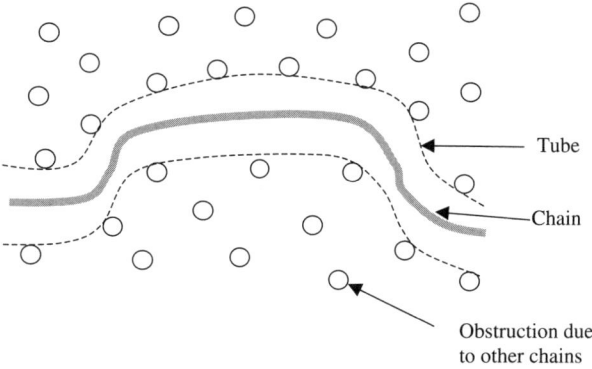

Figure 8.17 Schematic diagram of a polymer chain that reptates in a concentrated solution. The chain diffuses in a tube formed by the steric constraints of its neighbours

For a typical polymer solution the compliance resembles that shown in Figure 8.18. After a sharp rise the compliance reaches a plateau value (J_0). For times greater than the longest relaxation time (τ^*) of the solution the stress is no longer proportional to the strain, but to the rate at which the strain increases, i.e. the material becomes liquid-like and Newton's law of viscosity holds:

$$\sigma = J_1^{-1} \frac{d(\Delta l/l)}{d\tau} \qquad (8.75)$$

where $\Delta l/l$ is the strain. It is therefore possible to relate the viscosity to the compliance (J_1) on Figure 8.18:

$$J_1^{-1} = \eta \qquad (8.76)$$

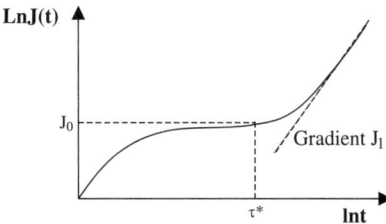

Figure 8.18 Schematic diagram of the compliance of a concentrated solution of entangled polymers as a function of time. There is a plateau (J_0) at intermediate times (τ^*) and at long times the gradient is J_1

In the microscopic picture the longest relaxation time of the system (τ^*) is the time for entangled cross-links to decay as the chains slither out of their tubes, i.e. the time for tube renewal. The Youngs modulus (E) for the entangled solution is given by the same equation as was derived for the rubber network (equation (8.56)):

$$E \sim \frac{kT}{N_e a^3} \quad (8.77)$$

where N_e is the average number of monomer units along a chain between two effective cross links, a is the Kuhn segment length and kT is the thermal energy. For the snake-like motion, the chain moves through a cylindrical tube (Figure 8.19) and for Gaussian chain statistics the tube diameter (d) is equal to the blob size (Section 8.2) given by:

$$d \sim a N_e^{\frac{1}{2}} \quad (8.78)$$

where N_e is the number of monomers in a blob and a is the step size. Similar to the calculation of the size of a chain in a capillary, the total contour length (Λ) of a tube is just the number of blobs in a chain (N/N_e) multiplied by the size of a blob:

$$\Lambda \sim \frac{N}{N_e} d \quad (8.79)$$

Therefore, since $d \sim a N_e^{\frac{1}{2}}$ the contour length can be expressed as:

$$\Lambda \sim a N N_e^{-\frac{1}{2}} \quad (8.80)$$

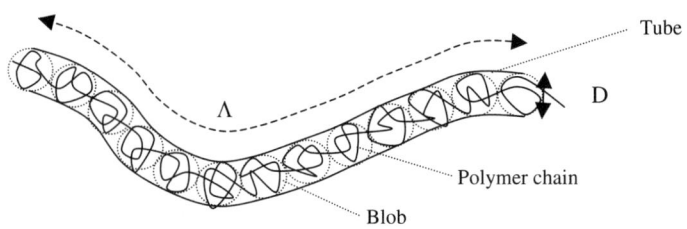

Figure 8.19 Schematic diagram of the conformation of blobs of a flexible polymer in a tube used in the scaling form of the reptation model. Λ is the length of the tube, and D is the blob diameter

DYNAMICS OF POLYMER CHAINS

For one dimensional diffusion the mean square fluctuations in the one dimensional displacement of the chain along the tube are related to the diffusion coefficient (D) from the standard relationship (Section 5.1):

$$\langle x^2 \rangle = 2Dt \tag{8.81}$$

where x is the distance along the tube. As before, the diffusion coefficient can be calculated for the chain from the fluctuation dissipation theory as:

$$D = \frac{kT}{\mu_t} \tag{8.82}$$

where kT is the thermal energy and μ_t is the total friction coefficient of a complete chain. The total friction coefficient experienced by the chain can be approximated by addition of the friction due to each subunit (μ) surrounded by its neighbours:

$$\mu_t = N\mu \tag{8.83}$$

where N is the number of chain subunits in a single chain. The longest relaxation time (τ^*) of the chain from the definition of the diffusion coefficient is:

$$\tau^* \sim \frac{\Lambda^2}{D_t} \sim N^3 a^{\frac{3}{2}} \frac{\mu}{N_e \rho kT} \tag{8.84}$$

The microscopic relaxation time (τ_m) of a molecular liquid is defined as $\tau_m \sim l^2/D$ (equation (8.70)), to give a slightly simpler form for the longest relaxation time:

$$\tau^* \sim \frac{N^3}{N_e} \tau_m \tag{8.85}$$

Two quantities can then be calculated from this reptation model that characterise the dynamics of entangled polymer solutions and are in good agreement with experiment, the viscosity and the diffusion coefficient. The viscosity (η) of the solution is approximately equal to the modulus (E) multiplied by the longest relaxation time of the fluid (a standard trick from rheology):

$$\eta \sim E\tau^* \sim \frac{\mu N^3}{aN_e^2} \tag{8.86}$$

The self-diffusion coefficient (D_s) of the polymer chains is the rate at which the chain fluctuates at a distance equal to the square of its size (R):

$$D_s \sim \frac{R^2}{\tau^*} \sim \frac{N_e T}{N^2 \mu} \qquad (8.87)$$

Experimental evidence for reptation is now very good. More sophisticated elaboration of the model provides a quantitative theory for the forced electrophoretic motion of DNA through polyacrylamide gels (Section 13.5). Quasi-elastic neutron scattering can explore the motion of labelled sections of polymers in a melt and is in agreement with the reptation model. Fluorescent recovery after photobleaching measurements of concentrated solutions provides data in agreement with the predicted self-diffusion coefficients. Reptation theories for the rheological response of polymer melts allow quantitative predictions for a range of mechanical spectroscopy experiments (Chapter 13). However, perhaps the strongest evidence to date is the microscopy images of a fluorescent DNA molecule in a concentrated solution of other untagged DNA chains. The DNA chain is shown to experience forced reptation when pulled with optical tweezers. The chain clearly has a memory of the position of its tube as it moves (Figure 8.20), it has experienced a process of reptation.

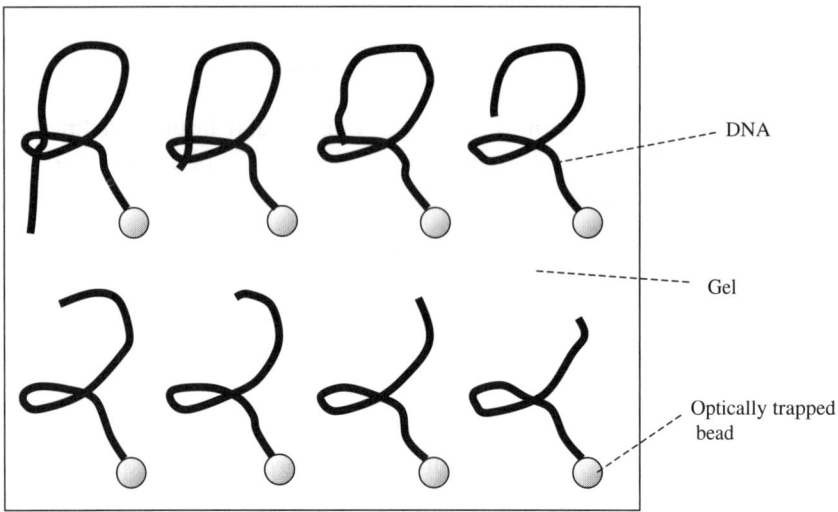

Figure 8.20 Reptation in entangled polymer chains can be clearly visualised using fluorescence microscopy and optical tweezers. A time sequence of images is shown for a DNA chain that moves in a polyacrylamide gel
[*Reprinted with permission from T.T. Perkins, D.E. Smith and S. Chu, Science, 264, 819, Copyright (1994) AAAS*]

8.6 TOPOLOGY OF POLYMER CHAINS – SUPER COILING

Many biologically important properties of duplex DNA and semi-flexible polymer chains are of a topological origin. The important characteristic of these chains is that they resist torsional distortion and can thus have a memory of their torsional state. The DNA of bacteria is circular and typically occurs in a compact supercoiled state in nature and this provides an important motivation for the study of super coiling (Figure 8.21).

It was discovered that DNA chains with the same molecular weight, but different values of the number of super twists (τ), separate in a well defined manner during agarose gel electrophoresis experiments. The superhelical state moves across a gel faster due to its more compact conformation. The state of a closed ring DNA chain is now known to be characterised by two topological invariants: the type of knot formed by the double helix as a whole and the *linking number* (L) of one strand with the other. The minimum of the energy of a closed ring DNA corresponds to the superhelical state; the number of twists in a superhelix depends on the degree of strand linking (L). Naturally occurring circular DNA chains are always negatively superspiralised. The axial *twisting* of strands around each other can differ from the magnitude of their linking number by the amount of *writhing*, which depends on the spatial form of the axis of the double helix. Viruses and bacteria can change the topology

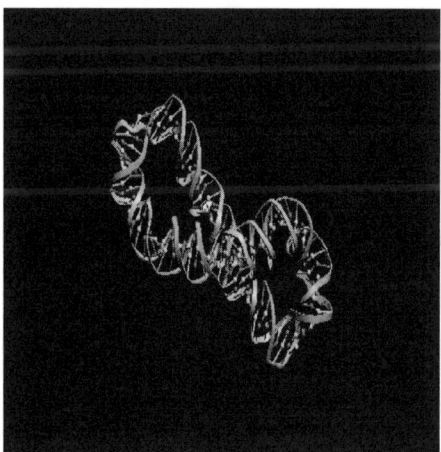

Figure 8.21 Molecular dynamics simulation of a super-helical circular DNA chain [*S. Harris, University of Leeds*]

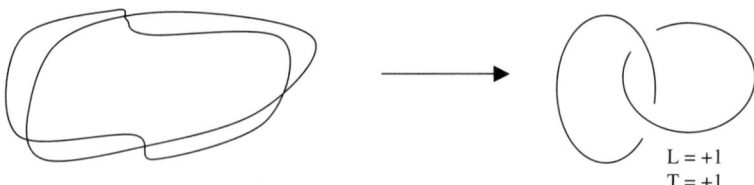

Figure 8.22 The linking number (L) is equal to the number of times one topologically connected chain passes over the other. The diagram illustrates the process by which the topology of the chain can be clearly identified by inspection

of their DNA during replication using toposiomerases. Topoisomerase I changes the linking number by $+1$ and topoisomerase II changes it by $+2$. Further enzymes introduce supercoils in DNA and are called the gyrases. Negative supercoils favour the unwinding of DNA and the subsequent processes of replication, transcription and recombination. Therefore, bacterial DNA is stored in a highly supercoiled, plectonomic structure (compact cylinders of DNA). The free energy of the superhelical state is proportional to the square of the density of superturns and the effective modulus of elasticity of a superhelical chain depends on both the bending and torsional stiffness of the polymer.

To be mathematically more precise, consider the topology of closed ribbons. Three numbers (topological invariants) characterise the closed ribbon formed by circular DNA: the *linking number* (L), the *twist number* (T) and the *writhing number* (W). The *linking number* (L) is the number of times the two edges are linked in space, and it is an integer. To find the linking number (L) a projection of the ribbon on to a plane is required and all points where a segment of one of the curves passes above the other need to be counted. This process is demonstrated in Figure 8.22. The *twist* (T) is the extent to which the normal vector (u) rotates around the ribbon direction (s). It can be calculated from the line integral along the path that describes a complete circuit around the chain (Figure 8.23):

$$T = \frac{1}{2\pi} \oint \omega ds \qquad (8.88)$$

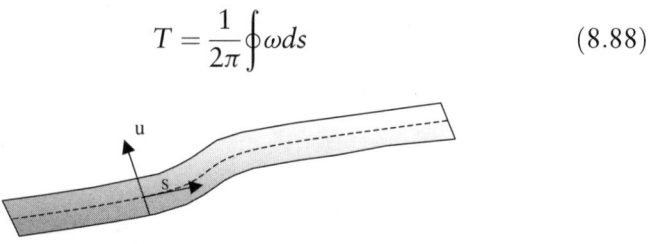

Figure 8.23 The twist number can be calculated from the extent that the normal vector to the ribbon (u), rotates around the tangent vector (s) that defines the ribbon direction

TOPOLOGY OF POLYMER CHAINS – SUPER COILING

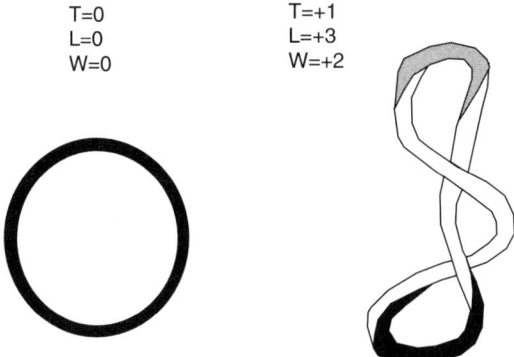

Figure 8.24 A rich variety of topological states are possible with duplex DNA and two alternatives are shown
[*Ref.: C.R. Cantor and P.R. Schimmel, Biophysical Chemistry III, Freeman, 1980*]

where ω is the angular rate per unit length, ds is the increment in arc length and T is the line integral. The *writhing number* is defined as:

$$W = L - T \qquad (8.89)$$

It is a measure of the extent to which the axis coils and folds in three dimensions. Figure 8.24 shows two possible configurations of a circular duplex DNA chain with the values of the twist and linking numbers.

FURTHER READING

A.Y. Grosberg and A.R. Khoklov, *Statistical physics of macromolecules*, Academic Press, 1995 and *Giant Molecules Here There Everywhere*, AIP, 1997. Two informative books on polymer physics.

J. Howard, *Mechanics of Motor proteins and the cytoskeleton*, Sinauer, 2003. Useful account of the mechanics of semi-flexible biopolymers.

C. R. Cantor and P. R. Schimmel, *Biophysical Chemistry Pt III. The behaviour of biological macromolecules*, Freeman, 1980. Old fashioned, but readable account of the statistical physics of biological molecules.

M. Rubinstein and R. H. Colby, *Polymer Physics*, Oxford University Press, 2004. Good exercises and a very clear account of the physical phenomena observed with synthetic polymers.

TUTORIAL QUESTION

8.1) The Young's moduli of elastin* and collagen are 1 MPa and 1.5 GPa respectively. What is the density of cross-links that would give rise to these moduli in a random Gaussian network? Assume that the sections of chain are completely random between cross-links. If the actual cross-linking density is on the order of 8×10^{25} m^{-3} in the two samples, can you explain the lack of agreement for the collagen data?

*Note that there is more ordering in the structure of elastin chains than this calculation might imply. The good agreement of the model with experiment is therefore to some extent fortuitous.

8.2) Consider the thermal fluctuation of a polymeric sickle cell haemoglobin aggregate observed with a fluorescence microscope. θ is the angle of the change in direction of a tangent to a filament along its length, s is the length of the particle and κ_f is the bending rigidity. The relationship between the bending energy and $\langle \theta^2 \rangle$, the average square fluctuation of the angle is given by:

$$E_{bend} = \frac{\kappa_f \langle \theta^2 \rangle}{2s}$$

For a virus of contour length 300 nm. Determine the value of $\langle \theta^2 \rangle^{\frac{1}{2}}$ that arises from thermal fluctuations, if the persistence length of the fibrous aggregate is 10^{-3} m.

8.3) A flexible titin molecule occurs in an approximately cylindrical hole inside striated muscle (Figure 8.25). If the molecule is unattached at either end what is the equilibrium length of the polymer ($R_\parallel$), assuming a Kuhn segment length of 30 nm, contour length of 750 nm, and a capillary diameter of 40 nm? The molecule is then attached to the sacromere at either end at a fixed

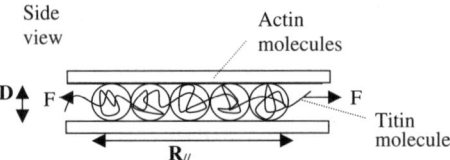

Figure 8.25 Schematic diagram of a titin chain compressed inside an action cage

distance of 0.6 μm. Will the loss in entropy of the chain due to it being confined in a capillary contribute to the elasticity of the chain?

8.4) A polymer chain in a good solvent is held between two optical traps. What would the force/extension curve look like? If the chain was gradually placed in a 'bad solvent' at constant separation how would the force exerted on the traps change?

8.5) Calculate the radius of gyration $\langle r^2 \rangle^{\frac{1}{2}}$ for phage DNA that has a persistence length of 450 Å and a contour length of 60 μm using a Kratky–Porod model.

9
Charged Ions and Polymers

Charged ions exist in biological systems in a wide range of forms. The ions can be small molecules (∼2 Å), protein nano-composites (10 nm) or giant linear aggregates (many centimetres for DNA with millions of charged groups). Examples of small ions found in biology include the carboxylic acid (COO^-) groups in aspartic and glutamic acid in proteins, the polar heads of fatty acids (COO^-–$(CH_2)_n$–CH_3), the positively charged amine groups in lysine, arginine and histdine, and the phosphate group (PO_4^-) in nucleic acid and phospholipids (Figures 9.1, 9.2, 9.3). All these ions are typically surrounded by a shell of associated water molecules and their interaction with this hydration shell has a dramatic effect on their physical properties.

Partial charges due to polarisation of covalent bonds also can exist and provide substantial interaction energies between neighbouring atoms. A particularly strong form of this polar interaction is observed in hydrogen bonded molecules.

There are a wide range of physiological uses for small ionised molecules. For example, charged molecules are used in signalling. Minute quantities of calcium (Ca^{2+}) ions regulate the control of molecular motors in striated muscle (μM) and a cocktail of sodium (Na^+), potassium (K^+) and calcium (Ca^{2+}) ions is required for the action of electrical impulses in nerve cells. There are therefore a large number of enzymes (ion pumps) designed to move and regulate the population of small ionic species in cells.

Many of the processes that involve charged ions in biological systems involve the process of *acid–base equilibria* (Section 1.1). A molecule

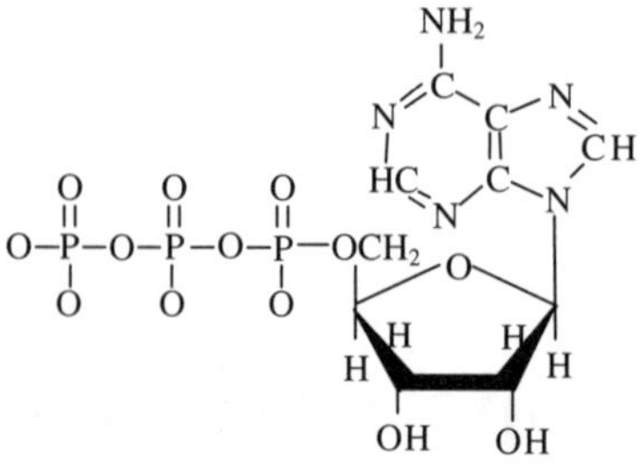

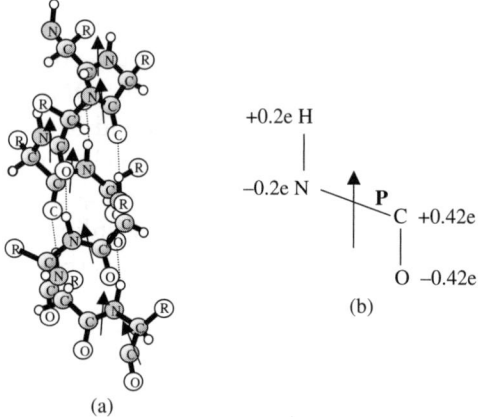

Figure 9.1 Example of a small charged biological molecule, adenosine triphosphate (ATP)

Figure 9.2 An alpha helix (a) can have a considerable dipole moment due to the alignment of the individual peptide dipoles along its backbone. (b) indicates a single peptide dipole (P). e is the electronic charge

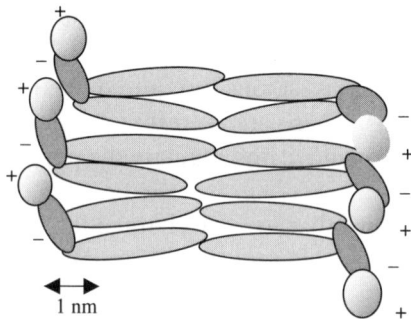

Figure 9.3 Charged lipid molecules in bilayers can produce one of the highest surface charge densities in nature with up to one electronic charge per $0.6\ \text{nm}^2$

(AH) is acidic if it can dissociate in water to provide a supply of hydrogen (H^+) ions, and conversely a molecule is basic if it can accept hydrogen ions. For an acid the process of dissociation can be described as:

$$AH \leftrightarrow A^- + H^+ \tag{9.1}$$

The study of acid–base equilibria is of central importance to aqueous physical chemistry and reference should be made to specialised texts for more details.

The behaviour of charged macromolecules, such as DNA, actin, carrageenan (a carbohydrate derived from seaweed) and chrondroitin sulfate (a glycosoaminoglycan found in the articulated joints of mammals), are highlighted here. The charged groups in the polymer are vitally important for both their solubility and correct biological functioning in a range of architectural, catalytic and information storage roles.

Simple electrostatic forces are a dominant factor that determines the structure of a large range of charged biological molecules and these forces are examined using some simple physical calculations. For example, the dipole moments of the constituent amino acids in a protein tend to line up in an alpha helix. These dipoles give the secondary structure a large cumulative dipole moment (Figure 9.2) and the forces between the dipoles help to determine the resultant morphology.

Although many useful theories exist for strongly charged biological systems it is necessary to emphasise some of their shortfalls. Electrostatic interactions between biomolecules are often long range and this can lead to an intractable many body problem. Thus a series of ingenious schemes and approximations need to be invoked to avoid some of the problems involved with long range interactions to allow tractable calculations to be completed in a variety of different biological scenarios.

9.1 ELECTROSTATICS

Some basic electrostatic phenomena are reviewed first. *Coulomb's law* gives the force ($\underline{f}$) between two point charges of magnitude q and q':

$$\underline{f} = \frac{qq'\underline{u}}{4\pi\varepsilon r^2} \tag{9.2}$$

where $\underline{u}$ is the unit vector directed between the two charges, ε is the dielectric constant of the material and r is the distance between the charges. This electrostatic force is thus directed along the line that joins the two point charges, and takes the direction in which a positive test charge would be translated. The local force (f) on a charge (q) is related to the electric field strength (E) it experiences due to the surrounding charge distribution:

$$\underline{f} = q\underline{E} \tag{9.3}$$

The energy (W) of an electric dipole ($\underline{p}$) in an electric field ($\underline{E}$) is the scalar product of the two vectors:

$$W = -\underline{p}.\underline{E} = -pE\cos\theta \tag{9.4}$$

where θ is the angle between the vectors $\underline{p}$ and $\underline{E}$. A torque (Γ) thus tends to orientate dipoles in an applied electric field. The torque tends to minimise the electric dipolar energy (W) and is equal to the vector product of the electric dipole and the electric field:

$$\underline{\Gamma} = \underline{p} \times \underline{E} \tag{9.5}$$

Gauss' theorem is a general relationship which relates the electric flux that crosses a closed surface to the amount of charge (q) contained within the surface. The electric flux (ϕ) is defined as the integral of the component of the electric field perpendicular to a surface, over the complete surface:

$$\phi = \oint \underline{E}.\underline{n}dS \tag{9.6}$$

where $\underline{n}$ is the unit vector normal to the surface (dS) and the integral ($\oint$) is taken over the complete closed surface. Gauss's theorem for an arbitrary charge distribution in a vacuum has the simple form:

$$\phi = \frac{q}{\varepsilon_0} \tag{9.7}$$

where ε_0 is the permittivity of free space and q is the charge contained within the surface. The total electric flux through a surface is proportional to the charge enclosed.

Gauss's theorem can be applied to a sphere that encloses a charge distribution (total charge q) of surface area $4\pi r^2$ which has a constant

ELECTROSTATICS

normal component of the electric field (E_r). It is thus easy to calculate the electric field around any *spherically symmetrical* charge distribution, e.g. a charged colloid, once the effect of the relative permittivity of water is understood. Equation (9.7) applied to the geometry of a sphere gives:

$$4\pi r^2 E_r = \frac{q}{\varepsilon_0} \tag{9.8}$$

Therefore, the radial electric field (E_r) around a point charge is:

$$E_r = \frac{q}{4\pi\varepsilon_0 r^2} \tag{9.9}$$

Similarly, Gauss's theorem can be applied to a *cylindrically symmetrical* charge distribution, which will later be used to model a polyelectrolyte. Consider a cylinder of known radius (r), length (L), with an electric field (E) normal to its surface, that contains a line of charge (charge per unit length λ), and ε_0 is the permittivity of free space. The electric flux crossing the cylinder's surface can be equated to the charge contained inside and Gauss's theorem gives:

$$2\pi r L E = \frac{\lambda L}{\varepsilon_0} \tag{9.10}$$

The electric field from a line charge (or cylindrically symmetrical object) is therefore:

$$E = \frac{\lambda}{2\pi\varepsilon_0 r} \tag{9.11}$$

Comparison with the expression for the electrical field around a sphere (equation (9.9)) shows that the electric field for a cylinder is radically different to that of a sphere ($E \sim 1/r^2$ for a sphere compared with $E \sim 1/r$ for a cylinder). These results have direct relevance to the relative electrostatic strength of the forces experienced by spherical colloids (e.g. globular proteins) and cylindrical polyelectrolytes (e.g. DNA).

For charge distributions with more complicated geometries vector calculus needs to be used to calculate the electric field. The equivalent vector calculus expression of Gauss's law relates the potential (ψ) to the charge density (ρ):

$$\underline{\nabla}.(\nabla\psi) = \frac{\rho}{\varepsilon_0} \tag{9.12}$$

By definition the gradient of the potential is equal to the electric field ($\underline{E}$):

$$\underline{E} = -\nabla \psi \qquad (9.13)$$

A combination of equations (9.12) and (9.13) provides the Poisson equation for the potential around an arbitrary charge distribution. This was solved in Chapter 2 to find the charge density near a plane.

When a material is placed in an electric field the mobile dipole moments line up due to the torque that they experience (equation (9.5)). It is found experimentally that for a polarised material in a vacuum the induced electric moment per unit volume (P) is proportional to the applied electric field (E):

$$\underline{P} = \varepsilon_0 \chi \underline{E} \qquad (9.14)$$

where χ is a constant of proportionality called the electric susceptibility of the material (Figure 9.4).

It is observed that the effect of a dielectric placed in an electric field is to reduce the electric field in proportion to the relative dielectric constant (ε). This observation explains why water is such a good solvent of ionic crystals such as sodium chloride (Na^+Cl^-). The relative permittivity of water (ε) is 80 at 20 °C and the electrostatic energy of the salt ions is reduced by this factor. However, the treatment of water as a dielectric can be complicated, since the dielectric constant varies from point to point that depends on the exact state of molecular polarisation. The polarisation is affected by the orientation of the water molecules in an electric field and the extent to which this is disturbed by their thermal agitation. Water is such an important solvent that a number of models have been developed to calculate its dielectric constant. One of the most successful of these is due to Lars Onsager, and the effective dielectric (ε) of the water molecules can be approximated (2–3% error) as:

$$\varepsilon = \frac{p'^2 g n}{2\varepsilon_0 k T} \qquad (9.18)$$

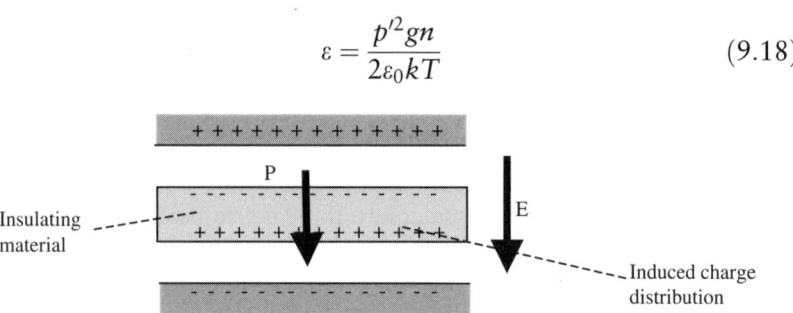

Figure 9.4 Schematic diagram that shows the polarisation (P) induced in an insulating material placed in an electric field (E)

where g is the correlation parameter that describes the relative orientation of the water molecules, p' is the mean dipole moment per molecule, n is the number of molecules per unit volume and kT is the thermal energy. Typical values of the correlation parameter and the mean dipole moment are 2.6 and 8.2×10^{-30} cm respectively.

An increase of the temperature of an aqueous system increases the rotational Brownian motion of the water molecules and shortens the lifetime of the tetrahedral hydrogen bonds that connect them together. The dielectric of water (ε) is therefore a function of both the temperature and the time scale at which an aqueous system is examined. On the assumption that the dielectric of water has a single dominant time scale, the dielectric as a function of frequency (ω) is given by:

$$\varepsilon = \varepsilon_\infty + \frac{\varepsilon_0 - \varepsilon_\infty}{[1 + (\omega \tau)^2]} \qquad (9.19)$$

where τ is a dielectric relaxation lifetime of the water dipoles and is in the order of 10^{-11} seconds. ε_0 and ε_∞ are two characteristic dielectric constants. It is found that the effective dipole moment ($\langle p \rangle$) of a dielectric induced by an electric field is inversely proportional to the temperature and this leads to the useful relationship:

$$\langle p \rangle = \frac{p^2 E}{3kT} \qquad (9.20)$$

where E is the applied electric field and p is the dipole moment of the constituent molecules. In dielectric spectroscopy experiments with larger charged molecules (e.g. DNA) a series of additional dynamic modes are measured that are shifted to longer time scales. These dynamic modes include both internal blob relaxations and whole chain motions.

Another useful electrostatic calculation is provided by a dielectric sphere placed in a uniform electric field. The relative magnitude of the dielectric of the sphere (ε_2) and the surrounding material are important (ε_1) (Figure 9.5) for the determination of the electric field that the sphere experiences. When the dielectric of the sphere is much bigger than the surrounding material ($\varepsilon_2 \gg \varepsilon_1$) the electric field inside the sphere is:

$$E_2 \approx \frac{3\varepsilon_1 E_0}{\varepsilon_2} \qquad (9.21)$$

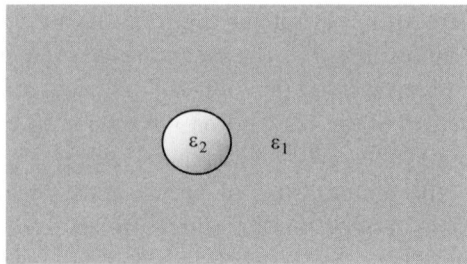

Figure 9.5 Schematic diagram of a globular particle (dielectric ε_2) immersed in a continuous material with a different dielectric constant (ε_1)

whereas when the dielectric of the sphere is much less that the surrounding material ($\varepsilon_2 \ll \varepsilon_1$) the electric field in the sphere is now:

$$E \approx \frac{3E_0}{2} \qquad (9.22)$$

A practical example of this situation is the electrophoresis of a globular protein in water that typically occurs with $\varepsilon_2 < \varepsilon_1$; the second of the two equations (9.22) must be used. An electric dipole (ψ_1) at the centre of a sphere is also affected by the ratio of the dielectric of the sphere to that of the surrounding material. The electric dipole moment of a sphere in this case is given by:

$$\psi_1 = 3\psi_0 \left(\frac{\varepsilon_2}{2\varepsilon_1 + \varepsilon_2} \right) \qquad (9.23)$$

where ψ_0 is the value of a dipole in a vacuum. With an alpha helix in a protein, this dipolar energy may have a role in specific internal interactions that affects the conformation of the protein.

It is also interesting to calculate the energy needed to move an ion between two different dielectric environments, e.g. from an oil (ε_1) to water (ε_2). It is given by:

$$\Delta W = \frac{q^2}{8\pi\varepsilon_0 r} \left(\frac{1}{\varepsilon_1} - \frac{1}{\varepsilon_2} \right) \qquad (9.24)$$

This is, however, only a crude approximation for the solvation energy of an aqueous ion, since it does not include the effects of the hydration shell (Section 9.3).

9.2 DEBYE–HUCKEL THEORY

A useful mean field theory (where fluctuations are neglected) of charge interactions is that due to Debye and Huckel. The starting point for the development of the Debye–Huckel theory for an ion in solution is the Poisson equation, which relates the potential (ψ) to the charge density (ρ). Equation (9.13) is taken and adapted for a material of relative permittivity (ε):

$$\nabla^2 \psi = -\frac{\rho}{\varepsilon \varepsilon_0} \qquad (9.26)$$

The condition of electrical neutrality is imposed on the solution; the quantity of charge in the system on the positively and negatively charged ions must balance:

$$\sum_i z_i n_i = 0 \qquad (9.27)$$

where z_i is the valence (size) of the charges and n_i is the number of charges. It is then assumed that around any ion the charge distribution is spherically symmetric (the arrangement of the charge depends only on the radius r) and the variation in charge follows the Boltzmann distribution, since it is in thermal equilibrium:

$$\rho_i(r) = \sum_i n_i z_i e = \sum_i n_i^0 z_i \exp[-w_{ij}(r)/kT] \qquad (9.28)$$

where $w_{ij}(r)$ is the potential energy corresponding to the mean electrostatic force exerted between ion i and j, and n_i^0 is the charge density at the origin of ion i.

The electrostatic environment of the ions that provides the interaction energy ($w_{ij}(r)$) is considered to a good approximation to be equal to the radially averaged Coulombic potential ($\psi_j(r)$) multiplied by the size of the charge on the ions ($z_i e$):

$$w_{ij}(r) \approx z_i e \psi_j(r) \qquad (9.29)$$

The spherical coordinates equation (9.26), combined with (9.28) and (9.29) gives the radially averaged Poisson–Boltzmann equation:

$$r^{-2} \frac{d}{dr}\left(r^2 \frac{d\psi}{dr}\right) = (\varepsilon \varepsilon_0)^{-1} \sum_i e z_i n_i^0 e^{-z_i e \psi(r)/kT} \qquad (9.30)$$

This equation is difficult to solve due to the exponential function on the right hand side, which stops analytic integration of the expression to solve for the potential (ψ). To make progress with the calculation the argument of the exponential is assumed to be small, i.e.:

$$\frac{z_i e \psi}{kT} \ll 1 \qquad (9.31)$$

the electrostatic potential energy is assumed to be much smaller than the thermal energy. The Poisson–Boltzmann equation (9.30) can then be expanded in the rescaled potential and only the first order terms are used to a first approximation:

$$r^{-2} \frac{d}{dr}\left(r^2 \frac{d\psi}{dr}\right) = \kappa^2 \psi \qquad (9.32)$$

Here an important constant has been introduced, the *Debye screening length* (κ^{-1}), defined to be:

$$\kappa^2 = e^2 \sum_i n_i^0 z_i^2 / \varepsilon \varepsilon_0 kT \qquad (9.33)$$

Equations (9.32) and (9.33) constitute the crucial components of the Debye–Huckel theory for simple ions. The Debye screening length (κ^{-1}) is the length over which the electrostatic forces decay when screened by salt ions in solution (Section 2.3). To find the exact potential and counterion profile, equation (9.32) has to be solved for a particular set of boundary conditions, but the Debye screening length can be used as an order of magnitude estimate for the range of the electrostatic interaction in many biological problems.

9.3 IONIC RADIUS

Stoke's law for the hydrodynamics of small spheres (Section 5.1) is not valid when their size is similar to that of the surrounding water molecules. Indeed ultrafast femtosecond laser experiments point to the importance of the viscoelasticity of water molecules in the determination of ion dynamics at small time scales, which is mirrored in the frequency dependence of the dielectric constant of water (equation (9.19)). The Stoke's radius (r_c) of an ion is found in a number of mobility studies (including

IONIC RADIUS

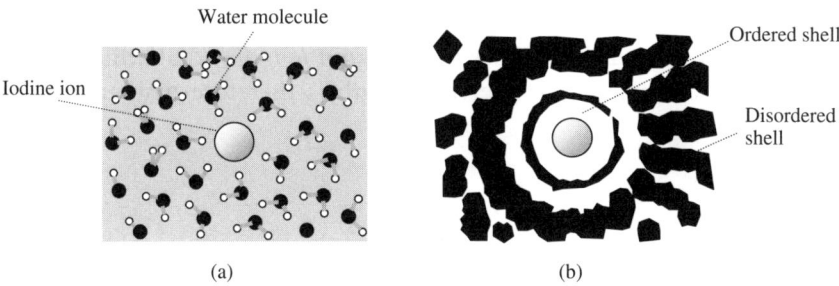

Figure 9.6 Schematic diagrams that show the results of molecular dynamics simulations of water molecules around an iodine ion at (a) 1 ps and (b) 30 ps
(The blurring in (b) indicates the time evolution of the water molecules and the more ordered shell of water close to the iodine molecule [*Reprinted with permission from Femtosecond Chemistry, Eds J. Manz and L. Woste, Copyright (1994) Wiley-VCH*])

laser and pulsed NMR) to be much larger than that expected from X-ray scattering experiments in the liquid state. This mismatch is interpreted as being due to a hydration shell that surrounds each of the ions in solution. Recent femtosecond laser experiments combined with molecular dynamics simulations demonstrate the lifetime of these cages of water is on the order of ~ 10 ps (Figures 9.6 and 9.7 for the hydration shell surrounding the aqueous iodine ions).

Bjerrum proposed a model in which the distance between ions in a concentrated solution can be small enough that transient associations are created between ions of opposite charge. An ion pair is formed when the distance between two elementary charges (e.g. $+e$ and $-e$) is such that the electrostatic energy of attraction is greater or equal to the thermal energy (kT). The thermal energy thus tends to disrupt these temporary ion pairs, which constantly associate and dissociate in solution in a

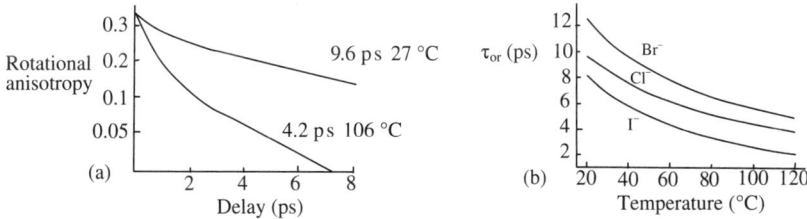

Figure 9.7 Pulsed laser experiments (a) can probe the rotational motion of halide ions in aqueous solution. The measured orientational diffusion time constants (τ_{or}) of the solvation shells of chlorine, bromine and iodine are shown as a function of temperature (b) [*Reprinted with permission from Femtosecond Chemistry, Eds J. Manz and L. Woste, Copyright (1994) Wiley-VCH*]

process of dynamic equilibrium. The thermal and electrostatic energies can be equated and this defines the *Bjerrum length* (l_B) at which stable ion pairs are formed. The Bjerrum length is therefore mathematically defined to be:

$$l_B = \frac{e^2}{4\pi\varepsilon\varepsilon_0 kT} \quad (9.34)$$

The dielectric constant of water (ε) is 80 at 20 °C and l_B is thus 7.12 Å at this temperature (these are useful numbers to remember). The ion pairs are short lived in normal simple electrolyte solutions due to the small Bjerrum length. However, ion pairs are an important effect in polyelectrolyte solutions and the Bjerrum length is a useful quantity to gauge the range of electrostatic interactions as a function of the dielectric of a solvent.

To understand the behaviour of charged ions in solution in more detail consider equation (9.4) for the orientational energy (W) of a dipole (p) in an electric field (E). The energy associated with the dipole moment of water is quite strong (10–20 kT) and charged ions are thus surrounded by layers of oriented water, which are not completely randomised by the thermal energies (Figure 9.6). Data from pulsed femtosecond laser experiments on the rotational motion of water molecules that surround halogen ions in aqueous solution are shown in Figure 9.7. The simulations are in good agreement with these experiments and show a restricted shell of water molecules around the ions in solution.

Detailed time averaged X-ray and neutron diffraction studies of liquid water indicate two regions of ordering around charged ions. In the first hydration shell there is ordered ice-like water and in the surrounding layer there is a second hydration shell of more disordered water.

The free energy of hydration (ΔG_H) for an ion is the work required to move an ion from a vacuum ($\varepsilon_1 = 1$) to a medium of relative dielectric ε_2. The free energy of hydration for an ion of charge q (a simplification of equation (9.24)) is given by:

$$\Delta G_H = \frac{q^2}{8\pi\varepsilon_0 r}\left(1 - \frac{1}{\varepsilon_2}\right) \quad (9.35)$$

It is assumed in this calculation that the ion is initially in a vacuum, the uncharged ion is transferred from a vacuum to the water and the ion is then recharged in the water. The ionic radii found from such calculations need some analysis. Values for the radii calculated are consistently bigger

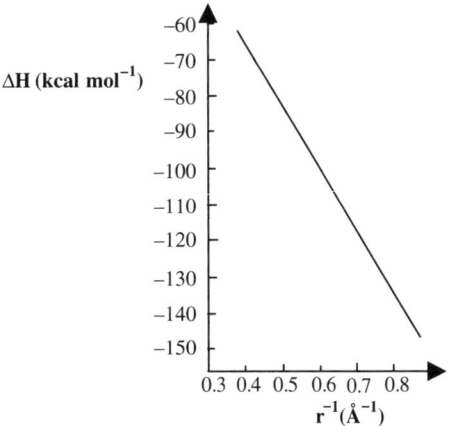

Figure 9.8 Trend of the variation of the hydration enthalpy (ΔH) with ionic radius (r) for small ions in solution
[Reprinted with permission from A.A. Rashin and B. Honig, *Journal of Physical Chemistry*, 89, 5588–5593, Copyright (1985) American Chemical Society]

than those measured in crystallography experiments (Figure 9.8). More sophisticated approaches that take into account the change in entropy due to the structure breaking properties of the ion on the dipole moments of the surrounding water molecules are therefore needed.

It is also possible to quantify the effects of ions on a solution from the change in viscosity (η) of a solution to which the ions are added. The reduced viscosity of the ionic solution is found to take a characteristic form:

$$\frac{\eta}{\eta_0} = 1 + Ac^{\frac{1}{2}} + Bc \qquad (9.36)$$

where A is a constant related to the electrostatics, B is a constant related to the degree of structural rearrangement of water, c is the ion concentration, and η_0 is the viscosity of the pure unperturbed water. Salty water (1 M NaCl ~ 5%) thus has a distinct increase in viscosity compared with a pure dialysed sample; the hydration shells that surround the water ions contribute to this increase.

The ability of an ion to restructure the water that surrounds it is related to its exact chemical nature. Hoffmeister classified salts in terms of their ability to precipitate proteins. The order of his series is explained by the effective charge density of ions in aqueous solutions when they are surrounded by their hydration shells and is a useful predictor in a wide range of hydrated biological reactions. Some typical anions and cations in solution are listed in equations (9.37) and (9.38):

Anions
Strongly hydrated Weakly hydrated
–stabilises protein solutions –destabilises protein solutions

$$SO_4^{2-} < CH_3COO^- < F^- < Cl^- < Br^- < NO_3^- < ClO_4^- < I^- < CNS^- \tag{9.37}$$

Cations
Weakly hydrated Strongly hydrated

$$(CH_3)_4N^+ < NH_4^+ < Cs^+ < Rb^+ < K^+ < Na^+ < H^+ < Ca^{2+} < Mg^{2+} < Al^{3+} \tag{9.38}$$

This is an approximate series and the specific chemistry of competing reaction schemes can sometimes reverse the order of close neighbours in the list.

The process of *salting out* that was used by Hoffmeister to classify ions often occurs practically in the purification and crystallisation of biological molecules (Section 3.4). Salt ions in a solution can cause a rearrangement of the water molecules, competing with the biological molecules for their hydration shell and hence can reduce their solubility. Thus the higher charge density ions in the Hoffmeister series are more effective for salting out proteins. For a specific protein empirical laws exist for the solubility as a function of the ionic strength of the form:

$$\log S = \log S_0 - KI \tag{9.39}$$

Where S is the solubility, S_0 is the value of the solubility at zero ionic strength ($I = 0$) and K is a constant proportional to the size of the protein. A fundamental derivation of such a formula in terms of a sticky protein–protein interaction potential is a current challenge for colloidal science. *Salting in* can also occur during the dissolution of proteins as electrostatic attractive forces between protein aggregates are disrupted by the addition of salt.

9.4 THE BEHAVIOUR OF POLYELECTROLYTES

The volume that surrounds a polyelectrolyte molecule in solution contains small ions of the opposite sign to the polyelectrolyte that have dissociated from the polymeric chain and that maintain global charge neutrality, e.g.

THE BEHAVIOUR OF POLYELECTROLYTES 219

Na^+, K^+ and Mg^{2+} for nucleic acids or acidic polysaccharides and OH^-, Cl^- anions for polyamines. Thus polyelectrolytes are surrounded by a cloud of counterions and the dissociation of the charged groups along the chain backbone is often the dominant contribution to the solubility of the molecules in water. Fluctuations can occur in the spatial distribution of the counterions, consistent with the requirements of thermal equilibrium, but globally charge neutrality must be maintained.

Polyampholytes (e.g. proteins) contain positive and negative charged groups bound to the same polymer chain. These charges can lead to anisotropic charge distributions within the protein. Hydrogen (H^+) and hydroxye (OH^-) are 'special' ions associated with polyacids and polybases in polyampholytes and polyelectrolytes, and make the charge fraction on these polymers particularly sensitive to changes in pH (Section 1.1). The neutralisation of the acidic or basic units with a corresponding alkali or acid removes this effect, i.e. a variation in the polymer concentration does not change the charge fraction on each polyelectrolyte chain; this is a useful simplification in many experiments on charged polyions.

The conformation of polyelectrolytes is strongly affected by the repulsion of charges along the backbone. Charge repulsion encourages extended conformations of chains at low ionic strengths. For example, the charged groups are important for the rigidity of semi-flexible polymers such as DNA and are a dominant factor with flexible polyelectrolytes such as alginates (a constituent of seaweed). Once polyelectrolytes overlap in more concentrated solutions the charge interaction becomes screened and the end-to-end distance of the molecules reduces.

The physics of charged macromolecules in solution is very rich. The chain conformation depends on the fraction of monomers that are charged, the concentration of monomers in the solution, the concentration of low molecular weight salt, the intrinsic rigidity of the polymer backbone and the quality of the solvent for the backbone chemistry (e.g. 'good', 'bad' and 'theta'; Section 8.2). Polyelectrolytes can be classified in terms of strongly and weakly charged behaviour. In strongly charged polyelectrolytes every monomer carries a charge. Therefore Coulomb monomer–monomer interactions are the dominant forces that determine the conformation of the chain if the chain backbone is flexible. If e is the charge held on a monomer and ε is the dielectric, the potential energy of the screened Coulomb interaction (potential $V(r_{ij})$) between charged links i and j separated by distance r_{ij} is given by:

$$V(r_{ij}) = \frac{e^2}{\varepsilon r_{ij}} e^{-r_{ij}\kappa} \qquad (9.40)$$

where κ^{-1} defines the Debye radius that determines the screening of the electrostatic interaction by other ions in solution (equation (9.33)).

The size of strongly charged polyelectrolyte macromolecules (L) in a dilute salt free solution is proportional to the number (N) of charged links, since a charged macromolecule is fully stretched ($L \sim Na$, where a is the monomer length). A weakly charged flexible polyelectrolyte macromolecule in dilute solution can be visualised as an extended chain of blobs (Figure 9.17). The length of the chain is again proportional to the number of monomers, but the chain size needs to be rescaled by the number of monomers in an electrostatic blob ($L \sim ND/g$; g is the number of monomers in a blob and D is the blob size).

In a polyelectrolyte solution of finite monomer concentration, the Coulomb interaction is screened by counterions, so that the chains become coiled on large length scales. The distance between the neighbouring chains in solution is of the order of the Debye screening radius. For flexible polyelectrolytes this charge screening causes a large contraction of the end-to-end distance upon coil overlap in semi-dilute solutions.

The *transport phenomena* involved with charged chains is an important but complicated subject. The counterion clouds associated with a polyelectrolyte must be dragged around when the polyelectrolyte moves and this process dissipates energy. The field of polyelectrolyte dynamics relates to the subject of electrophoresis, the driven motion of a charged molecule in an electric field, which will be investigated in Section 13.5 due to its importance in DNA sequencing.

The addition of charged groups to a polymeric molecule has a dramatic effect on its osmotic pressure in solution. The osmotic pressure of the solution is dominated by the contribution of the counterions. Consider a negatively charged polyelectrolyte in solution. The number of cations (n_+) in the solution is provided by both any additional small molecule salts in the mixture and the counterions associated with the polyelectrolytes. However, the number of negatively charged anions (n_-) is predominantly due to the salt, since the number of polyelectrolyte chains is negligibly small compared with the number of small salt ions. The number of positive (n_+) and negative (n_-) charged units in the solution can therefore be written as:

$$n_+ = n_s + n_p v \qquad (9.41)$$

$$n_- = n_s + n_p \approx n_s \qquad (9.42)$$

where n_s is the number of cations and anions per unit volume and n_p is the number of molecules of a polyion per unit volume when negatively charged polyions of valence v are dissolved. Each independent unit in the solution contributes kT per unit volume to the osmotic pressure, much like the molecular contribution to the pressure of an ideal gas as the molecules bounce off the walls of their container. The total *osmotic pressure* of the solution is thus:

$$\pi = kT\phi(n_+ + n_-) = kT\phi(n_p v + 2n_s) \quad (9.43)$$

where ϕ is the fraction of polyelectrolyte ions that are dissociated per unit volume. Thus the osmotic pressure of charged polymers is much larger than their neutral counterparts, by a factor of $\phi n_p v$, and this has a range of associated phenomena, e.g. the swelling behaviour of nappies and the shape of the cornea in the eye.

The calculation that gives equation (9.43) does not assume that all the counterions are dissociated. To decide on the nature of *counterion binding* in polyelectrolytes the osmotic coefficient (ϕ) is introduced. The osmotic coefficient is determined experimentally as the ratio of the osmotic pressure (π) to a reference value π_0 ($\phi = \pi/\pi_0$) defined for full dissociation:

$$\pi_0 = (2n_s + n_p v)kT \quad (9.44)$$

Thermodynamically the osmotic pressure can be shown to be the negative rate of change of free energy with the volume (V) of the solution. The difference between the osmotic pressure and the reference value for full dissociation ($\pi - \pi_0$) is due to the additional electric free energy in the system that causes the counterions to bind to the polyions (charge condensation).

9.5 DONNAN EQUILIBRIA

The process by which the osmotic pressure of a system is regulated by the affinity of molecules for their counterions is called *Donnan equilibrium* which is an important phenomenon in a wide range of biological systems. Striated muscle fibres (actin and myosin assemblies) are held apart by a Donnan pressure. The cornea in the eye is osmotically stressed and the Donnan pressure is important to maintain the interfibrillar spacing (Figure 9.9). In mammalian cells ion pumps are required to regulate the Donnan pressure to provide the correct working environment of the cell. Malfunctioning of the Donnan equilibrium in any of these examples would be catastrophic for the organism involved.

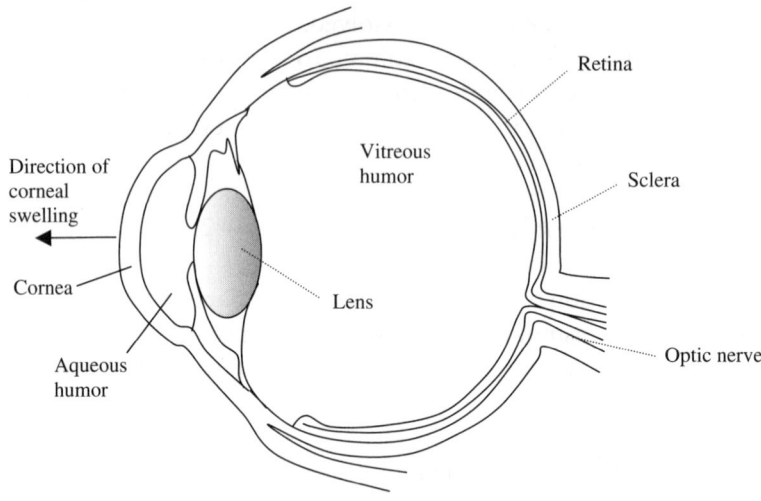

Figure 9.9 Donnan equilibrium controls the swelling of the human eye and can effect the cornea, aqueous humor, lens and vitreous humor

Donnan equilibrium describes the extent to which small counterions travel through a partition between two ionic environments and this is a useful simple scenario with which to approach the subject. Consider two compartments, A and B, separated by a membrane through which only small molecules such as water and ions can pass (Figure 9.10). Compartment B contains polyions and salt whereas compartment A only has small salt molecules. The chemical potentials of the salt are identical on the two sides of the membrane (μ_s^A and μ_s^B), since there is no change in the

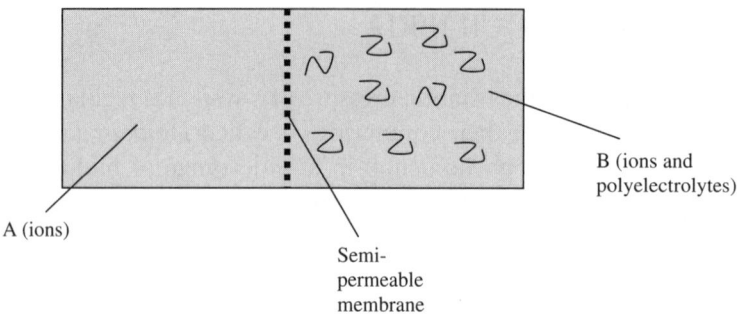

Figure 9.10 Schematic diagram of Donnan equilibrium of the concentration of ions and polyions maintained between two compartments (A and B) separated by a semi-permeable membrane

free energy of each salt molecule as it passes through the membrane when in thermal equilibrium:

$$\mu_s^A = \mu_s^B \qquad (9.45)$$

Electrical neutrality must be maintained in each compartment to avoid an excessive electrostatic energy penalty. n is the molar concentration of a particular particle species. The salt is assumed monovalent and the polyions each carry v charges. The number of positive (n_+) and negative (n_-) charges in each compartment must balance

$$\text{In } A: n_+^A = n_-^A \qquad (9.46)$$
$$\text{In } B: n_-^B + \phi v n_p = n_+^B \qquad (9.47)$$

where n_p is the molar concentration of the polyelectrolyte and ϕ is the fraction of charges that dissociate from the polyelectrolytes (the osmotic coefficient). When the system is in full equilibrium the cation concentration is lower in A than B whereas the anion concentration is higher. This process of Donnan equilibirum is characterised by a coefficient (Γ) defined as:

$$\Gamma = \lim_{n_p \to 0} \frac{(n_+^B - n_+^A)}{v n_p} = \frac{\phi}{2} \qquad (9.48)$$

For the case where no counterions are bound to the polyion ($\phi = 1$) the Donnan coefficient (Γ) is $\frac{1}{2}$. For DNA in a low ionic strength medium a typical value of the Donnan coefficient (Γ) is -0.1. This implies that 80% of the counterions behave as bound to the polyelectrolyte, in agreement with that expected from the discussion of counterion condensation in Section 9.8. In the intracellular environment there is a complex process of Donnan equilibrium due to the interchange of ions between the mixed cocktail of polyionic species.

9.6 TITRATION CURVES

There is an important difference between the two possible mechanisms through which charges can be placed on a polyion, denoted the *annealed* and *quenched* mechanisms. When a weakly charged polyelectrolyte is obtained by copolymerisation of neutral and charged monomers the total number of charges and their position is fixed. This is called a *quenched polyelectrolyte*.

Polyacids or polybases are polymers in which the monomers can dissociate and acquire a charge that depends on the pH of the solution. The dissociation of a hydrogen ion (H^+) ion from an oppositely charged polymer (e.g. in a COOH group) gives a negative charge (COO^-. This is an *annealed polyelectrolyte*, the total number of charges on a given chain is not fixed, but the chemical potential of the hydrogen ion and the chemical potential of the charges is imposed by the pH of the solution.

Acid–base equilibria apply to biological polyelectrolyte molecules (Section 1.1), which predominantly fall within the annealed category. For example, hydrogen ions (protons) can bind to a basic unit on a polyelectrolyte (A^-) to give an acid AH, as described in equation (9.1). The association constant (K_a) for the acid–base equilibria is defined as:

$$K_a = \frac{c_{AH}}{c_{A^-} c_{H^+}} \tag{9.49}$$

where c indicates the concentration of the species in moles. θ is defined as the fraction of acid monomers (A^-) with a bound proton:

$$\theta = \frac{c_{AH}}{c_{A^-} + c_{AH}} \tag{9.50}$$

This can be rearranged to give:

$$\frac{\theta}{1-\theta} = K_a c_{H^+} \tag{9.51}$$

Titration is the experimental process in which a well characterised acid or base is added to a solution of polyions to determine their degree of dissociation. The progress of a titration reaction can be determined using the conductivity of the solution or by the use of pH sensitive dyes. A titration curve is shown in Figure 9.11. v is defined as the mean number of bound protons and n is the total number of ions, so the fraction of dissociated ions is $\theta = v/n$. Therefore equation (9.51) can be reexpressed as:

$$\frac{v}{n-v} = K_a c_{H^+} \tag{9.52}$$

TITRATION CURVES

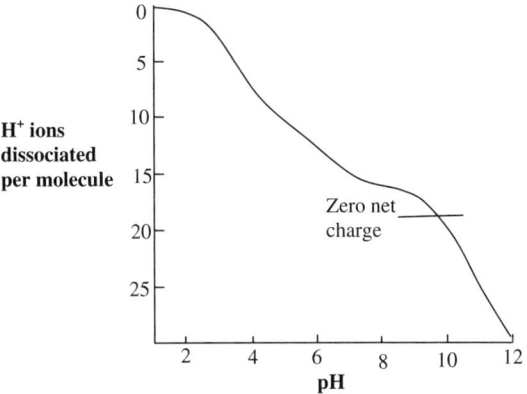

Figure 9.11 Titration of a polyacid
(The concentration of hydrogen ions dissociated per molecule is shown as a function of the solution pH [*Reprinted with permission from C. Tanford and J.D. Hauenstein, J. Am. Chem. Soc, 77, 5287–5291, Copyright (1956) American Chemical Society*])

The degree of dissociation (α) is one minus the degree of association:

$$\alpha = \frac{c_{A^-}}{c_{A^-} + c_{AH}} = 1 - \theta \tag{9.53}$$

The dissociation constant (K_d) is equal to the inverse of the association constant:

$$K_d = K_a^{-1} \tag{9.54}$$

Equation (9.52) can therefore be rewritten:

$$\frac{(1-\alpha)}{\alpha} = K_a^{-1} c_{H^+} \tag{9.55}$$

It is useful to define two important parameters that relate to the strength of the hydrogen ion concentration (pH) and the degree to which a polyacid can associate (pK_a):

$$\mathrm{pH} = -\log c_{H^+} \tag{9.56}$$
$$pK_a = -\log K_a \tag{9.57}$$

The logarithm is introduced to facilitate calculations with concentrations that can vary by many orders of magnitude. The exponential nature of the electrostatic interaction makes solutions sensitive to a vast range of hydrogen ion concentrations from nM to M, and there are a correspondingly vast range of hydrogen ion concentrations and equilibrium constants.

The logarithm of equation (9.55) can be taken to give the Henderson–Hasslebach equation. The equation is an extremely useful relationship which relates the pH of a solution and the intrinsic pK_a value of an ionisable group to the charge fraction (α) of ionisable groups in the solution (Section 1.1):

$$\text{pH} = pK_a + \log\left(\frac{\alpha}{1-\alpha}\right) \quad (9.58)$$

The Henderson–Hasslebach equation requires the assumption that values of the association constant (K_a) are independent of the charge on the polyion. This is not a completely reasonable assumption as will be seen. The change in free energy (ΔG_0 upon the association of the polyions with their counterions) can be related to the association (K_a) and dissociation (K_d) equilibrium constants:

$$\Delta G_0 = -RT \ln K_a = RT \ln K_d \quad (9.59)$$

From the definition of the pK_a value and equation (9.57) this gives:

$$\Delta G_0 = -2.3RTpK_d \quad (9.60)$$

The factor of 2.3 occurs due to the change in base from 10 to e, i.e. the log becomes a ln. pK_d is defined in an analogous fashion to equation (9.57). W_{el} is defined as the work required to bind hydrogen ions onto a polyion and is the work done against the polyion potential provided by all the charged groups. Therefore an expression for the work of binding is:

$$W_{el} = e\psi(a) \quad (9.61)$$

where $\psi(a)$ is the potential at the surface of the polyion and e is the electronic charge. The total change in free energy of a polyion with N groups is therefore:

$$\Delta G = \Delta G_0 + NW_{el} \quad (9.62)$$

Thus there is an effective pK_a value (pK'_a) measured in an experiment given by:

$$pK'_a = pK_a + 0.43\frac{e\psi}{k_B T} \quad (9.63)$$

This expression allows the effect of neighbouring charged groups on the pK_a value of a polyion to be quantified to a first approximation.

9.7 POISSON–BOLTZMANN THEORY FOR CYLINDRICAL CHARGE DISTRIBUTIONS

The Poisson equation for the potential (ψ) as a function of the charge density is given by equation (9.26). Substitution of a Boltzmann distribution for the energies of the counterions in equation (9.26) for a monovalent salt gives the *Poisson–Boltzmann equation*:

$$\nabla^2 \psi = 2n_s e \frac{\sinh(e\psi/kT)}{\varepsilon\varepsilon_0} \qquad (9.64)$$

where n_s is the density of salt ions. Through a similar process to that with which the Poisson–Boltzmann equation was solved in spherical geometry in Section (9.2), numerical solutions of the Poisson–Boltzmann equation for a line charge with cylindrical symmetry are shown in Figure 9.12.

There are three separate charged regions that surround a polyion and they can be classified in terms of their distance from the surface of the polymer. At long distances, in the *Debye–Huckel* region, the ions are treated as point charges and form a double layer around the polymer screening its electrostatic force. The Poisson–Boltzmann equation (9.64) can be linearised for the polyion in a similar spirit to that of a simple ion in equation (9.32). At intermediate distances, in the *Gouy region*, the

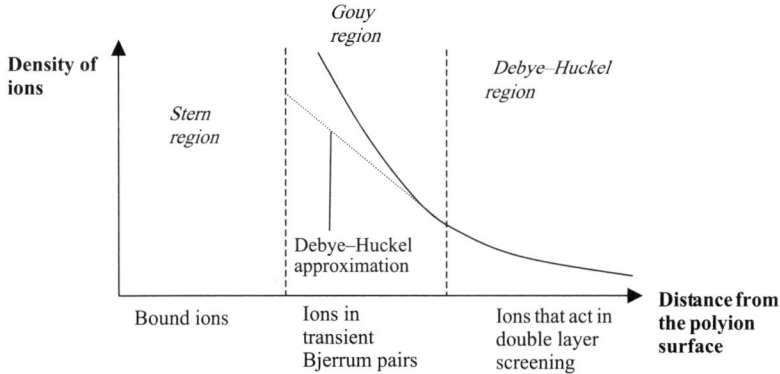

Figure 9.12 Density of ions as a function of the distance from the surface of the polyion showing Stern, Guoy and Debye regions

cylindrical symmetry can be used with the point charge model, but the Poisson–Boltzmann equation must be solved without any approximation since the potential is large and this is computationally more intensive. Numerical solutions show that a condensed phase of counterions occurs near the polyelectrolyte. At short distances, in the *Stern region*, the cylindrical symmetry of the chain disappears at very close distances to the charged groups along the polyion, and the structural and geometrical factors of the specific chain chemistry have to be considered.

9.8 CHARGE CONDENSATION

In a solution of a sufficiently strongly charged polyelectrolyte, a fraction of counterions stays in the immediate vicinity of the polymer chain, effectively neutralising some of the chains' charge. This phenomenon is referred to as *counterion condensation* and was encountered already in the discussions of Donnan equilibrium and osmotic pressure. Manning devised a simple model for understanding charge condensation with polyelectrolytes and the model is a useful starting point for developing an explanation of the interactions between charged biological polymers. The effective potential that determines the forces between polyions is regulated by the condensation of the counterions that surround the chains; the effective charge fraction of a chain is often much smaller than expected from the chain chemistry.

The assumptions required in the Manning continuous model are that the solvent is a continuum with a uniform dielectric constant (ε), the ions are represented by a continuous charge density ($\rho(r)$) and the polyion is modelled by a line charge of infinite length characterised by a charge density parameter (ξ). Using Gauss' theorem for a line charge, the electric field (E) can be calculated around a polyelectrolyte chain in water (compare with equation (9.11)):

$$E = \frac{e}{2\pi\varepsilon\varepsilon_0 br} \tag{9.65}$$

where b is the linear charge density (units of Coulomb m^{-1}), e is the electronic charge and r is the equipotential radius for a cylinder around the line of charge (Figure 9.13). The radial electric field (E) is related to the potential (ψ) by the spatial derivative:

$$E = -\frac{d\psi}{dr} \tag{9.66}$$

CHARGE CONDENSATION

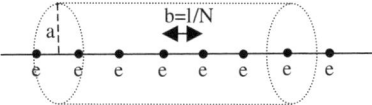

Figure 9.13 Geometry of a cylindrical polyelectrolyte used in the calculation of Manning charge condensation
(b is the distance between two adjacent charged groups (charge e) along the backbone equal to the length of the chain (*l*) divided by the total number of charged units (*N*). a is the radius of a cylinder surrounding the charged groups along the chain backbone.)

Equation (9.66) can be integrated to provide the potential as a function of radius:

$$\psi = A - 2e \ln\left(\frac{r}{4\pi\varepsilon\varepsilon_0 b}\right) \tag{9.67}$$

A monovalent counterion with a unit charge (e) at a certain radial distance (r) away from the polyion acquires a potential energy $E_p(r)$ (Figure 9.14):

$$E_p(r) = e\psi(r) \tag{9.68}$$

A Boltzmann distribution for the thermalised energies of the counterions is assumed and this can be written in terms of a charge parameter (ξ):

$$e^{-E_p/kT} = W_0 r^{-2\xi} \tag{9.69}$$

where $W_0 = e^{eA/kT}$ and the charge parameter (ξ) is defined as the ratio between the Bjerrum length (l_B) and the linear charge density (b):

$$\xi = \frac{l_B}{b} \tag{9.70}$$

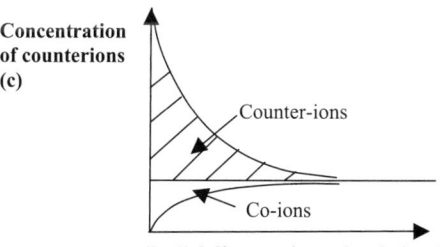

Figure 9.14 Radial distribution of the concentration of counterions and co-ions that surround a polyelectrolyte

The number of counterions inside a cylinder (radius r_0) of unit length is proportional to the integral:

$$\int_0^{r_0} W_0 r^{-2\xi} 2\pi r \, dr = 2\pi W_0 \int_0^{r_0} r^{1-2\xi} dr \qquad (9.71)$$

It is important to note that this integral diverges at the origin ($r = 0$) if the charge density parameter is greater than one ($\xi > 1$), which is an unphysical result. A condensed layer of counterions is invoked to avoid this problem, which maintains the charge parameter at a value of one and a finite energy for the system. The fraction of charge neutralised is simply expressed in terms of the charge density parameter:

$$\frac{(\xi - 1)}{\xi} = 1 - \frac{1}{\xi} \qquad (9.72)$$

This analysis can be extended to the case of multivalent counterions. For an ion of charge ze the neutralised charge fraction is:

$$1 - \frac{1}{z\xi} \qquad (9.73)$$

The condensation process is explained physically in terms of the formation of ion pairs (Bjerrum pairs) on the surface of the polymer. The effect of Manning condensation above a critical charge fraction on the effective charge fraction of the polyelectrolyte chain is illustrated in Figure 9.15. The effective charge fraction reaches a constant value above the Manning threshold.

For example, consider a double stranded chain of DNA at 20 °C. The distance between phosphate groups ($b = 1.7$ Å) is taken for the B form of

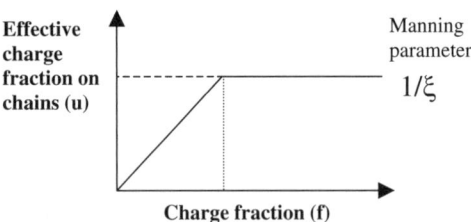

Figure 9.15 The effective charge fraction (u) on a polymer chain as a function of the charge fraction expected from the number of chargeable groups (f). At high charge fractions the effective charge fraction saturates at the Manning value ($1/\xi$).

CHARGE CONDENSATION

DNA and hence the charge parameter (ξ) is 4.2. The fraction of charge neutralised is given by equation (9.73) and on average three quarters of the phosphate groups are neutralized!

The *Osmotic coefficient* (ϕ, equation (9.43)) can be measured experimentally. The osmotic pressure ($\pi_{\exp}$) found using osmometry on a polyelectrolyte in distilled water is proportional to the concentration of counterions (c_e), since this is the dominant contributing species, so:

$$\pi_{\exp} = kT\phi c_e \qquad (9.74)$$

and the osmotic coefficient can be identified with the fraction of neutralised charge:

$$\phi = 1 - \frac{\xi}{2} \qquad (9.75)$$

Such a theory is in reasonable agreement with experiment. Manning condensation can also be used to explain *association phenomena* in highly charged biological molecules, e.g. drug binding on to DNA.

In summary, the continuum Manning model predicts there is an excess of counterions in the vicinity of a polyion and the charge density variable (ξ) is important for describing the effective charge fraction of such a linear polyelectrolyte. However, there are two limits on the validity of this type of Manning model. The polyelectrolytes are required to be of infinite length and there is assumed to be a vanishingly small free ion concentration. A more satisfactory approach is to solve the full Poisson–Boltzmann (PB) equation numerically for the cylindrical geometry (equation (9.64)), and use this to determine the degree of counterion association. The PB equation is valid at physiological salt concentrations (0.15 M), which is not true for the Manning approach. The Manning model is, however, conceptually much simpler to use than the Poisson–Boltzmann equation and provides a useful starting point for learning about charge condensation.

Recent anomalous X-ray scattering experiments provide direct structural information on the morphology of counterion clouds around polyions. The concentration of rubidium counterions on the surface of a DNA chain as a function of the amount of divalent magnesium in the solution is shown in Figure 9.16. Such data are found to be in reasonable agreement with a Poisson–Boltzmann model.

Charge condensation is also found with spherical colloids. Solution of the Poisson–Boltzmann equation in dilute solutions with spherical symmetry

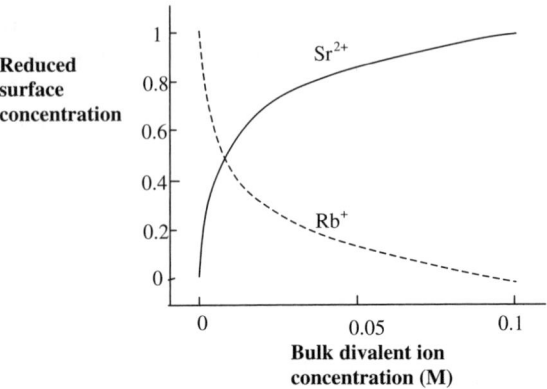

Figure 9.16 Anomalous small angle X-ray scattering data for DNA molecules associated with rubidium (Rb^+) and strontium (Sr^{2+}) counterions
(Rubidium ions are displaced from the surface of the DNA as the bulk divalent counterion concentration is increased in accord with Poisson–Boltzmann theory [*Reprinted with permission from K. Andresen, R. Das, H.Y. Park et al, Physical Reviews Letters, 93, 248103, Copyright (2004) American Physical Society*]

(equation (9.32)) can provide an effective charge to renormalise the interaction strength in a DLVO treatment (equation (2.18)) for the interparticle potential. Charge renormalisation due to counterion condensation is required for highly charged colloids (large values of the unscreened structural charge $Q_{str}e$) that have small radii of curvature (R), when the following inequality for monovalent counterions is obeyed:

$$\frac{Q_{str}l_B}{4R} \geq 1 \tag{9.76}$$

And for these highly charged and curved colloids the charge is pinned at a value (the effective colloidal charge, Q_{eff}) given by:

$$Q_{eff} = \frac{4R_{eff}}{l_B} \tag{9.77}$$

Where R_{eff} is the radius of an effective smaller colloid.

9.9 OTHER POLYELECTROLYTE PHENOMENA

For weakly charged polyelectrolytes pronounced counterion condensation only occurs in a poor solvent (where the blobs are globular) and in this case it constitutes an avalanche-like process that results in the condensation of nearly all the counterions on the macromolecule. This can provide an important contribution to the folding process in highly charged globular proteins.

The Coulombic interactions of a strongly charged polyelectrolyte tend to stiffen the chain and lead to an increase in its persistence length (l_p). The contribution to the total persistence length due to the electrostatics (l_e) is called the electrostatic persistence length. A useful theory that predicts the persistence length of charged semi-flexible chains is due to Odjik, Skolnick and Fixman. The OSF theory is applicable to semi-flexible biopolymer chains such as actin and DNA. The electrostatic component (l_e) is added on to the intrinsic rigidity due to the backbone chemistry:

$$l_T = l_p + (4l_b\kappa^2)^{-1} = l_p + l_e \tag{9.78}$$

where l_b is the Bjerrum length, κ^{-1} is the Debye screening length and l_p is the intrinsic persistence length. A similar behaviour (separation of the persistence length into two components) is expected for flexible polyelectrolytes whose conformation consists of a bayonet of blobs, but there continues to be some dispute as to how the blob size renormalises the effective length of the charged chains (Figure 9.17). Both κ^{-1} and κ^{-2} dependences of the electrostatic persistence length (l_e) are predicted theoretically for flexible polyelectrolytes. Some caution is therefore required when applying equation (9.78) to chains with flexible architectures.

When a small fraction of the links of a cross-linked polymer network are charged its collapse in a poor solvent proceeds (as the solvent quality deteriorates) as a discrete first order phase transition. The abrupt change in the size of the network is associated with the additional osmotic pressure of the gas of counterions in the charged network (Figure 9.18). This process of collapse is analogous to the globule-coil transition of a single

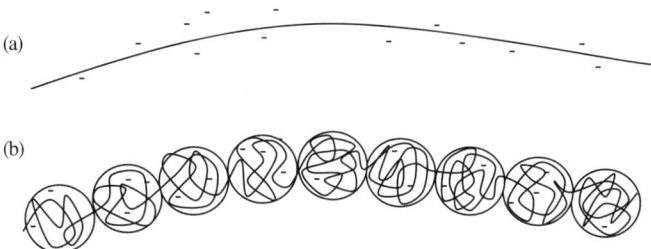

Figure 9.17 The OSF type approach for the calculation of the electrostatic contribution to the persistence length is possible in both semi-flexible (a) and flexible polyelectrolyte solutions (b).

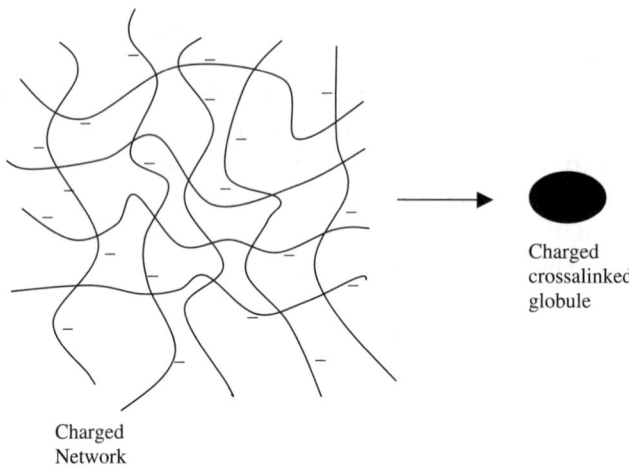

Figure 9.18 Charged cross-linked network contracts into a cross-linked globule with a first order transition as the quality of the solvent for the polymer chains deteriorates (The positively charged counterions are not shown for simplicity.)

polyelectrolyte chain (Section 3.3) and has been used to produce biomimetic muscular contraction of synthetic polyelectrolyte gels.

The compatibility of a mixture of two polymers improves substantially after one of the components is weakly charged. Many of the extreme phase separation phenomena observed in Section 3.5 are thus avoided in vivo in the intracellular environment due to this mechanism.

An area of hot debate over many years is that of an attractive force between like-charged polyelectrolytes. The current consensus is that an attractive force does exist due to shared counterion effects between adjacent polyions, but an accurate quantitative prediction of the associated experimental phenomena is yet to be made.

FURTHER READING

A.Y. Grosberg and A.R. Khoklov, *Statistical physics of macromolecules*, Academic Press, 1994. Clear compact account of the properties of polyelectrolytes.

M. Daune, *Molecular Biophysics*, Oxford University Press, 1999. Useful introduction to the physical properties of ions and polyions.

G.B. Benedek and F. Villars, *Physics with illustrative examples from Medicine and Biology*, Springer, 2000. Well written expansive account of biophysical applications of electromagnetism.

TUTORIAL QUESTIONS

9.1) The dissociation pK_a value for lysine is 10 and the pK_a value for polylysine is 9.5. Can you account for the shift?

9.2) What is the effective charge fraction on a polymer of hyaluronic acid according to the Manning model, given that the distance between charged groups along the polymer backbone is 5 Å? Each charged group is assumed to have a single unscreened electronic charge before condensation.

9.3) An amyloid fibre associated with a pathogenic misfolded protein has a charge fraction (f) of 0.5, and a repeat unit of 1 nm. Using the Odjik, Skolnick and Fixman calculation estimate the electrostatic contribution to the persistence length of the fibre in a buffer solution with a Debye screening length (κ^{-1}) of 4 nm.

9.4) A very big industrial application for polyelectrolytes is in the gels that fill disposable nappies (or diapers in the USA). What physical properties of polyelectrolytes do you think make them especially suitable for this critical technology? The osmotic pressure of a charged polymer is dominated by the counterions. Compare the osmotic pressure of a charged PAMPS gel (poly acryl amides methyl propane sulfonic acid) with that of its neutral counterpart at a polymeric monomer concentration of 1 mM. State any assumptions used.

9.5) How big is a statistical blob of uncharged polylysine (a polypeptide)? If you charge up the polylysine by a change in pH how big is the charged blob? What happens to the conformation of the chain during the process of charging?

10
Membranes

Every living cell is surrounded by an outer membrane (Figure 10.1). The membrane acts as a partition that divides the cell between its interior and extracellular environment. It is the interface through which a cell communicates with the external world. Biological membranes are involved in a wide range of cellular activities. The membrane participates in simple mechanical functions such as motility, food entrapment and transport. Also, highly specific biochemical processes are made possible by the membrane's structure, including energy transduction, nerve conduction and biosynthesis. Adhesion between cellular membranes is thought to be a critical factor in the determination of the morphology and development of organisms (morphogenesis) from the initial ball of dividing cells (the blastula).

Biological lipids in solution self-assemble into thin bilayer membranes that can compartmentalise different regions within a cell and protect the inside of the cell from the external environment. The membrane remains intact even when the bathing medium is extremely depleted of lipids due to the lipids' extremely low critical micelle concentration (Section 6.1). As a result of unsaturation or branching of the constituent lipids, membranes are in a fluid state at physiological temperatures, with rapid two dimensional rearrangements possible of the neighbouring lipids.

Long-chained polypeptide polymers are often embedded in membranes and consist of long strings of amino acid residues ($\sim$500 000). The polypeptides are relatively rigid when compared with the lipids in the surrounding cell membrane and they are amphiphilic with their surface exposed to both hydrophobic and hydrophilic regions of the

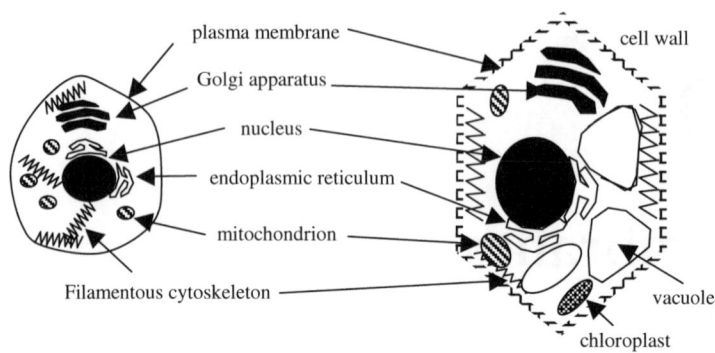

Figure 10.1 Schematic diagrams of an animal and a plant cell that show the compartmentalised structures, and some individual pieces of intracellular machinery [Ref.: adapted from B. Alberts, A. Johnson, J. Lewis et al., The Cell, Garland Science, 2002]

membrane. The membrane proteins induce stress in the surrounding lipids and take part in a range of physiological functions including adhesion and signalling.

The basic structure of all cells is the same. Fluid sheets surround the cell and its internal compartments, while semi-flexible filaments form a more rigid internal scaffolding within the cell and contribute to its mechanical integrity. Deformations of the cell boundary are due to a number of processes; both compositional inhomogeneity of the bilayers (phase separation) and anisotropic structuration of the cell walls under lateral stress and pressure from structural elements, e.g. microtubules.

10.1 UNDULATIONS

Membranes are two dimensional objects. Fluctuations in their shape (undulations) are specific to their dimensionality and are of primary importance to their physical properties. There is an important difference in the undulations of fluid bilayers in which there is no shear resistance (Figure 10.2) and those that can sustain an in-plane shear stress due to ionic or covalent bonds between neighbouring atoms or molecules. Both varieties of membrane can occur naturally.

From standard continuum mechanics, the compression modulus (κ_V) in three dimensions is related to the change in volume (V) with pressure (P) (Section 11.1) via:

$$\kappa_V^{-1} = -\frac{1}{V}\left(\frac{\partial V}{\partial P}\right)_T = \frac{\beta \langle (\Delta V)^2 \rangle}{V_0} \qquad (10.1)$$

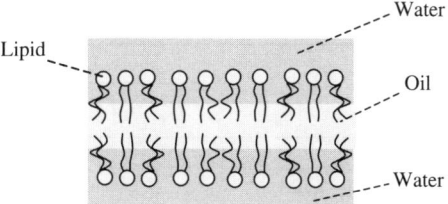

Figure 10.2 The structure of a fluid lipid bilayer at the oil/water interface

where $\partial V/\partial P$ is calculated at constant temperature, ΔV is the fluctuations in the volume, V_0 is the volume at zero temperature and $\beta = 1/kT$ (k is Boltzmann's constant and T is the temperature). For membranes an analogous two dimensional compressibility can be related to the fluctuations in the area of the membrane (ΔA):

$$\frac{1}{\beta \kappa_{V2D}} = \frac{\langle (\Delta A)^2 \rangle}{A_0} \qquad (10.2)$$

where A_0 is the area at zero temperature. A membrane with a large areal compressibility (κ_{v2D}) only experiences small fluctuations in its area at fixed pressure (Figure 10.3).

The mechanism through which undulations affect the size of membranes is still an open area of research. In a similar manner to that in which the size of a polymer is dependent on the quality of the solvent (Section 8.2), the average size and area of a membrane is related to the interplay between the solvent–membrane interaction and the excluded volume of the two dimensional surface. The scaling behaviour of the radius of gyration and the surface area of a range of model closed bag

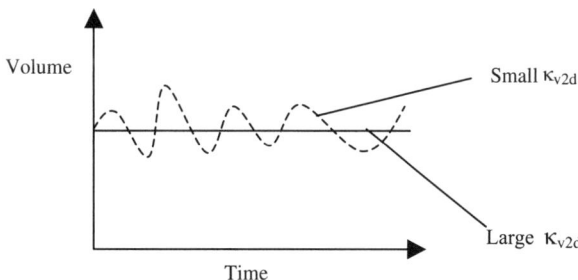

Figure 10.3 Schematic diagram that shows the fluctuations in volume as a function of time for a membrane. A large compressibility suppresses volume fluctuations and vice versa

Table 10.1 Scaling behaviour of closed bags in three dimensions

| | Scaling law | |
Configuration	$\langle R_g^2 \rangle \propto L_c^{2\nu}$	$\langle A \rangle \propto L_c^{2\eta}$
Inflated (good solvent)	$\nu = 1$	$\eta = 3/2$
Flory-type (theta solvent)	$\nu = \frac{4}{5}$	
Branched polymer	$\nu = 1$	$\eta = 1$
Dense (poor solvent)	$\nu = \frac{2}{3}$	$\eta = 1$

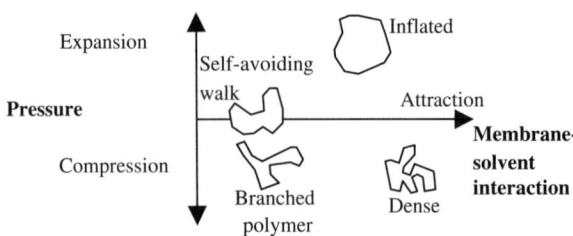

Figure 10.4 Phase diagram for membranes as a function of pressure and the attractive membrane–solvent interaction
[Ref.: D.H. Boal, Phys. Rev. A, 1991, 43, 6771–6777]

morphologies are described in Table 10.1. The different scenarios for membrane–solvent interactions are shown in Figure 10.4.

10.2 BENDING RESISTANCE

At zero temperature a membrane minimises its bending energy and adopts a shape that is flat or uniformly curved. At finite temperature, the spatial decorrelation of the normals to the membrane surface, introduces a finite roughness to the surface and provides a measure of its elasticity in response to thermal fluctuations. This spatial decorrelation of the normals is characterised by a persistence length (ξ_P, Figure 10.5); the two dimensional analogue of that found for a semi-flexible polymer (Section 8.1).

The normal to a membrane surface at a point in space ($\underline{r}$) is defined as $\underline{n}(\underline{r})$ and the correlation of the normals at an average separation (Δr) is found to decay exponentially:

$$\langle \underline{n}(\underline{r_1}).\underline{n}(\underline{r_2}) \rangle = e^{-\Delta r/\xi_p} \qquad (10.3)$$

$$\underline{\Delta r} \equiv \underline{r_1} - \underline{r_2} \qquad (10.4)$$

BENDING RESISTANCE

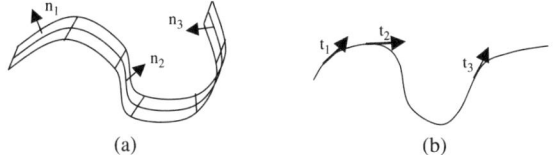

Figure 10.5 The correlation of the normals (n_i) to a membrane (a) provides a definition for the persistence length of a membrane and the correlation of tangent vectors (t_i) to a polymer (b) is related to the persistence length of a polymer

where r_1 and r_2 are two points on the membrane. In contrast to the case of a polymer, where the persistence length (l_p) was found to be proportional to the bending modulus ($l_p \sim \kappa/kT$, Section 8.1), the persistence length of a membrane (ξ_p) depends exponentially on the bending modulus and the temperature:

$$\xi_p \sim b \exp(2\pi\kappa_b/kT) \qquad (10.5)$$

where b is a characteristic step length (in metres) along the membrane. The membrane persistence length is thus a much more sensitive function of the modulus and temperature than the case with a polymer. A flat zero temperature membrane obeys $R_g^2 \sim L^2$; the area (R_g^2) of a membrane in three dimensions is approximately equal to its contour area (L^2, where L is the contour length of the sides of the membrane). When the membrane is subject to thermal fluctuations, the size of the membrane without self-avoidance grows very slowly with a contour area $R_g^2 \sim lnL$, which can be shown both by simulation and analytical calculation using Fourier decomposition of the surface profile.

The scattering of X-rays and light from a stack of membranes can be used as a sensitive measure of their undulations (Figure 10.6). The stack of membranes are used to reinforce the scattering signal through the process of constructive interference of the scattered waves from the periodic structure. Typically elastic scattering experiments measure the structure factor ($S(q)$) as a function of momentum transfer (q, the inverse length scale, Section 13.1):

$$S(q) = N^{-2} \left\langle \sum_{m,n} e^{iq \cdot (r_m - r_n)} \right\rangle \qquad (10.6)$$

where N is the number of membranes in a stack and the summation is carried out over all the separate pairs of molecules in the stack. There are no true Bragg peaks at finite temperature in such scattering experiments

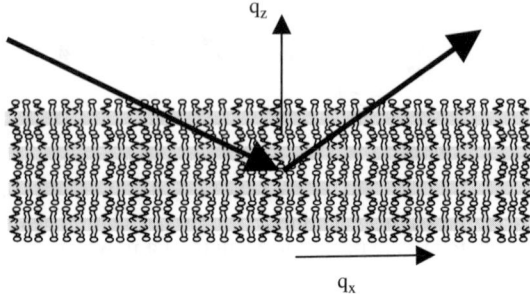

Figure 10.6 Schematic diagram of the scattering geometry of radiation from a stack of membranes
(q_z is the momentum transfer perpendicular to the membrane and q_x is that lateral to the membrane)

due to the undulations of the membrane stack, and the scattering profiles exhibit power law singularities (Figure 10.7) in a similar manner to that observed with smectic liquid crystals. In the direction perpendicular to the surface (z) the scattered intensity ($I(0, 0, q_z)$) takes the form:

$$I(0, 0, q_z) \propto (q_z - q_m)^{-2+\eta_m} \quad (10.7)$$

where q_m is the value of the momentum transfer centred on the mth order peak of the constructive interference of the scattered waves ($q_m = 2\pi/md$ and d is the membrane spacing). Parallel to the surface the scattered intensity has the form:

$$I(q_\|, 0, q_m) \propto q_\|^{-4+2\eta_m} \quad (10.8)$$

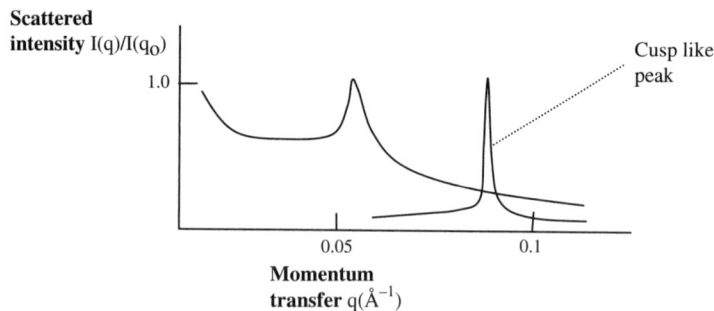

Figure 10.7 Cusp like peaks are found in X-ray experiments due to the constructive interference of X-rays scattered from stacks of lipid membranes. The scattered intensity is shown as a function of the momentum transfer of the X-rays

ELASTICITY

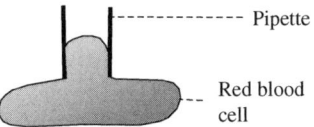

Figure 10.8 Micropipette apparatus can be used to examine the elasticity of red blood cells using digital analysis of the resultant membrane profile

where $q_{\parallel}$ is the momentum transfer parallel to the surface. There are no peaks in the data in this case and a characteristic continuous decrease from a singularity at zero q is measured. The exponent η_m found in equations (10.7) and (10.8) is related to the elastic modulus of the membrane:

$$\eta_m = \frac{m^2 \eta_1^2 kT}{8\pi\sqrt{BK}} \quad (10.9)$$

where $K = K_c/d$, d is the interlamellar spacing, K_c is the elasticity modulus, η_1 has different forms that depend on the dominant force of interaction between the layers (e.g. undulation forces or electrostatics), B is the bulk modulus and kT is the thermal energy. It is also possible to examine single membranes using a reflectivity geometry by means of the total internal reflection of X-rays, neutrons and light radiation. The calculation of the reflectivity equivalent of the structure factor given in equations (10.6) and (10.7) for transmission experiments is slightly more complicated, but quantitative measurements of single membrane fluctuations can be made from the profile of the reflected radiation. Image analysis of the curvature of membranes aspirated with a micropipette also provides a useful method for calculating membrane bending elasticity (Figure 10.8).

10.3 ELASTICITY

The Possion ratio (v) is a measure of how a material contracts in the transverse direction when it is stretched longitudinally (Section 11.1). It is defined in two dimensions as the ratio of the strains parallel (u_{xx}) and perpendicular (u_{yy}) to the direction of extension:

$$v = -\frac{u_{yy}}{u_{xx}} \quad (10.10)$$

when the stress is applied along the x-axis. The negative sign is included so that most 'normal' materials have positive values of the Poisson ratio. However, unusually, crumpled membranes have negative Poisson ratios. This can be simply demonstrated by crumpling a piece of paper into a ball. Extending the paper between your hands then causes it to expand laterally, it has a negative Poisson ratio. Another unusual example of a negative Poisson ratio is found in the Section 11.3 on foams in three dimensions, which have openly connected membraneous structures.

The shear modulus (μ) for a two dimensional elastic fibrous network is given by the expression:

$$\mu \approx \rho kT \qquad (10.11)$$

where ρ is the density of cross-links and kT is the thermal energy. This is a very similar result to the three dimensional case of cross-linked polymers (equation (8.56)), and has a similar derivation. The exact prefactor in equation (10.11) depends on the co-ordination number of the network (Section 8.3).

The stress and internal pressure in an elastic spherical membrane are related to its curvature (Figure 10.9). The resultant tensile force due to the wall stress is equal to the average area of the wall times the average stress ($\langle\sigma\rangle$):

$$\pi(r_o^2 - r_i^2)\langle\sigma\rangle \qquad (10.12)$$

where r_0 is the external radius and r_i is the internal radius. In equilibrium, the stresses in the wall of the membrane are balanced by the internal pressure (p_i):

$$\pi(r_o^2 - r_i^2)\langle\sigma\rangle = \pi r_i^2 p_i \qquad (10.13)$$

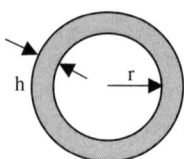

Figure 10.9 The geometry of a spherical membrane of thickness (h) and internal radius (r)

Equation (10.13) is rearranged and the shell thickness (h) is defined as $r_o - r_i$ to give:

$$\langle \sigma \rangle = p_i \frac{r_i^2}{r_o^2 - r_i^2} = p_i \frac{r_i^2}{h(r_o + r_i)} \tag{10.14}$$

If the elastic membrane on the sphere is thin the internal radius is approximately equal to the external radius ($r_o \approx r_i = r$) and equation (10.14) can be further simplified:

$$\langle \sigma \rangle = \frac{force}{area} = \frac{rp}{2h} \tag{10.15}$$

where p is the interior pressure, h is the shell thickness and r is the radius of the sphere. This explains why balloons are initially difficult to expand, since the applied shear stress is small for balloons with small radii (r).

The stresses in the cell wall of a *cylindrical membrane* (e.g. a cylindrical bacterium) are a little different to those of a sphere and have important medical applications (Figure 10.10), e.g. the mechanical properties of a range of tubular organs. For a cylinder, $\langle \sigma_\theta \rangle (r_o - r_i) L$ is the resultant force on the cross section and is balanced by the pressure that acts on the inside of the cylinder $2 r_i L p_i$:

$$2 \langle \sigma_\theta \rangle (r_o - r_i) L = 2 r_i L p_i \tag{10.16}$$

where $\langle \sigma_\theta \rangle$ is the average value of the stress (σ_θ) as a function of the angle θ over the cross section and L is the length of the cylinder. Therefore, defining the thickness of the membrane as h ($h = r_o - r_i$), the stress in the hoop direction is:

$$\langle \sigma_\theta \rangle = \frac{rP}{h} \tag{10.17}$$

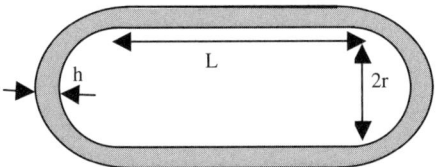

Figure 10.10 The geometry of a cylindrical membrane, length (L), thickness (h) and end cap radius of curvature (r)

The stress in the axial direction (σ_z) is similar to the case of a sphere:

$$\langle \sigma_z \rangle = \frac{rP}{2h} \tag{10.18}$$

where P is the pressure and r is the cylinder radius. The hoop stress is thus twice that in the axial direction. This explains why sausages and internal organs burst in the direction of the hoop, i.e. the break runs parallel to the long axis of the cylinder.

More generally it can be shown that for two principal curvatures ($1/R_1$ and $1/R_2$), pressure (P) and two line tensions τ_1 and τ_2:

$$\frac{\tau_1}{R_1} + \frac{\tau_2}{R_2} = P \tag{10.19}$$

For fluid sheets the tension is isotropic and equation (10.19) becomes the Young–Laplace equation that was encountered earlier (Section 7.4):

$$\tau \left(\frac{1}{R_1} + \frac{1}{R_2} \right) = P \tag{10.20}$$

Mathematically a unit tangent vector (t) to the membrane can be defined:

$$\underline{t} = \frac{\partial \underline{r}}{\partial s} \tag{10.21}$$

where $\underline{r}$ is a point on the membrane and s is a distance along the membrane surface. The curvature (c) can be defined:

$$c = \underline{n} \cdot \frac{\partial \underline{t}}{\partial s} = \underline{n} \cdot \left(\frac{\partial^2 \underline{r}}{\partial s^2} \right) \tag{10.22}$$

where n is the normal to the curvature at position r. The simplest model for the free energy density (F) of a membrane is:

$$F = \frac{\kappa_b}{2}(c_1 + c_2 - c_0)^2 + \kappa_G c_1 c_2 \tag{10.23}$$

where c_0 represents the spontaneous curvature, c_1 and c_2 are quadratic terms in the Taylor expansion of the surface around a point (the principal curvatures), κ_b is the bending modulus introduced earlier and κ_G is the

ELASTICITY

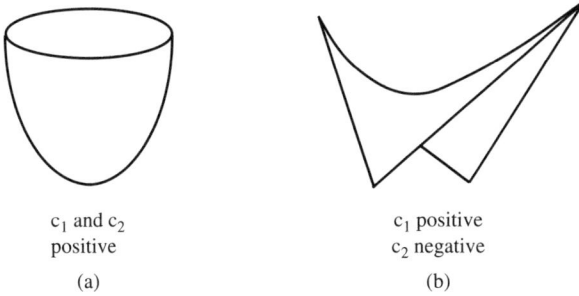

c_1 and c_2 positive
(a)

c_1 positive
c_2 negative
(b)

Figure 10.11 Schematic diagram of surfaces with curvature (C_1 and C_2) of the same sign (a) and opposite sign (b)

saddle-splay modulus. For a stable film κ_b in the free energy density (equation (10.23)) must be positive. The product $c_1 c_2$ is called the Gaussian curvature and has a large impact on the morphology that a membrane adopts (Figure 10.11). The integral of the Gaussian curvature over a closed surface is an invariant:

$$\oint c_1 c_2 dA = 4\pi(n_c - n_h) \qquad (10.24)$$

where $(n_c - n_h)$ is the difference between the number of connected components and the number of saddles, and $\oint$ is an integral over the complete surface. For a bilayer membrane with no spontaneous curvature and topology conserving fluctuations, the bending energy is simpler than equation (10.23)

$$E_{el} = \oint \left[\frac{\kappa_b}{2}(c_1 + c_2)^2\right] dA \qquad (10.25)$$

where κ_b is the bending modulus.

Biologically, biochemical products must be transported from their point of manufacture to the sites of usage. The products must be packaged to prevent their loss during transport and this process involves small membrane-bound entities (vesicles). The packages are labelled so they can be recognised at their point of destination and they need to be shipped along efficient transportation routes (Figure 10.12). Equation (10.25) for the bending energy of a membrane can be used to calculate the sizes observed experimentally for simple spherical vesicles (they are normally less complicated in their construction than a complete cell) and their mechanism of formation.

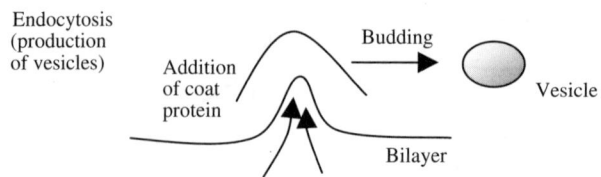

Figure 10.12 The formation of a vesicle used for transportation occurs through a process called endocytosis

10.4 INTERMEMBRANE FORCES

A range of mesoscopic forces are important in the interactions of membranes. At small distances the membranes attract due to the interaction of the induced electric dipole moments (Van der Waals force). The balance between van der Waals and electrostatic forces occurs at intermediate lengths and often defines an average inter-membrane distance (Figure 10.13). Futhermore, there is often a furry coat of polymers attached to the exterior of the plasma membrane that impedes cell adhesion (Figure 10.14) due to the induced steric repulsive force (Section 2.4). Water also interacts with the structure of membranes and can give rise to a long range hydrophobic interaction which sensitively depends on the ionic environment. Non-covalent binding is possible between specific molecules attached to membranes at very close distances of approach. There are a range of proteins that are designed to connect cells together with ionic bonds and are a major determining factor in the cohesion of cells into tissues.

The measurement of line tensions is possible by measuring the size of disk-like vesicles (an indirect method of measurement), by osmotic swelling measurements, or through the production of holes in membranes (Figure 10.15) and the observation of the critical hole radius.

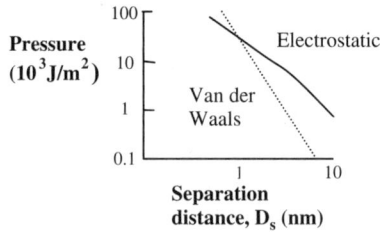

Figure 10.13 Pressure between two rigid charged plates as a function of their degree of separation
[*Reprinted with permission from D. Boal, Mechanics of the Cell, Copyright (2002) Cambridge University Press*]

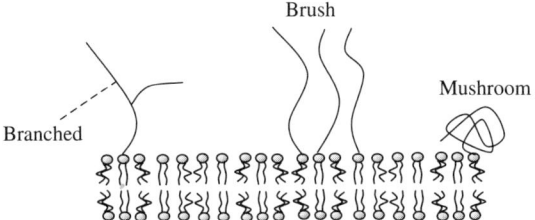

Figure 10.14 The possible conformations of the polysaccharides in the glycoproteins of cell membranes can include branched, parallel brush and mushroom morphologies

At zero temperature a membraneous system in which a hole has been introduced (energy U) acts to minimise its enthalpy (H):

$$H = U - \tau A \tag{10.26}$$

where τ is the tension of the membrane in two dimensions and A is the area. The energy of a circular hole (U) formed in a bilayer is given by:

$$U = 2\pi R \lambda \tag{10.27}$$

where λ is the line tension and R is the hole radius. The difference in area between the sheet and the hole system with respect to the intact sheet is πR^2. From equation (10.26) the change in enthalpy (ΔH) on the production of a hole is therefore given by:

$$\Delta H = 2\pi R \lambda - \tau \pi R^2 \tag{10.28}$$

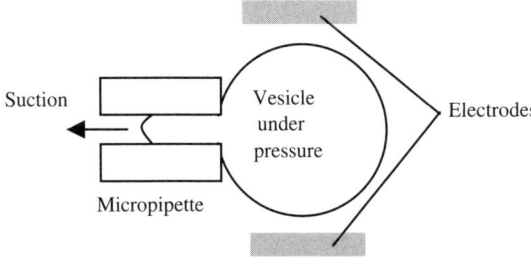

Figure 10.15 A technique for the study of the surface tension of cells. Holes are formed by the action of an electric field

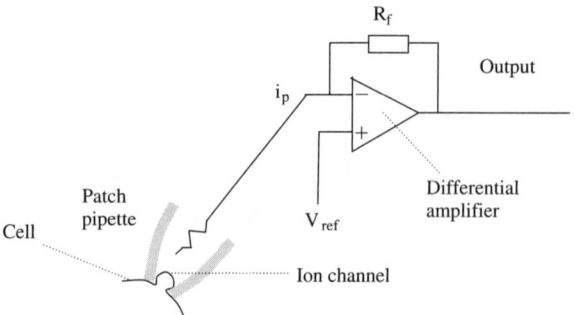

Figure 10.16 The patch clamp method can be used to examine the electrical properties of membrane proteins. A pipette is attached to a cell and the electrical activity from an ion channel is detected with a differential amplifier

The extrema of this expression can be calculated and the maximum value of the enthalpy (ΔH) occurs at a critical hole radius (R^*) of:

$$R^* = \frac{\tau}{\lambda} \qquad (10.29)$$

The line tension (λ) can then be calculated experimentally using this expression.

Proteins embedded in a cell membrane can provide it with electrochemical activity. Patch clamps are used to measure electrical properties of membranes that are induced by such proteins. A very narrow clean capillary pipette is taken and a small degree of suction is applied to a membrane. The tiny electrical potentials typically formed by cells are measured with respect to a reference voltage (Figure 10.16). A very high degree of amplification is needed to measure the voltage produced by a membrane protein; the trick is to separate the current due to the cellular events from the background thermal noise. Such electrical data provides information on the conduction of membranes and has even enabled the kinetics of individual channel opening and closing events to be followed.

Anomalous X-ray experiments have recently provided detailed information on the distribution of counterion clouds near membranes (Figure 10.17). These clouds determine the electrostatic potential experienced by the membranes.

FURTHER READING

D. Boal, *Mechanics of the cell*, Cambridge University Press, 2002. Thorough, well explained account of the physics of membranes.

G. Forgar, S.A. Newman, *The biological physics of developing embryos*, Cambridge University Press, 2005. Interesting discussion of the role of membrane adhesion in cellular morphogenesis.

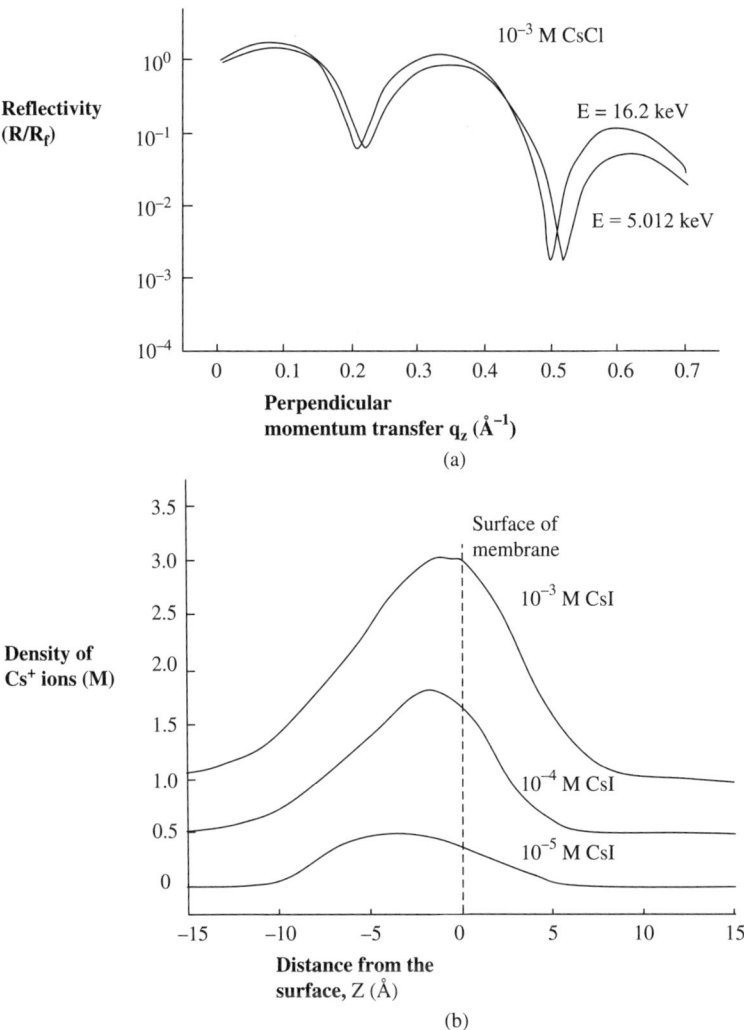

Figure 10.17 Anomalous X-ray reflectivity data from phospolipid bilayers (a) Reflectivity curve that shows the reflected intensity as a function of momentum transfer (q_z); (b) is the calculated Poission–Boltzmann counterion profile that shows the density of cesium counterions as a function of the perpendicular distance from the bilayer surface [*Reprinted with permission from W. Bu, D. Vaknin, and A. Travesset, Physical Rev. E, 72, 060501, Copyright (2005)American Physical Society*]

TUTORIAL QUESTIONS

10.1) What is the line tension of a membrane if the critical hole radius in it surface is 2.6 nm and its surface tension is $0.03 \, \text{Jm}^{-2}$?

10.2) What are the axial and hoop stresses on a cylindrical bacterium if its internal pressure is 1×10^5 Pa, the membrane thickness is 1 nm and the radius is 1 µm?

10.3) By what factor does the persistence length of a membrane ($\kappa/kT = 5$) change if its bending modulus is doubled? By what factor would the exponent change, measured in an ideal X-ray scattering experiment, for the power law cusp from a stack of such membranes? Assume all the other parameters of the membrane remain unchanged.

10.4) A membrane is moved from a good solvent to a poor solvent, and its contour length is L_c. According to the scaling laws for closed bags, by what factor does its radius of gyration and area change? Assume the prefactors are the same in both cases.

11
Continuum Mechanics

Two architectural themes occur again and again in the continuum mechanics of naturally occurring biomaterials: *fibrous composites* and *cellular solids* (and combinations of the two). *Fibrous composites* consist of stiff rigid rods (e.g. collagen or cellulose) combined with a highly viscous dissipative filler (e.g. proteoglycans or lignins). The stiff rods resist extension and compression providing the composite with its strength, whereas the dissipative filler increases the material's toughness by many orders of magnitude. Composite materials are now widely used in a range of synthetic products (e.g. skis, the fuselage of aircraft etc) but biological composites continue to out perform many of their synthetic counterparts due to their well optimised structures on the nanometre length scale.

Cellular materials are widespread in biology. Cellular solids have reasonable strengthes and toughnesses, but they only use a fraction of the structural component to achieve these properties. Thus cellular materials provide a mechanical solution with greatly improved stiffness/weight ratios and consequently occur in a wide range of biological tissues that include bones and woods.

A list of some characteristic properties commonly encountered in the mechanics of materials is shown in Table 11.1. All of these properties are important in biological situations and their extension to anisotropic nanostructured materials needs to be carefully considered.

One of the simplest experiments that can be performed on a biomaterial concerns the application of a force (stress) that causes it to extend (become strained). A wide range of stress–strain properties available

Table 11.1 Important material properties found in the solid mechanics of biomaterials

Functional attribute	Material property	Units
Stiffness	Modulus of elasticity, E_{init}	Nm^{-2}
Strength	Stress at fracture, σ_{max}	Nm^{-2}
Toughness	Energy to break at fracture	Jm^{-3}
Extensibility	Strain at failure, ε_{max}	No units
Spring efficiency	Resilience	%
Durability	Fatigue lifetime	s to failure
Spring capacity	Energy storage capacity, W_{out}	Jkg^{-1}

with elastic proteins is shown in Figure 11.1. Typically, the stress–strain curves are linear for small deformations and Hooke's law is obeyed, with the force (and corresponding stress) proportional to the extension (and corresponding strain). Numerical values of the material properties of some proteins and standard synthetic materials are tabulated in Table 11.2.

11.1 STRUCTURAL MECHANICS

For an isotropic material there is a simple relationship between the stress ($\sigma = F/A$, the force (F) divided by the area (A)) and the strain ($e = \Delta l/l$,

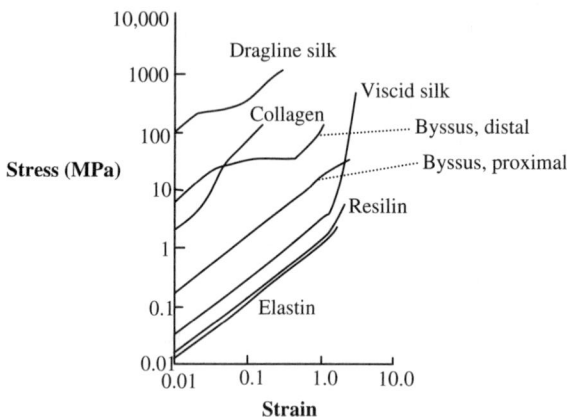

Figure 11.1 Stress–strain curves for a range of structural proteins
Hooke's law, with stress proportional to strain, is observed with all these materials at small degrees of strain [Ref.: adapted from J. Gosline, M. Lillie, E. Carrington et al., Phil. Trans. R. Soc. Lond. B, 2002, 357, 121–132]

Table 11.2 Mechanical properties of a range of structural proteins commonly encountered in biophysics and some synthetic equivalents
[Ref.: Adapted from J. Gosline, M. Lillie, E. Carrington et al., Phil. Trans. R. Soc. Lond. B, 2002, 357, 121–132]

Material	Modulus (GPa)	Strength (GPa)	Extensibility	Toughness (MJm^{-3})	Resilience (%)
Elastin	0.0011	0.002	1.5	1.6	90
Resilin	0.002	0.004	1.9	4	92
Collagen	1.2	0.12	0.13	6	90
Synthetic rubber	0.0016	0.0021	5	10	90
Mussel byssus proximal	0.016	0.035	2.0	35	53
Dragline silk	10	1.1	0.3	160	35
Viscid silk	0.003	0.5	2.7	150	35%
Kevlar(e.g. in bullet proof jackets)	130	3.6	0.027	50	—
Carbon fibre	300	4	0.013	25	—
High tensile steel	200	1.5	0.008	6	—

the change in length (Δl) divided by the original length (l)). For a linear material the stress/strain relationship is:

$$\sigma = Ee \qquad (11.1)$$

where the constant of proportionality is Young's modulus (E, units: Pascals).

The *Poisson ratio* (v) for an isotropic material is a dimensionless number given by the ratio of the perpendicular strain (e_{perp}) to the longitudinal strain (e) (Figure 11.2):

$$v = -\frac{e_{perp}}{e} \qquad (11.2)$$

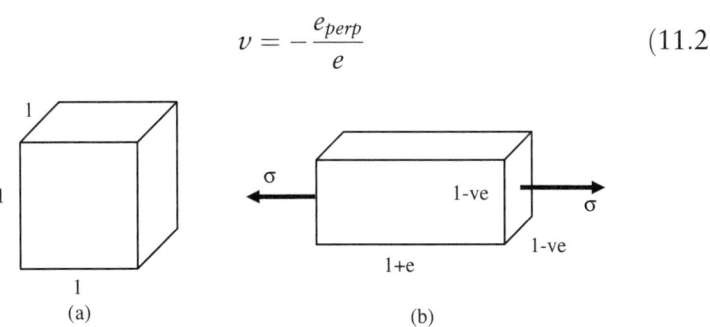

Figure 11.2 The geometry of a stressed cube is used to illustrate the Poisson ratio (v), which quantifies the reduction in size perpendicular to the direction of the applied stress. (a) Unstressed cube of material and (b) the material with an extension (e) in the direction of the uniaxial stress (σ)

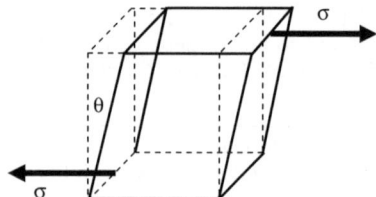

Figure 11.3 Geometry used to define the shear modulus (*G*) of a material in terms of the shear stress (σ) and the angle of deformation (θ)

It describes how the change in morphology of a material under stress is coupled in the directions parallel and perpendicular to the direction of the stress. The Poisson ratio is a half for an incompressible material in uniaxial extension.

The *shear modulus* (*G*) describes the resistance of a material to a shearing motion, e.g. the motion of two parallel faces of a material relative to one another (Figure 11.3). For small angles of deformation the shear modulus is equal to the stress (σ) divided by the angle of deformation (θ) and it therefore is measured in Pascals:

$$G = \frac{\sigma}{\theta} \tag{11.3}$$

The *bulk modulus* (*K*, units: Pascals) quantifies the variation in volume (*dV*) of a material when the pressure is changed (*dp*) (Figure 11.4). It is defined through the equation:

$$\frac{1}{K} = -\left(\frac{1}{V}\right)\frac{dV}{dp} \tag{11.4}$$

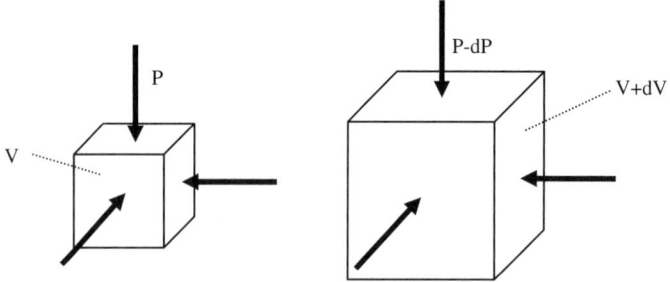

Figure 11.4 The bulk modulus of a solid (*K*) defines the change in volume (*dV*) of a material in response to a change in pressure (*dP*)

STRUCTURAL MECHANICS

The factor $\frac{1}{V}$ makes the bulk modulus independent of the volume of material considered; and the negative sign allows positive compressibilities to be used to describe the typical case in which an increase in pressure causes a decrease in volume.

For an isotropic elastic material the bulk modulus (K), Young's modulus (E), shear modulus (G) and Poisson ratio (v) are interrelated by two simple relationships:

$$K = \frac{E}{3(1-2v)} \quad (11.5)$$

$$G = \frac{E}{2(1+v)} \quad (11.6)$$

Thus knowing two of the characteristic constants (from E, K, v and G) for an isotropic elastic material allows the other two to be calculated.

For anisotropic materials (the predominant moiety among biological structures) a full tensorial analysis is required to describe the stress and strain of a material (Figure 11.5). In three dimensions the stress tensor (σ_{ij}) simultaneously desribes the extensive, compressive, dilatational and shear forces on a small volume element. In Cartesian coordinates (x, y, z) the stress tensor is given by:

$$\sigma_{ij} = \begin{bmatrix} \sigma_{xx} & \sigma_{xy} & \sigma_{xz} \\ \sigma_{xy} & \sigma_{yy} & \sigma_{yz} \\ \sigma_{xz} & \sigma_{yz} & \sigma_{zz} \end{bmatrix} \quad (11.7)$$

Similarly, there are nine constants that describe the strain (e_{ij}) of the material in response to the stress in three dimensional Cartesian

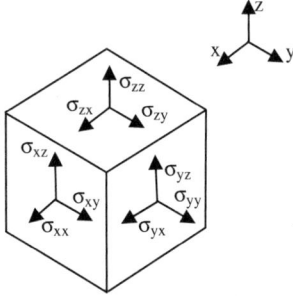

Figure 11.5 Tensorial nature of the stress (σ_{ij}) on a cubic volume element of a material

coordinates. For linear elastic materials the stress is related to the strain by a tensorial equation:

$$e_{ij} = s_{ijkl}\sigma_{kl} \tag{11.8}$$

where s_{ijkl} are the compliance constants that characterize the elastic response of a particular material. Thus eighty one ($3^4 = 81$) compliance constants are required to describe an anisotropic material that exhibits no symmetry in its morphology. Fortunately, in most practical situations the number of independent constants is reduced due to the symmetry both in the molecular structure of the sample and the morphology of the particular specimen chosen.

11.2 COMPOSITES

Composite materials are constructed from a mixture of discrete rigid units combined with a dissipative matrix. A range of composite morphologies commonly occur in nature. These include two dimensional laminates (the rigid units are planar) and one dimensional fibres (the rigid units are one dimensional, Figure 11.6 and Figure 11.7). Furthermore, a wide range of tessellations are possible for the rigid embedded units and these provide important consequences for the mechanical properties. For fibrous composites the length of the rigid units embedded is a critically important parameter for the resultant mechanical properties, and separate models have been developed to describe the stresses that are

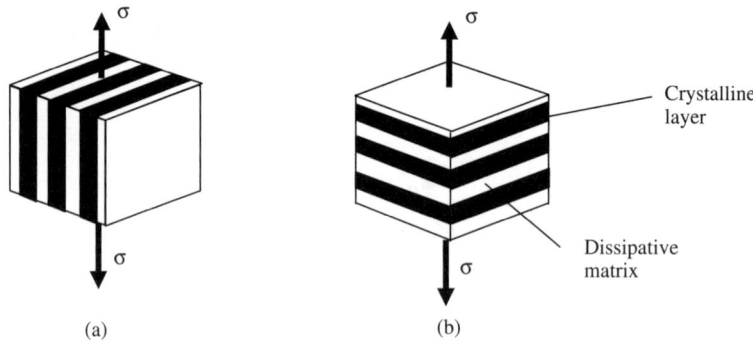

Figure 11.6 The modulus of laminated composites depends on the direction of the applied stress (σ), which is shown parallel to the layers (a) and perpendicular to the layers (b)

experienced by *short* and *long composites*. Examples of biocomposites include elastin/collagen in heart walls (fibrous composites) and nacre/protein in sea shells (laminar composites, Section 15.6). There is a direct link between the physical properties of fibrous composites and solid liquid crystals (Section 4.4); often the two descriptions are used interchangeably. For example, with polymeric liquid crystals the amorphous material in the flexible linkers can act as a dissipative filler, and the nematic inserts provide resistance to extension (a fibrous composite).

Inorganic/organic biocomposites such as dentine, the hard surface layer in teeth, often have only small amounts of organic material (~1%), but this dissipative filler has an extremely large impact on the resultant physical properties. In particular, the toughness of the material (e.g. a combination of hydroxyapatite and protein in dentine) is many orders of magnitude larger than the toughness of a single crystal.

To a reasonably good first approximation it is possible to calculate the effective Young's modulus for a *layered composite material*. For a force applied *parallel* to the composite (Figure 11.6(a)) the force applied to this mixed material (F_m) is equal to the sum of the forces on the crystal (F_c) and that on the amorphous phase (F_a), e.g. an adhesive protein:

$$F_m = F_c + F_a \tag{11.9}$$

These forces can be reexpressed in terms of average stresses and areas:

$$\sigma_m A_m = \sigma_c A_c + \sigma_a A_a \tag{11.10}$$

where σ_m, σ_c and σ_a are the stresses on the mixture, the crystal and the protein respectively. A_m, A_c and A_a are the corresponding areas. Equation (11.10) can be divided by the total volume of the mixture. This allows the effective Young's modulus of the mixture (E_m) to be calculated in terms of the Young's modulus of the two constituent phases (E_c and E_a) and the volume fraction of the crystalline phase (ϕ_c):

$$E_m = E_c \phi_c + E_a(1 - \phi_c) \tag{11.11}$$

For a force applied *perpendicular* to a layered composite (Figure 11.6(b)) the total extension (δl_m) is the sum of the extension of each individual phase (δl_c and δl_a for the crystal and amorphous phases respectively):

$$\delta l_m = \delta l_c + \delta l_a \tag{11.12}$$

The extension can be converted into strains (ε) using the definition:

$$\varepsilon = \frac{\delta l}{l} \quad (11.13)$$

where δl is the change in length and l is the initial length. If l_m, l_c and l_a are the lengths of the mixture, crystal, and amorphous components respectively, ε_m, ε_p and ε_c are the corresponding strains. Therefore, equation (11.12) can be reexpressed as:

$$\varepsilon_m l_m = \varepsilon_c l_c + \varepsilon_a l_a \quad (11.14)$$

The strains can be calculated in terms of the Young's modulus of the individual components if it is noted that the stresses (σ) on each component are identical:

$$\frac{\sigma l_m}{E_m} = \frac{\sigma l_c}{E_c} + \frac{\sigma l_a}{E_a} \quad (11.15)$$

The stresses cancel out from this expression and the lengths can be reexpressed in terms of the volume fraction of the components. The final result for the Young's modulus of a layered composite strained in a direction perpendicular to the layers is:

$$E_m = \frac{E_a E_c}{E_a(1 - \phi) + E_c \phi} \quad (11.16)$$

For a *fibre composite* the modulus *parallel* to the direction of stress is:

$$E = \eta E_f \phi_f + E_a(1 - \phi_f) \quad (11.17)$$

Anomalous X-ray experiments have recently provided detailed information on the distribution of counterion clouds near membranes (Figure 10.17). These clouds determine the electrostatic potential experienced by the membranes.

where E_f and E_a are the Young's modulus of the fibres and amorphous matrix respectively (Figure 11.7). ϕ_f is the volume fraction of the fibres. This expression looks similar to that for a layered composite (equation (11.18)) with the inclusion of the scale factor (η). The scale factor is given by:

$$\eta = 1 - \frac{\tanh ax}{ax} \quad (11.18)$$

FOAMS

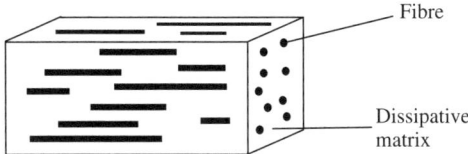

Figure 11.7 The arrangement of fibres in a composite material with uniaxial orientation

where a is the aspect ratio (length (l) of the fibres divided by the diameter ($2r$), $l/2r$) and x is given by the expression

$$x = \left[\frac{2G_a}{E_f \ln(R/r)}\right] \quad (11.19)$$

where G_a is the shear modulus of the matrix in which the fibres are embedded, E_f is Young's modulus of the fibres, R is the distance of separation between the fibres and r is the fibre radius.

Anisotropic composites also occur in nature. For example, helicoidal materials are often observed in biological materials (elastin/collagen in heart walls) and the orientation of the fibres provides additional resistance against torsional rotation and tearing.

11.3 FOAMS

Foams (or equivalently cellular solids) are a morphology optimised for strength and weight. A range of mechanical properties for some standard biological cellular solids are included in Table 11.3.

Table 11.3 Mechanical properties of some biological cellular solids

	Youngs Modulus (E)	Fracture stress σ_f (MPa)	Volume fraction of solid (%)	Density (ρ) (kgm^{-3})	Poisson ratio (v)	Toughness
Soft wood	10 GPa	180	0.96	10^2–10^3	0.3–0.68	12 kJm^{-2}
Hard wood	~17.5 GPa	240	0.73	10^2–10^3	0.01–0.78	11 kJm^{-2}
Cork	~20 MPa	15	0.15	170	0–0.1	60–130 Jm^{-2}
Bone	12 GPa	105	0.05–0.7	10^2–10^3	0.36	600–3000 Jm^{-2}
Carrot	7 MPa	1	0.03–0.38	10^3	0.21–0.49	200 Jm^{-2}

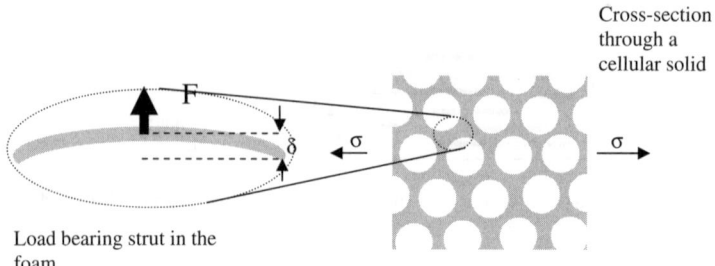

Figure 11.8 The mechanical strength of a foam is related to the elasticity of the struts in the material
(The average force on the struts is F and their deformation is δ. σ is the average stress applied to the foam.)

It is possible to motivate the elasticity of foams using a simple scaling theory. For the compression of a foam, the stress is resisted by strut-like sections of the foam morphology (Figure 11.8). The deformation of a single strut (δ) at its midpoint by a force (F) is related to the Young's modulus of the strut (E), the length of the strut (a) and the thickness (t):

$$\delta \sim \frac{Fa^3}{Et^4} \qquad (11.20)$$

The force required to deform the strut (F) is related to the compressive stress (σ) on the strut ($F \sim \sigma a^2$), so:

$$\sigma \sim \frac{\delta t^4 E}{a^5} \qquad (11.21)$$

The Young's modulus of an unfilled foam (E_f) is approximately the stress (σ) divided by the strain (γ). The strain (γ) for a given displacement is inversely proportional to the length of a fibre ($\gamma \sim a^{-1}$) and therefore equation (11.21) can be rewritten as:

$$E_f \sim \frac{\sigma}{\gamma} \approx \left(\frac{t}{a}\right)^4 E \qquad (11.22)$$

This Young's modulus of an unfilled foam can be rewritten in terms of the relative density of the material:

$$\frac{E_f}{E} = C_1 \left(\frac{\rho*}{\rho_s}\right)^2 \qquad (11.23)$$

FRACTURE

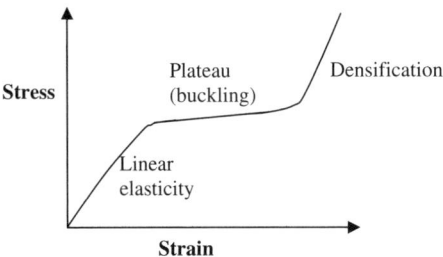

Figure 11.9 Compressional behaviour of an open celled foam
(The stress is shown as a function of strain. As the strain increases there is a linear Hookean region, followed by a plateau due to buckling of the struts, and finally the structure densifies before failure.)

where C_1 is a constant of proportionality, ρ^* is the density of the foam and ρ_s is the density of the material in the struts. The formulation is in reasonable agreement with experiment for a wide range of unfilled foam morphologies. The behaviour described in this calculation is also close to that expected with cross-linked semi-flexible polymers such as actin and fibrin, and there are many mechanical features held in common between these types of polymers and open celled foams.

Filled foams such as biological tissues predominantly contain water. They cannot be compressed and have markedly different mechanical properties to unfilled foams. Mechanical failure in this case corresponds to the filled cells bursting, and the strain is experienced by the stretching of the cellular walls, not bending of the materials. The theoretical prediction for filled foams is a lower dependence of the Young's modulus of the foam on the thickness and the length of the struts, $E_f \sim (t/a)^2 E$ compared with $(t/a)^4 E$ for the unfilled case.

A schematic diagram of the compressional behaviour expected for an open celled foam, e.g. cork or bone, is shown in Figure 11.9. The compaction of the foam under compressive stress and its densification at medium and high strains is characteristic of a cellular solid. An Euler buckling transition is associated with the collapse of struts at high strain.

The Poisson ratio for unfilled cellular solids can have an unusual behaviour. It can be negative, with the classic example of cork stoppers used to seal wine bottles. Cork stoppers expand as they are stretched in the opening of a bottle and seal the contents.

11.4 FRACTURE

The mechanism through which biological materials break and fracture is of vital concern in the lives of many organisms. Mechanical materials are

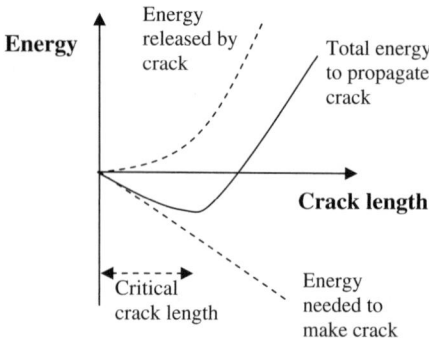

Figure 11.10 The energy released by the propagation of a crack in a one component solid is a function of its length. There is a critical crack length due to the interplay between the energy needed to nucleate the crack and the energy released by the crack growth
[Reprinted with permission from S. Vogel, Comparative Biomechanics, Copyright (2003) Princeton University Press]

often nanostructured in order to terminate cracks as they form; this greatly increases the energy they can absorb before failure. Such termination of cracks in soft viscoelastic fillers contributes to the toughness of nacre and bone.

The energy released by the propagation of a crack in a one component solid has two competing terms, the energy released by the advance of the crack and the energy absorbed to make the crack surface (Figure 11.10). There is thus a critical length (L_c) above which a crack begins to propagate. The critical crack length is related to the work of failure per unit area (W_f) and the strain energy stored per unit volume (W_s):

$$L_c = \frac{W_f}{\pi W_s} \quad (11.24)$$

Typically, for Hookean materials, the strain energy stored (W_s) by an elastic material is half the stress (σ) times the strain (ε i.e. $\sigma\varepsilon/2$):

$$L_c = \frac{2W_f}{\pi\sigma\varepsilon} \quad (11.25)$$

To avoid the propagation of cracks biological materials use a series of different mechanisms; a high work of fracture (the work done during the fracture process), a low strain energy at fracture extension, limitation of the material to low values of stress and strain, division of the material transversely to the direction of load and the use of materials that develop blunt cracks so stresses are minimised at their tips. Cellular solids profit from their ability to terminate cracks in voids in this respect, and the effect greatly increases their toughness.

11.5 MORPHOLOGY

As any architect of a macroscopic building will testify, the geometry of a structure has a significant impact on its mechanical properties. *Tubes*, *struts*, *braces* and *helicoids* are standard motifs that occur in biology and optimise the performance of biological structures for their required roles, e.g. the modification of the flexural stiffness to provide additional torsional stiffnesses. *Tubes* have high flexural and torsional stiffness with a minimum of structural material. *Struts* and *braces* have the same bending resistance as a single continous piece of material, with a large reduction in weight (foams have a random array of struts and braces). Active truces (arrays of motile struts) occur in the spines of many vertebrates, with soft ligaments providing active modification of the bending rigidity. *Helicoidal* structures in materials can improve fracture and tear energies by an order of magnitude, as linear fractures must occur across the strong fibre axis for at least part of their path length.

FURTHER READING

J.D. Humphrey and S.L. Delange, *An Introduction to Biomechanics: Solids and fluids, analysis and design*, Springer-Verlag, 2004. Detailed engineering approach to biomechanics.

J. Vincent, *Structural Biomaterials*, Princeton, 1990. Very good introductory course on biological materials.

S. Vogel, *Comparative Biomechanics: Life's Physical Worlds*, Princeton, 2003. Detailed introductory text.

L.J. Gibson and M.F. Ashby, *Cellular Solids*, Cambridge University Press, 1997. Classic text on the mechanics of foams.

J. Gosline, M. Lillie, E. Carrington et al., *Elastic proteins; biological roles and mechanical properties*, Phil. Trans. R Soc. London B, 2002, 357, 121–132.

B.K. Ahlborn, *Zoological Physics*, Springer, 2004. Some simple zoological examples of the physics of materials are covered.

TUTORIAL QUESTIONS

11.1) What is the ratio of the Young's modulus parallel and perpendicular to a layered biocomposite if the volume fraction of the crystalline material is 0.9 and the Young's modulus of the crystalline and amorphous fractions are 50 MPa and 50 GPa respectively?

11.2) Estimate the Young's modulus of a dried cellular solid if the Young's modulus of the walls is 9 GPa, the diameter of the struts is 1 μm and the length of the struts is 20 μm. Estimate the change in the Young's modulus if water now fills the pores. Assume that the mechanical properties of the struts are unchanged in the hydrated environment.

12
Biorheology

Traditionally experimental rheology has consisted of placing a sample in the scientific equivalent of a food blender (a *rheometer*). The response of the material in the blender provides a probe of the materials *viscoelasticity*. Rheology thus considers the measurement of the viscoelasticity of materials. A classic example of a viscoelastic substance is the childs toy, silly putty. This material can flow when it is drawn in the hand and bounces when subjected to a rapid collision with the floor. Silly putty therefore acts as a viscous fluid (a liquid) at long times and an elastic solid at short times; thus the viscoelasticity of the material is seen to be a time dependent phenomenon. All materials are viscoelastic to some degree, but many biological materials have very carefully tailored mechanical behaviour that simultaneously exhibit finely tuned viscosities and elasticities over physiologically important time windows.

In terms of a fundamental understanding of condensed matter, the field of rheology introduces the key concept of irreversible dissipative behaviour. Dissipative behaviour needs to be included in the development of realistic statistical models of biological processes. Thus experimental rheology allows a quantitative method of testing models from theories of irreversible thermodynamics.

The mechanisms by which biological materials store and dissipate energy are of prime importance in many biological functions, e.g. shock absorbers formed from cartilage (Section 15.1), the contraction of striated muscle (Section 14.1), the resilience of skin to impacts and how bloods cells are pumped through arteries (Figure 12.1). Rheological

---

Applied Biophysics: A Molecular Approach for Physical Scientists   Tom A. Waigh
© 2007 John Wiley & Sons, Ltd

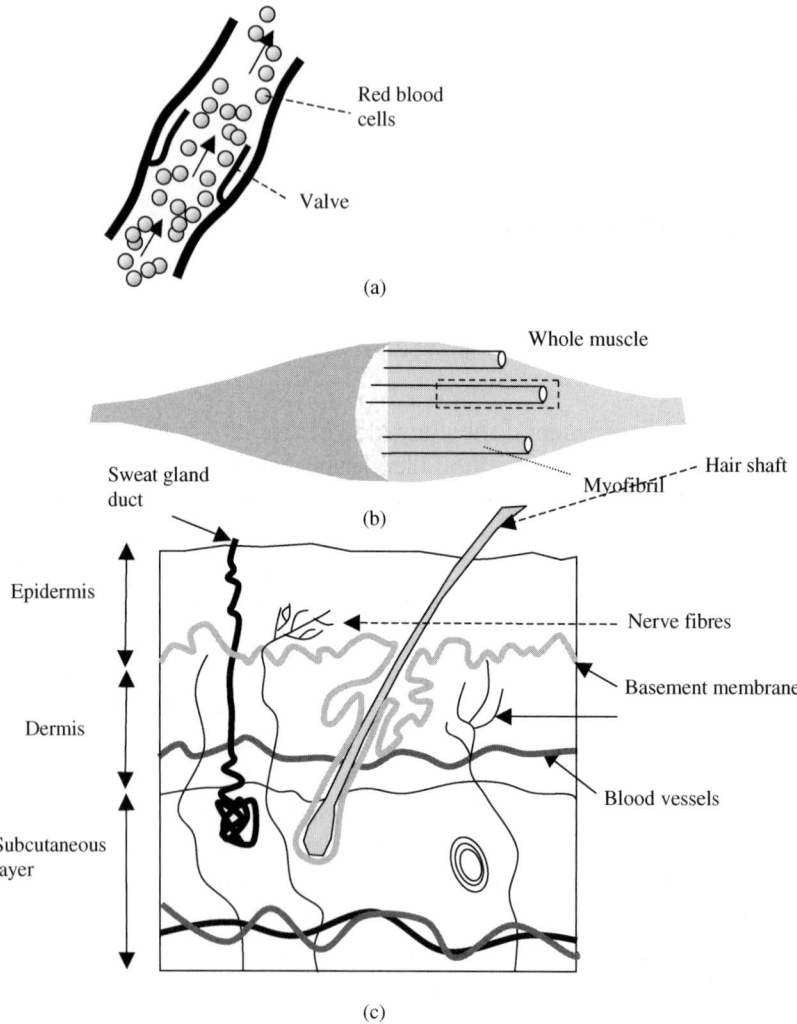

Figure 12.1 Examples of biological materials in which the viscoelasticity is vital to their function
((a) blood pumped through veins (microrheology, thixotropy etc), (b) the action of striated muscle (active stresses) and (c) the complex viscoelastic composite that is skin (the viscoelasticity of mixtures).)

measurements are thus of key importance in the determination of the behaviour of biomaterials in such bioengineering problems.

The rheological behaviour of biological materials can be exceedingly complicated. However, as a first step, two *dimensionless numbers* are introduced to qualitatively understand the flow behaviour: the *Peclet number* and the *Deborah number*.

The *Peclet number* determines when the stress applied to a material substantially deforms the microstructure. For example, consider a colloidal solution that consists of spherical particles (radius, a). The Stoke's–Einstein relationship (equation (5.11)) gives an expression for the diffusion coefficient (D) of the constituent particles as:

$$D = \frac{kT}{6\pi\eta a} \qquad (12.1)$$

where η is the viscosity of the solvent and kT is the thermal energy. From equation (5.39), the relaxation time associated with the diffusive motion (t_a) of the particles (the time for the colloids to diffuse their own radius) is then:

$$t_a = \frac{a^2}{D} = \frac{6\pi\eta a^3}{kT} \qquad (12.2)$$

These spherical particles experience the applied stress (σ) and the characteristic time for shear flow (t_s) is the reciprocal of the shear rate ($t_s = \dot{\gamma}^{-1}$). The Peclet number (Pe) is simply the ratio of the time for diffusive rearrangement (t_a) to the time associated with the rate of shear flow (t_s):

$$Pe = \frac{t_a}{t_s} = \frac{6\pi\eta a^3 \dot{\gamma}}{kT} < 1 \qquad (12.3)$$

where the inequality follows from the requirement that the microstructure is undisturbed.

For a predominantly viscous liquid the stress is directly proportional to the shear rate (Newton's law of viscosity):

$$\sigma = \eta\dot{\gamma} \qquad (12.4)$$

Therefore, the expression for the Peclet number (equation (12.3)) for a colloidal solution can be reexpressed as:

$$Pe = \frac{6\pi a^3 \sigma}{kT} < 1 \qquad (12.5)$$

where $kT/6\pi a^3$ is the thermal stress and σ is the mechanical or rheological stress. The inequality follows from the requirement that the convective motion due to the applied stress must be less than that due to Brownian motion or the microstructure will be significantly disturbed.

A further condition in order for an experiment to measure the linear viscoelasticity of a sample is that the structural relaxation by diffusion must occur on a time scale (τ) comparable to the measurement time (t). The *Deborah number* must be less than one ($De < 1$) for linear experiments and by definition:

$$De = \frac{\tau}{t} \quad (12.6)$$

When $De \gg 1$ the material is solid-like; when De is of the order of one or below the material is liquid-like. Both the conditions on the Peclet number and the Deborah number need to be satisfied in order to observe linear viscoelasticity experimentally.

Rheometers can normally function in both oscillatory and unidirectional shear modes. Steady state shear experiments typically measure non-linear rheology, whereas oscillatory measurements are sensitive to the linear viscoelasticity in the limit of small amplitude oscillations.

12.1 STORAGE AND LOSS MODULI

A fundamental aim of models developed for the rheology of a material is to relate the applied force (stress) to the resultant deformation (strain); this is provided by a *constitutive equation*. A simple general constitutive equation for a viscoelastic material will first be constructed to illustrate the key concepts involved.

For an elastic Hookean solid the stress (σ) is proportional to the strain (γ) with an elastic constant (μ):

$$\sigma = \mu\gamma \quad (12.6)$$

This expression can then be extended into three dimensions using tensorial notation for an isotropic (with a single elastic constant, μ) material as:

$$\sigma_{xy} = \mu\gamma_{xy} \quad (12.7)$$

For a viscoelastic solid the mathematical expression that relates the stress to the strain is more complicated, with the stress slowly reducing to zero

STORAGE AND LOSS MODULI

with time as the material relaxes. The solid will be modelled in one dimension for simplicity and it is assumed that the viscoelastic response of the system is additive in time. The stress can then be expressed as an integral over the relaxation modulus:

$$\sigma = \int_{-\infty}^{t} G(t-t')\left(\frac{d\gamma}{dt'}\right) dt' \qquad (12.8)$$

where $G(t-t')$ is the relaxation modulus (the stress per unit applied strain) and $d\gamma/dt'$ is the shear rate.

Now consider the system subject to an oscillatory strain (e.g. $\gamma(t)$ due to mechanical vibration of the sample) of the form:

$$\gamma(t) = \gamma_0 \sin \omega t \qquad (12.9)$$

where γ_0 is the amplitude of displacement, t is the time and ω is the frequency of the oscillation. The corresponding strain rate is simply the differential with respect to time of equation (12.9) and is therefore:

$$\frac{d\gamma}{dt} = \omega \gamma_0 \cos \omega t \qquad (12.10)$$

Through substitution of this expression for the strain rate in equation (12.8) it can be shown that an alternative functional form for the stress as a function of time (t) is:

$$\sigma(t) = G'(\omega)\gamma(t) + \frac{G''(\omega)}{\omega} \frac{d\gamma(t)}{dt} \qquad (12.11)$$

where $G'(\omega)$ is a frequency dependent shear modulus, a measure of the elastic energy stored in the network (the elastic storage modulus), and G'' is the component of the shear modulus that corresponds to the energy dissipated in the material (the dissipative or loss modulus). $G''(\omega)/\omega$ can be defined as the frequency dependent *dynamic viscosity*. Newton's law of viscosity relates the shear stress (σ) to the rate of shear $(\dot{\gamma})$ and is:

$$\sigma = \eta \frac{d\gamma}{dt} \qquad (12.12)$$

Therefore, it can be seen that the conventional viscosity of the material is given by the limit of the dynamic viscosity $(\eta = G''/\omega)$ as the

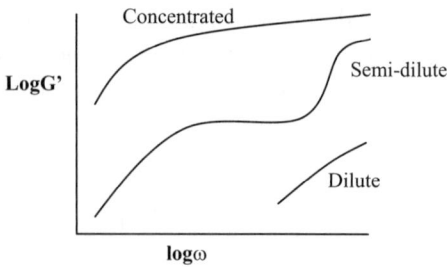

Figure 12.2 Schematic diagram of the elastic shear modulus (G') as a function of frequency (ω) for solutions of flexible polymers. The frequency dependence of the three dominant regimes of polymer concentration is illustrated

frequency tends to zero ($\omega \rightarrow 0$) for a material with no elasticity ($G' = 0$).

As an example of the response of a viscoelastic material to an oscillatory strain the storage modulus of some polymer solutions is shown as a function of frequency in Figure 12.2. Obviously the modulus of a viscoelastic polymer solution is expected to increase with the polymer concentration, since there are more large structure forming molecules that resist the shear. However, the expectation for the frequency dependence of the storage modulus needs much more careful thought and it will be considered in detail in Section 12.3.

A useful function that characterises the linear viscoelasticity of a material is the ratio of the elastic modulus to the dissipative modulus; it is defined to be equal to the tangent of the phase angle (δ):

$$\tan \delta = \frac{G'}{G''} \quad (12.13)$$

It is then found that the resultant stress on a linear viscoelastic material for an applied sinuisoidal strain (equation (12.9)) is also sinuisoidal, but has a phase lag (δ):

$$\sigma = \sigma_0 \sin(\omega t + \delta) \quad (12.14)$$

This phase lag helps quantify the linear viscoelasticity of the material. Evidently a material that is predominantly elastic will have a large value for tanδ from equation (12.13) (e.g. elastin) whereas a predominantly fluid material (e.g. blood plasma) will have a very low value.

To begin to understand the time response of biological materials simple mechanical models can be useful. The simplest models are those

STORAGE AND LOSS MODULI

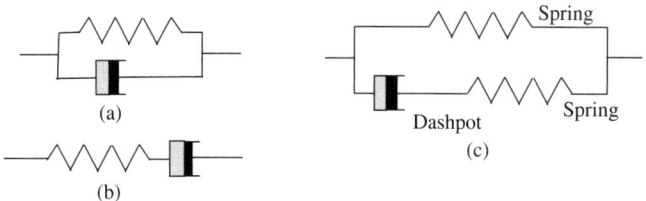

Figure 12.3 Standard mechanical constitutive equations can be derived for (a) the Maxwell model, (b) the Kelvin model and (c) the standard linear solid

due to *Kelvin* and *Maxwell*, which are constructed from a dashpot and a spring in parallel and series respectively (Figure 12.3). The solution of these models provides some simple constitutive equations that can then be used to model the viscoelastic response of real biological materials. Ideally the connection between the viscoelasticity and dynamics of the constituent biological molecules is required (what is the molecular origin of the spring and dashpot?), but simple constitutive equations are a useful first step towards this goal.

For an *elastic spring* (an ideal elastic solid) the shear modulus (G) is given by:

$$\sigma = G\gamma \tag{12.15}$$

and the stress is proportional to the strain. In contrast, for a *viscous dashpot* (an ideal viscous fluid), a totally dissipative structure, the stress is proportional to the rate of shear as per Newton's law, equation (12.12):

$$\sigma = \eta\dot{\gamma} \tag{12.16}$$

where η is the viscosity.

For the *Kelvin model* (a dashpot and an elastic spring in parallel, Figure 12.3(a)) the stresses in both elements can be added, since they resist the imposed stress in concert:

$$\sigma = G\gamma + \eta\dot{\gamma} \tag{12.17}$$

For the *Maxwell model* (a dashpot and an elastic spring in series, Figure 12.3(b)) the strain rates in both elements add linearly, since the total strain is just the sum of that in each component. Equation (12.15) can be differentiated and the result added to equation (12.16) to give:

$$\dot{\gamma} = \frac{\dot{\sigma}}{G} + \frac{\sigma}{\eta} \quad (12.18)$$

The decay of stress with time after a rapid step increase in strain can be calculated in the Maxwell model from a solution to equation (12.18) and is given by:

$$\sigma(t) = \sigma_0 e^{-t/\tau_m} \quad (12.19)$$

where the time constant is given by $\tau_m = \eta/G$ and σ_0 is the initial stress. The time dependent relaxation modulus is given by $G(t) = \sigma(t)/\gamma$ and a simple expression for it can then be calculated for a Maxwell material:

$$G(t) = \frac{\sigma_0}{\gamma} e^{-t/\tau_m} \quad (12.20)$$

Similarly, the decay of strain (γ) with time after a constant stress (σ_0) has been applied can be found using the Kelvin model through the solution of equation (12.17). It is:

$$\gamma = \frac{\sigma_0}{G}[1 - e^{-Gt/\eta}] \quad (12.21)$$

More sophisticated mechanical models can be created that capture both the stress and strain relaxation simultaneously such as the standard linear solid (Figure 12.3(c)). However, these models are still very much the domain of bioengineering, a molecular biophysicist hopes to make the connection between the viscoelasticity and the molecular structure to develop a detailed reductionist understanding of the phenomena (Section 12.3).

12.2 RHEOLOGICAL FUNCTIONS

Experiments that study linear viscoelasticity are a form of mechanical spectroscopy. The sample is struck with a mechanical perturbation (either a stress or a strain) and its response is measured as a function

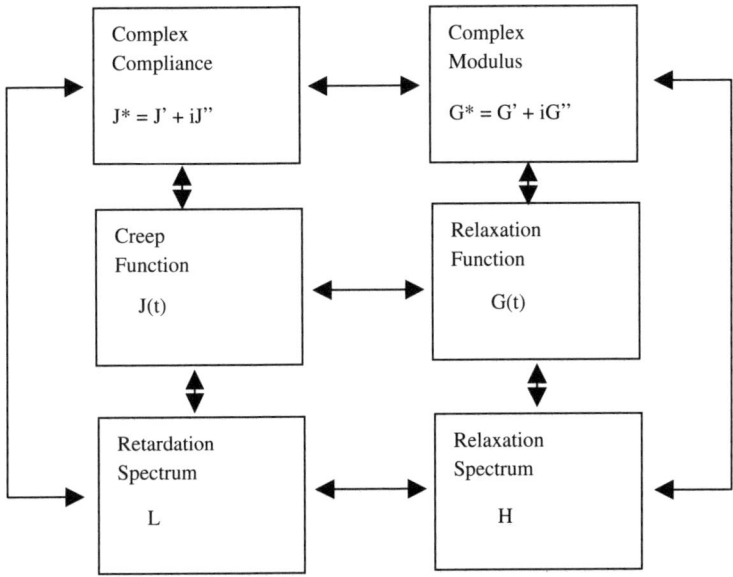

Figure 12.4 Interrelationship between linear rheological measures (complex compliance (J^*), complex modulus (G^*), creep function ($J(t)$), relaxation function ($G(t)$), retardation spectrum (L) and the relaxation spectrum (H) [Reprinted with permission from J.W. Goodwin and R.W. Hughes, *Rheology for Chemists*, Copyright (2000) Royal Society of Chemistry])

of time. A series of equivalent rheological measures can be used to quantify the experiments (Figure 12.4); these include the complex compliance ($J^* = J' + iJ''$), complex modulus ($G^* = G' + iG''$), creep function ($J(t)$), relaxation function ($G(t)$), retardation spectrum (L) and relaxation spectrum (H). This range of measures is convenient for a number of reasons. Often an experimentalist needs to impose a low stress or strain to obtain a linear response, and this is facilitated with the measurement of a particular rheological function. The data can then be mathematically translated to a different function for comparison with an actual process at a larger stress or strain. The use of linear transformations can also allow the prediction of viscoelastic behaviour over a wide range of time scales, e.g. outside the experimental time window for a single rheological function.

A flavour of some of the mathematical inter-relationships between the rheological functions is shown in equations (12.22–12.25). All of these transformations are linear and well determined. The data can be transformed back and forth between these functions with no loss of information:

$$G(t) = \int_{-\infty}^{\infty} He^{-t/\tau} d\ln\tau = \frac{2}{\pi} \int_{-\infty}^{\infty} \frac{G''(\omega)}{\omega} \cos\omega t \, d\omega \quad (12.22)$$

$$G(t) = \frac{2}{\pi} \int_{-\infty}^{\infty} \frac{G'(\omega)}{\omega} \sin\omega t \, d\omega \quad G'(\omega) = \frac{J'(\omega)}{J'(\omega)^2 + J''(\omega)^2} \quad (12.23)$$

$$G''(\omega) = \frac{J''(\omega)}{J'(\omega)^2 + J''(\omega)^2} \quad J'(\omega) = \omega \int_{0}^{\infty} J(t) \sin\omega t \, dt \quad (12.24)$$

$$J''(\omega) = -\omega \int_{0}^{\infty} J(t) \cos\omega t \, dt \quad J(t) = \int_{-\infty}^{\infty} L[1 - e^{-t/\tau}] d\ln\tau \quad (12.25)$$

These numerical transformations are very standard and a number of software packages are available that provide robustly implemented algorithms for their calculation.

12.3 EXAMPLES FROM BIOLOGY

12.3.1 Neutral Polymer Solutions

Many polymeric solutions exist in biological systems. Even solid polymers such as collagen and spider silk pass through a concentrated lyotropic phase during their synthesis. The dynamics of flexible neutral polymers is much simpler than the scenario with charged polymers (polyelectrolytes), and this is the initial focus of the discussion. There are three basic concentration regimes that affect the dynamics of neutral flexible polymeric molecules: *dilute* (the chains do not overlap), *semi-dilute* (the chains form an overlapping mesh) and *concentrated* (the thermal blob size is equal to the correlation length and the chains are Gaussian on all length scales). The range of dynamic behaviour exhibited by a neutral flexible polymer chain as a function of the number of monomers in a chain (N) and the concentration of monomers (c) is

EXAMPLES FROM BIOLOGY

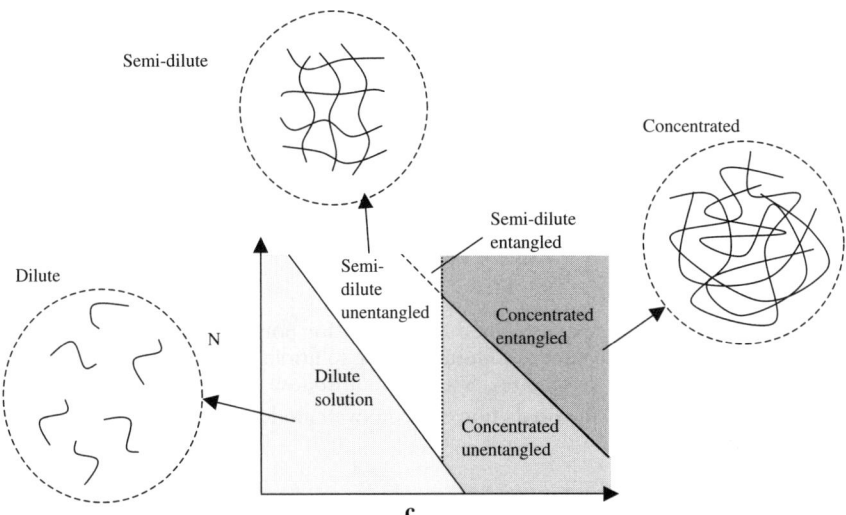

Figure 12.5 The phase diagram for the dynamics of flexible polymer solutions (The number of monomers in the polymer chains (N) is shown versus the polymer concentration (c).)

shown in Figure 12.5. In the dilute regime polymer solutions are predominantly viscous. Above the overlap concentration (c^*) the chain dynamics are strongly coupled with an interplay of Zimm and Rouse modes due to screening of the hydrodynamics of one chain by another. The elastic component of the viscoelasticity becomes important at lower and lower frequencies as the polymer concentration is increased and its absolute value increases. Polymers can reptate above a critical polymer length and monomer concentration (Section 8.5), which has a dramatic effect on their viscoelasticity; they demonstrate entangled rheology. The concentration at which the dynamic transition to reptation occurs is called the *entanglement concentration* (c_e).

The overlap concentration (c^*) is defined for the transition between the dilute and semi-dilute regimes (Figure 12.5). This concentration is given by:

$$c^* = \frac{1}{4\pi R_g^3/3} \quad (12.26)$$

where R_g is the radius of gyration of a chain and c^* has units of the number of chains per unit volume. In the case of rigid molecules the radius of gyration is equal to the long axis of the molecular rod (the contour length, L).

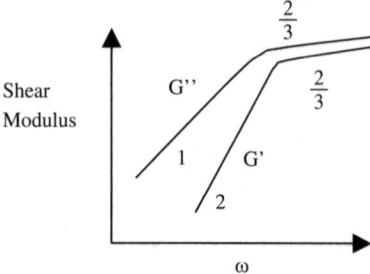

Figure 12.6 The complex shear moduli (G', G'') for polymer chains that obey the Zimm model for their dynamics (dilute polymer solutions and semi-dilute chains at length scales below the mesh size). Shear moduli follow $G'' \sim \omega^1$, $G'' \sim \omega^2$ at low frequencies (identical to the prediction of a Maxwell model) and $G' \sim G''/1.73 \sim \omega^{\frac{2}{3}}$ at high frequencies

The *Rouse and Zimm models* for polymer dynamics were encountered previously (Section 8.5) and can be used to describe the viscoelasticity of flexible polymer solutions (Figures 12.6 and 12.7). At low frequencies both models predict power laws of one and two for the frequency (ω) dependence of G' and G'' respectively, equivalent to a Maxwell model. At high frequencies a $\omega^{\frac{2}{3}}$ dependence of the shear moduli on frequency is observed for the Zimm model ($G' \sim G''/1.73$), whereas there is a $\omega^{\frac{1}{2}}$ dependence for Rouse modes ($G' \sim G''$). A detailed analysis of the linear viscoelastic spectrum of flexible polymers that uses the whole spectrum of dynamic modes can become quite involved, so only a few key results are quoted here. The relaxation modulus ($G(t)$) of semi-dilute polymer

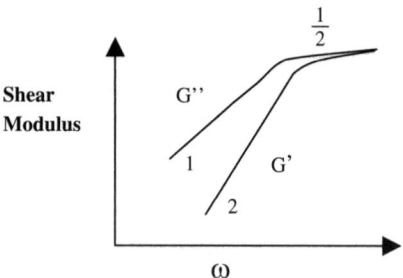

Figure 12.7 The complex shear modulus for polymer chains that obey the Rouse model for their dynamics (concentrated polymer solutions and semi-dilute solutions at lengths above the mesh size). The shear moduli follow $G'' \sim \omega^1$, $G' \sim \omega^2$ at low frequencies (identical to the prediction of a Maxwell model) and $G' \sim G'' \sim \omega^{\frac{1}{2}}$ at high frequencies

solutions of relatively short chains can be calculated with the Rouse model. For long chains the reptation concept is also required. Stresses in the Rouse model are determined by the orientation of the chains under the influence of an external force. The steady state shear viscosity (η) and steady state compliance ($J(t)$) of Rouse macromolecules are proportional to the number of links (N) in the chain. Solutions of long polymer chains exhibit unusual viscoelastic properties due to their reptative motion. The reptation model presented in Section 8.5 provides a calculation of the viscosity (η) of polymeric solutions and the viscosity is found to be a strong function of the degree of polymerisation ($\eta \sim N^3$). However, experiments indicate that the viscosity follows a 3.4 power law dependence on the degree of polymerisation ($\eta \sim N^{3.4}$) in real polymeric solutions (Figure 12.8). This deviation from the prediction of the reptation model is thought to be associated with fluctuations of the contour length of the tubes and the tube renewal process. The simple picture of a stationary tube of constraints provided in Section 8.5 thus needs to be modified. Reptative dynamics in polymeric solutions can be induced by a range of factors: an increase in the degree of polymerisation of the chains (the viscosity varies from a N^1 to $N^{3.4}$ dependence as the material starts to reptate), an increase in the persistence length of the chains, the addition of bulky side groups to the polymeric molecules and an increase in the polymer concentration (Figure 12.8).

When the persistence length of a polymer chain is significant compared to its contour length, *semi-flexible* models are required to describe the contributions of the internal modes of the polymer chain to the viscoelasticity. There is a wide range of biological materials that consist of networks of such rods and ropes. These materials conform to the

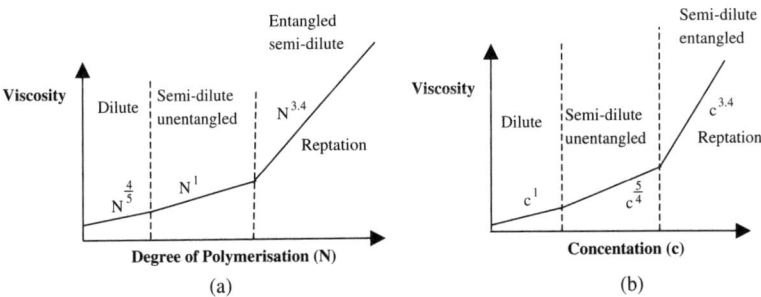

Figure 12.8 The dependence of the viscosity of a neutral flexible polymer solution on (a) the degree of polymerisation of the chains (N) and (b) the monomer concentration (c)
[Ref.: R. Colby and M. Rubinstein, Macromolecules, 1990, 23, 2753–2757]

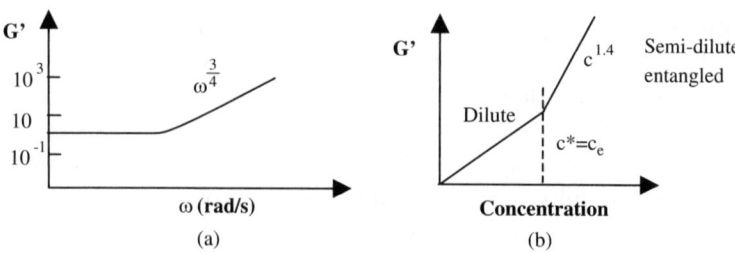

Figure 12.9 The elastic shear modulus (G') of long semi-flexible polymer networks has a number of unique experimental signatures. (a) The high frequency viscoelasticity follows $G' \sim \omega^{\frac{3}{4}}$ and (b) the modulus as a function of polymer concentration follows a $G' \sim c^{1.4}$ scaling above the overlap concentration [*Ref.*: D. Morse, *Macromolecules*, 1998, 31, 7044–7067]

predictions of relatively new theoretical models for semi-flexible chain dynamics, e.g. intermediate filaments, actin, peptide fibrils and microtubule solutions. Two experimental effects that are well predicted by the semi-flexible models are the high frequency viscoelasticity and the polymer concentration dependence of the shear modulus. A schematic diagram of the shear storage moduli of semi-dilute solutions of semi-flexible actin filaments determined using high frequency microrheology techniques such as diffusing wave spectroscopy, are shown in Figure 12.9. At very high frequencies the elastic modulus for semi-flexible chains differs from the Rousse and Zimm models. The shear modulus follows a power law dependence on the frequency with an exponent of 0.75 ($\omega^{\frac{3}{4}}$), which is thought to be characteristic of the transverse fluctuations of the semi-flexible chains.

12.3.2 Polyelectrolytes

The viscoelasticity of polyelectrolytes is a tricky field to approach both experimentally and theoretically, and is still under development. Examples of biological polyelectrolytes include nucleic acids, seaweed extracts, hyaluronic acid, proteoglycans and muscle proteins. Hyaluronic acid is contained in articulated joints and the rheology of this polyelectrolyte in synovial fluid is important for the mobility of the joint. Proteoglycans (comb polyelectrolytes) are used in a wide range of roles, as shock absorbers, for the reduction of friction and as barrier materials, e.g. protecting the stomach from self-digestion and minimising the adhesion of eye balls to eyelids. The rheology of DNA has a range of

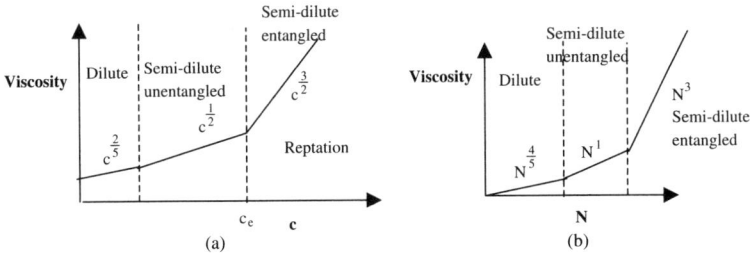

Figure 12.10 The scaling dependence of the viscosity of a solution of flexible polyelectrolytes on (a) the polymer concentration and (b) degree of polymerisation (N) [Ref.: A.V. Dobrynin, R.H. Colby and M. Rubinstein, Macromolecules, 1995, 28, 1859–1871]

biotechnological applications, e.g. to provide a detailed understanding of the results of electrophoresis experiments. Due to the relatively large size of DNA molecules they can be examined on a single molecule basis and this allows a range of single biomolecular polyelectrolyte rheology experiments to be performed.

Polyelectrolytes can be subdivided into flexible, semi-flexible and rigid classifications that depend on the persistence length of the backbone, in a similar manner to neutral polymers. However, the inclusion of charges along the backbone of a polyelectrolyte tends to increase the chains' persistence length when compared with their neutral counterparts (Section 9.9). Semi-flexible and rigid polyelectrolyte chains can be modelled by direct adaptation of the neutral results. Indeed, the predictions are identical to those of neutral polymers in the case of high salt concentrations where most of the charge interactions are screened. To predict the viscoelasticity of flexible polyelectrolytes current models adjust the results for neutral flexible polymers through the inclusion of the statistics of charged blob chain conformations. Examples of the predictions of scaling models are shown in Figure 12.10 for the viscosity on the concentration and degree of polymerisation. The entangled regime is highlighted by a change of the power law dependence on the concentration of the viscosity from $\frac{1}{2}$ to $\frac{3}{2}$ ($c^{\frac{1}{2}}$ to $c^{\frac{3}{2}}$) (Figure 12.10) and the unentangled semi-dilute regime (with a $c^{\frac{1}{2}}$ of the viscosity) is anomalously large. This behaviour is clearly observed experimentally for flexible polyelectrolytes. Polyelectrolytes also shear thin at very low shear rates which makes the measurement of intrinsic viscosities more challenging with these materials. This shear thinning property is useful practically in articulated joints, since the resistive forces decrease as the shear rate is increased (Figure 12.11).

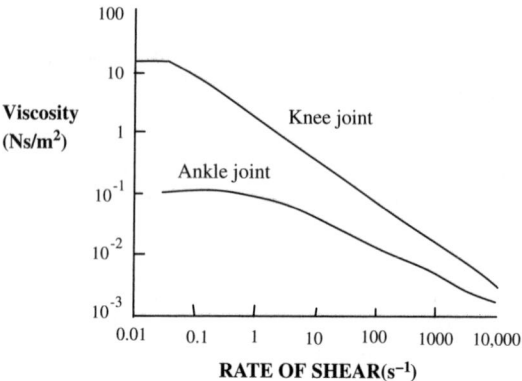

Figure 12.11 Non-linear shear thinning rheology of synovial fluids (flexible polyelectrolytes) from the ankle and knee joints as a function of shear rate
[*Reprinted with permission from R.G. King, Rheol. Acta, 5, 41–44, Copyright (1966) Springer Science and Business Media*]

A complete understanding of the linear viscoelasticity of flexible polyelectrolytes is still being developed. An example of an entangled flexible polyelectrolyte solution is shown in Figure 12.12, synovial fluids from umbilical cords and articulated joints.

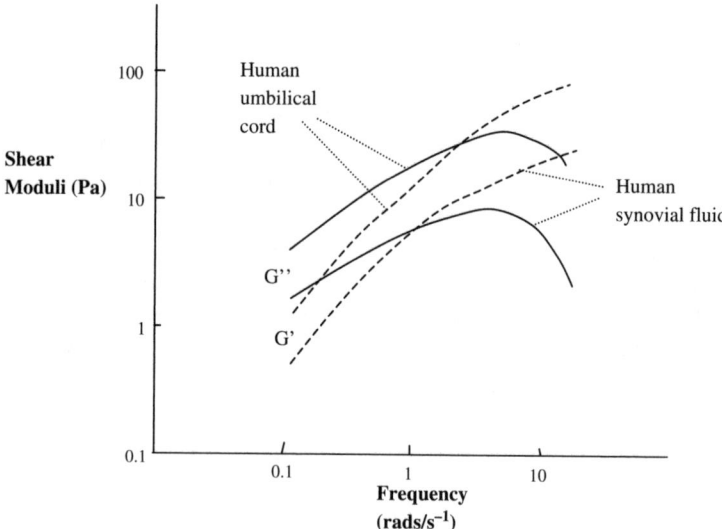

Figure 12.12 The shear moduli of fluids from the human umbilical cord and synovial fluid (both materials contain large quantities of hyaluronic acid, a flexible polyelectrolyte) as a function of frequency. G'' is indictaed by the continuous lines and G' the dashed lines
[*Reprinted with permission from D.A. Gibbs, Biopolymers, 6, 777–791, Copyright (1968) John Wiley & Sons, Inc.*]

12.3.3 Gels

A gel is a material in which the polymeric components are cross-linked to form a network (Figure 12.13). The crucial difference between an elastomer (Section 8.3) and a gel is that there is a large amount of solvent associated with the components of a gel, swelling its microstructure and altering the mobility of the chains. Gelled phases of matter are realised by a range of biological molecules, e.g. many foods are gelled biopolymers such as starches (custard and Turkish delight), pectins (jams) and denatured collagen (table jelly).

Cross-links in biopolymers are separately categorised as *physical* (of a weak nature, e.g. electrostatic or hydrogen bonds) or *chemical* (strong in nature, e.g. covalent/ionic bonds). With physical gels the moduli become low when the cross-links (e.g. helical, or egg box connections between chains) are melted at high temperatures. Chemical gels in contrast cannot be reformed, since permanent chemical cross-links exist between the subunits (often disulfide links in collagenous systems). The heat treatment of chemical gels at high temperatures results in a complete irreversible breakdown of the chemical structure.

In many biopolymer networks the degree of cross-linking can be switched on or off by varying the temperature, solvent quality, electrostatics or number of specialised biomolecular cross-linkers. At the point where the degree of cross-linking is sufficiently large for one complete aggregate to span the sample volume (called the *percolation* or the *gel*

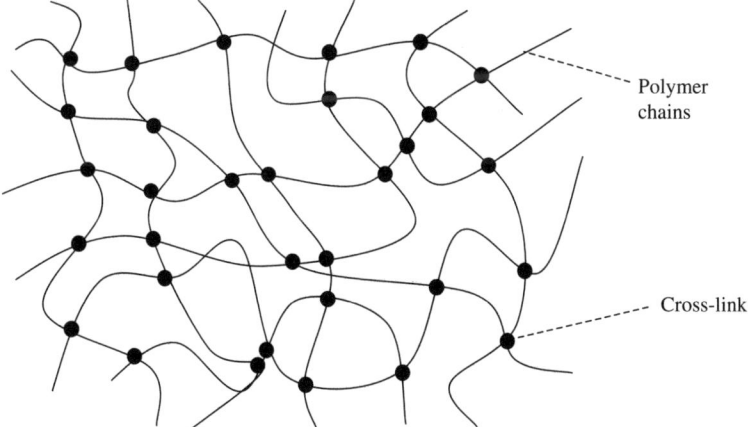

Figure 12.13 Schematic diagram of a polymeric gel showing the chemical cross-links (The solvent molecules that swell the network and help determine the chain conformations are not shown for simplicity.)

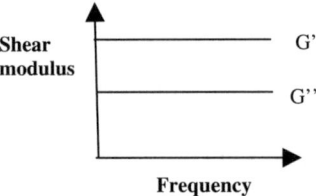

Figure 12.14 Schematic diagram of the complex shear moduli of a gelled flexible polymeric network at low frequencies; $G' > G''$ and the frequency dependence is fairly weak
(Note the similarity with Figure 12.9(a) for a semi-flexible network at low frequencies for G'.)

point), the relaxation modulus ($G(t)$) is described by a simple power law (the gel equation):

$$G(t) = St^{-n} \qquad (12.27)$$

where S is the strength of the gel and t is the time. Using equations (12.22–12.25) to transform between the different measures of linear viscoelasticity, the complex shear moduli (G^*) can be constructed from this simple power law for the relaxation function ($G(t)$), and then compared with the results of an oscillatory experiment. Above the percolation threshold and at low frequencies where the internal modes cannot be observed, the shear moduli of a gel has a weak frequency dependence (Figure 12.14). Characteristic rubbery type behaviour occurs for gels with the elastic modulus much higher than the dissipative modulus ($G' \gg G''$). To a first approximation the elastic modulus of flexible polymer gels is proportional to the density of cross-links, in a similar manner to that seen with the behaviour of elastomers (Section 8.3).

The model of *sticky reptation* has been developed to describe the rheology of associating physical gels (Figure 12.15). The stickers that

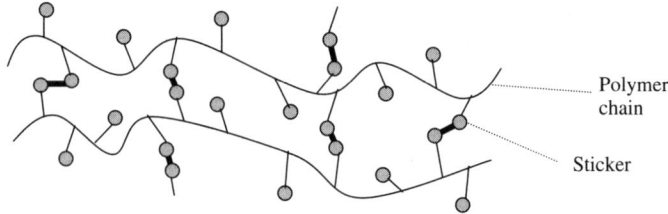

Figure 12.15 The entangled rheology of associating physical gels can be described using the theory of sticky reptation. An additional time scale is introduced into the chain dynamics due to the sticker life time, that radically slows down the motion of the chains

EXAMPLES FROM BIOLOGY   285

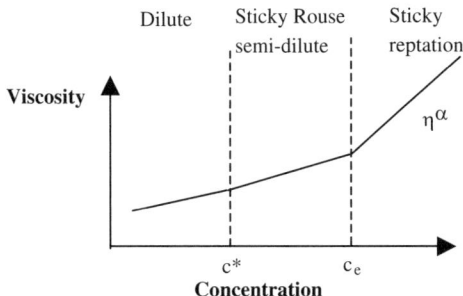

Figure 12.16 The viscosity of an associating polymeric solution is a very strong function of the polymer concentration at high concentrations, with α a large exponent. The dynamic behaviour can be very rich. A simplified phase diagram is shown and the sticker life times modify the dynamics in both the semi-dilute (Rouse, $c* < c < c_e$) and concentrated (Reptation, $c > c_e$) regimes
[*Ref.*: M. Rubinstein and A.N. Semenov, *Macromolecules*, 1998, 31, 1386–1397]

form the cross-links between the polymer chains introduce a second time scale to the dynamics of the chain motion. Above the entanglement concentration the chains move like centipedes through their reptation tubes with stickers for legs. The viscosity therefore becomes a very strong function of the polymer concentration (Figure 12.16) compared with unassociating polymer solutions, due to the dramatic slowing down of the chain motion with the introduction of more stickers. Other unusual phenomena can also occur with sticky polymers such as *shear thickening* (polymers normally shear thin), with the resistance of the network (the shear modulus) increasing with increasing shear rate.

Many naturally occurring gels are formed from polyelectrolytes. The osmotic pressure of the counterions associated with the polymer chains in these materials has a series of dramatic consequences for the physical properties of charged gels. The free counterions increase the swelling of the gel by a large degree and modify the elastic moduli. The mechanical properties of charged gels are an important consideration in a range of biological problems, e.g. for the viscoelasticity of cartilage, the cornea and striated muscle.

Many common biological polymers are semi-flexible and their cross-linked networks have a range of unique properties related to the rigidity of the chains (Figure 12.17). One interesting phenomena is that of *strain hardening*, the elastic modulus increases as the samples are strained. The semi-flexible networks of fibrin in blood clots have a virtually perfect signature of elasticity in the linear shear moduli over a wide range of frequencies (seven orders of magnitude, Figure 12.18). However, the

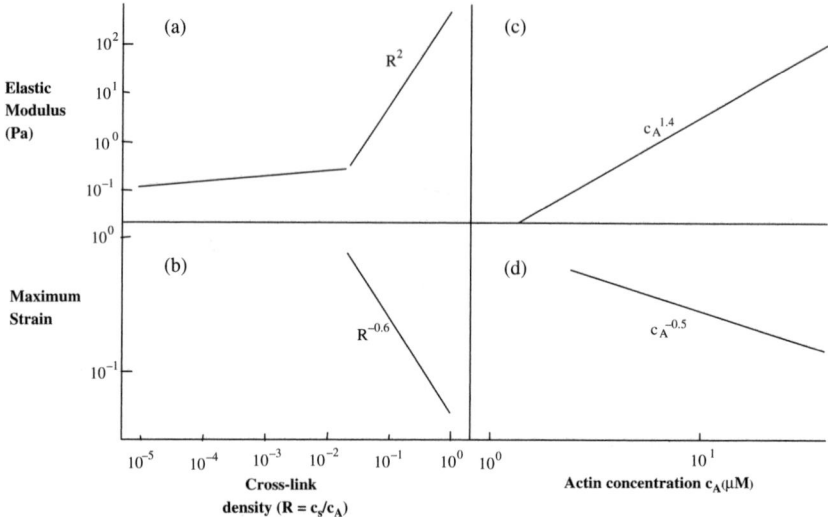

Figure 12.17 Trends found experimentally in the linear viscoelasticity of cross-linked actin, a well defined semi-flexible polymeric gel
((a) Elastic modulus as a function of cross-linking density; (b) maximum strain resisted by the gels as a function of cross-linking density; (c) elastic modulus as a function of actin concentration (cross-linking density = 0.13); (d) maximum strain as a function of actin concentration at fixed cross-linking density (cross-linking density = 0.13)
[Reprinted with permission from M.L. Gardel, J.H. Shin, F.C. MacKintosh et al., Science, 304, 1301–1305, Copyright (2004) AAAS])

semi-flexibility of the fibrin chains is clearly observed in the strain hardening phenomena at high strains (the shear modulus increase with strain). Perfect flexible chain elastomeric gels (e.g. polyacrylamide) do not show such behaviour (Figure 12.19).

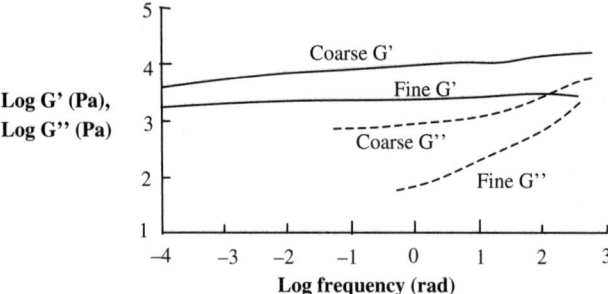

Figure 12.18 Linear viscoelasticity of the fibrin clots responsible for the clotting process in blood. Semi-flexible fibres occur in the clots that contain a large number of cross-links and demonstrate almost perfect elasticity over six orders of magnitude in frequency
[Reprinted with permission from W.W. Roberts, O. Kramer, R.W. Rosser et al., Biophy. Chemist, 152–160, Copyright (1974) Elsevier]

EXAMPLES FROM BIOLOGY 287

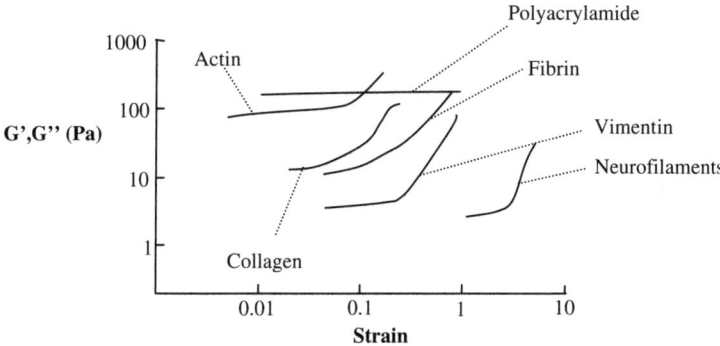

Figure 12.19 Strain hardening of semi-flexible networks is shown by the increase in the storage modulus with strain for a range of semi-flexible polymers
(The polyacrlamide gel is shown for comparison as it is a synthetic flexible polymer network that does not strain harden [*Reprinted with permission from C. Storm, J.J. Pastore, F.C. MacKintosh et al., Nature, 435, 191–194, Copyright (2005) Macmillan Publishing Ltd*])

12.3.4 Colloids

Colloidal science encompasses the large field of research that relates to particulate dispersions. One of the simplest examples of a colloidal system is a monodisperse suspension of identical particles; important biological examples of this scenario are provided by globular proteins and icosohedral viruses in solution.

For dilute dispersions of spheres the flow field is shown in Figure 12.20. Einstein solved the Navier–Stokes equations at low Reynold's number and found a surprisingly simple expression that relates to the dispersion's viscosity (η) to the volume fraction (φ) of colloids:

$$\eta = \eta_0 \left(1 + \frac{5\varphi}{2} + O(\varphi^2) \right) \qquad (12.28)$$

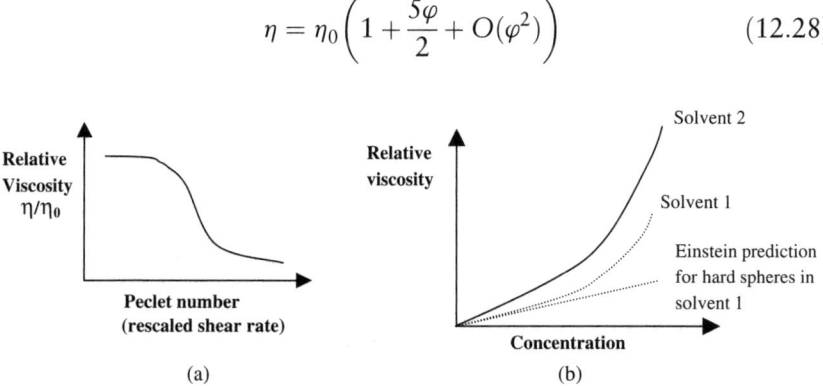

Figure 12.20 The relative viscosity (viscosity/viscosity of solvent) of colloidal suspensions as a function of (a) shear rate and (b) concentration of the colloids

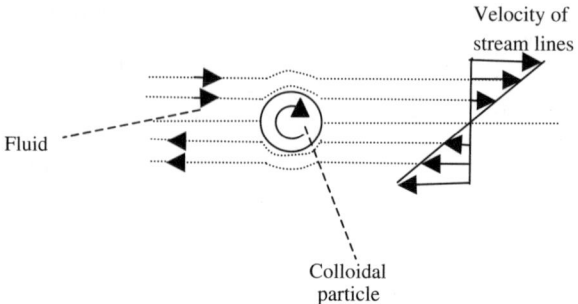

Figure 12.21 Steady flow field in a fluid around a spherical colloidal particle. The shear field has a vorticity and the colloidal particle rotates with a constant angular velocity

where $O(\varphi^2)$ is a term of order φ^2 and η_0 is the viscosity of the solvent. This Einstein equation can only model the behaviour of fairly dilute solutions of colloids (Figure 12.21) at low shear rates. At higher shear rates dilute colloidal solutions typically shear thin and the effects of a number of competing hydrodynamic effects need to be calculated.

The phase behaviour of a colloidal dispersion has a strong effect on the viscosity of a solution. Increasing the volume fraction of colloids above 2–3% also can cause a break down in equation (12.28) due to the change in the colloidal microstructure and a wide range of gel, fluid, or jammed phases are possible. Novel phenomena occur when high density colloidal solutions are sheared. The shear thinning phenomena that occur in suspensions of blood cells is illustrated in Figure 12.22. At low shear rates the blood cells aggregate into 'rouleaux' structures. Further increases in the shear rate cause a break up of these structures and a decrease in the viscosity. At even higher shear rates, there is a further reduction in viscosity associated with the formation of strings of blood cells, as these regular structures slide more easily past one another in the shear field.

12.3.5 Liquid Crystalline Polymers

Many biological polymers are liquid crystalline, e.g. DNA, cellulose, carrageenan and α helical peptides. These liquid crystalline polymers strongly shear thin in ways that depend on their defect structures (Section 4.3). It is thus important to measure the orientation of the

EXAMPLES FROM BIOLOGY

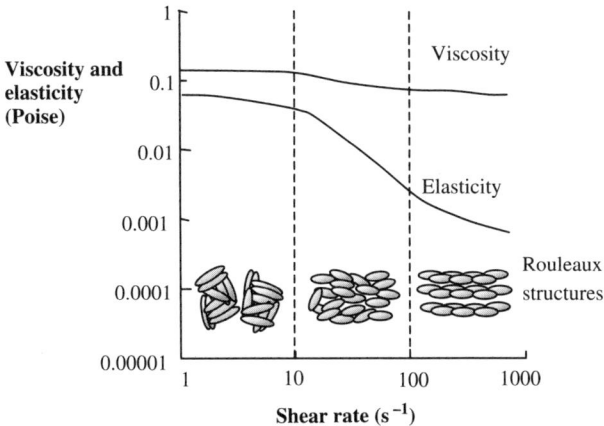

Figure 12.22 The non-linear rheology of blood cells and colloidal gels can demonstrate shear thinning (and the related phenomenon of thixotropy due to transients in the structural rearrangement of the colloids)

nematic director (with additional chiral and smectic order parameters if necessary) with respect to the direction of shear to understand the viscoelastic properties of these materials. An example of the non-linear flow properties of nematic liquid crystalline polymers is shown in Figure 12.23. Different degrees of shear thinning occur with shear rate. Often three power law behaviours are observed experimentally, a type I

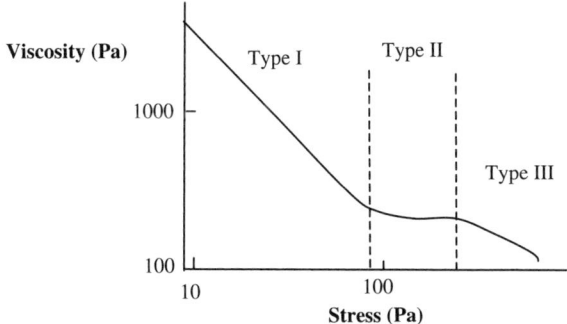

Figure 12.23 The flow behaviour of liquid crystalline polymers as a function of shear rate often has three characteristic regimes; Type I shear thinning, the Type II Newtonian plateau and Type III shear thinning (data from a cellulose derivative)
[*Ref.: S. Onogi and T. Asada, in Rheology, Proceedings of the Eighth International Congress on Rheology, Naples, Italy, 1980, Plenum*]

shear thinning $\eta \sim \dot{\gamma}^{-\frac{1}{2}}$ regime, followed by a type II Newtonian plateau and, finally, a further type III shear thinning regime.

12.3.6 Glassy Materials

Many amorphous biological materials exhibit glassy phenomena (e.g. resilin at 10 Hz). Ergodicity (the exploration of all the molecules' energy states) is lost in these materials. Glasses result from, in the simplest picture, undercooled liquids in which the viscosity increases so rapidly that it prevents the formation of a crystalline phase. Everyday examples of biological glasses are boiled sweet (sugars) and a wide range of biopolymers in lowly plasticized states, e.g. starches in dehydrated foods.

Glasses are further categorised as *strong* and *weak*. In strong glasses the short range order tends to persist above the glass transition, whereas fragile glasses have no such memory. The viscosity of glasses has a characteristic dependence on the temperature, with the viscosity changing by up to sixteen orders of magnitude during the transition (Figure 12.24).

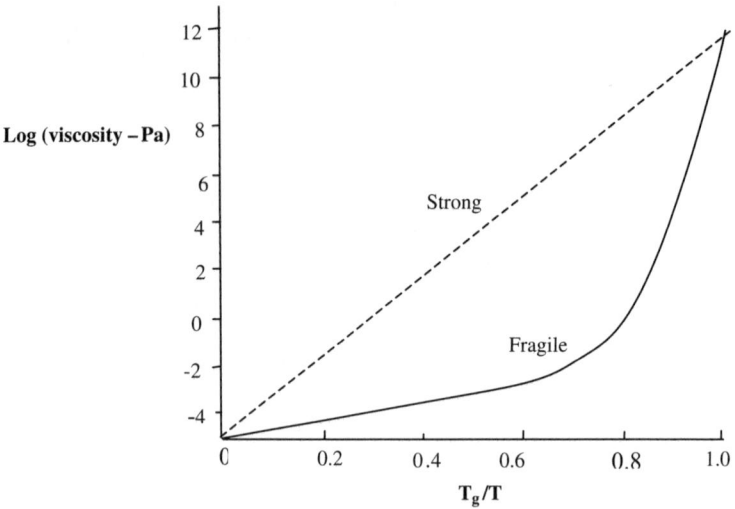

Figure 12.24 The viscoelasticity of glasses (fragile/strong) can be characterised by plots of the viscosity as a function of the reciprocal of the temperature (rescaled by the glass transition temperature, T_g)
[Ref.: J. Angell, J. Non-Cryst. Sol, 1988, 102, 205]

12.3.7 Microfluidics in Channels

There is a large current research effort in the area of microfluidics and microrheology. These studies consider the rheology of viscoelastic fluids as a function of length scale. Examples include how blood cells squeeze through the micron sized capillaries in arteries (the effective viscosity is much smaller than expected from a simple Pouseille flow calculation, Figure 12.1(a)), the interaction between mucins and cilia in the lungs and the motion of DNA on templated surfaces.

FURTHER READING

R.G.Larson, *The Structure and Rheology of Complex Fluids*, Oxford University Press, 1999. Advanced level text, an extensive range of materials is covered.

J.W.Goodwin and R.W.Hughes, *Rheology for Chemists*, Royal Society of Chemisty, 2000. A very useful simple introduction to the key concepts of rheology.

TUTORIAL QUESTIONS

12.1) In a running synovial joint the cartilage surfaces move relative to one another at a velocity of 1 cm s^{-1}. The distance between the surfaces in the cartilage can be on the order of 10 μm. What is the shear rate experienced by the synovial fluid? What is the approximate Peclet number of a hyaluronic acid chain in the fluid if its radius of gyration is 1 μm and the viscosity of the surrounding fluid is 0.001 Pa.s?

12.2) The longest time scale important in the viscoelasticity of a boiled sweet (glassy sugar molecules) can be described using a Maxwell model. What is the time scale for stress relaxation if the viscosity is 1 MPa.s and the elastic modulus is 10^{-3} Pa?

12.3) It is assumed that there is no interaction between blood particles in a suspension. Estimate the viscosity of a dilute blood suspension. The viscosity of the surrounding buffer is 10^{-3} Pa.s, and the volume fraction of the blood cells is 2%.

13
Experimental Techniques

A vast range of experimental techniques is used to analyse the structure and dynamics of biomolecules. A subset of methods that emphasise the physical behaviour of biological molecules is examined here and reference should be made to more specialised texts for detailed descriptions of analytical biochemical methods. The discussion of nuclear magnetic resonance, terahertz, ultraviolet, infra-red, mass and Raman spectroscopies is avoided, since they require too much space to be covered satisfactorily. Only the mechanical form of spectroscopy (*rheology*)– where the sample is hit with a mechanical perturbation and its response in time is observed–is examined in detail here.

Historically, the primary methods for structural determination of biological molecules have been high resolution probes such as scattering (neutrons, X-ray, light, elastic/inelastic) and microscopy (light and electron). There are many good accounts of standard scattering techniques in the biophysical literature and microscopy is well described in undergraduate optics textbooks. It is hard to beat the discussion in Cantor and Schimmel for detail on the process of scattering from biomolecules, so instead, after a brief introduction, some modern developments in the field of *scattering*, such as quasi-elastic scattering, microfocus scattering and coherent diffraction microscopy, are covered. In addition, methods of *single molecule force measurement*, *osmometry*, *sedimentation*, *tribology*, *solid mechanics* and *electrophoresis* are explained to give a modern emphasis to the subject.

13.1 STATIC SCATTERING TECHNIQUES

The field of scattering encompasses a vast range of fundamental physical processes and techniques. The basic geometry of a scattering experiment is shown in Figure 13.1. Incident radiation or particles interact with the sample and are deflected through an angle (θ). The *momentum transfer* (q) of the scattering process is simply related to the reciprocal of the length scale (d) probed:

$$q = \frac{2\pi}{d} \tag{13.1}$$

For a particular form of radiation the momentum transfer for an elastic (energy is conserved) scattering process can be calculated from:

$$q = \frac{4\pi}{\lambda} \sin\left(\frac{\theta}{2}\right) \tag{13.2}$$

where λ is the wavelength. The use of momentum transfers rather than scattering angles allows the results of experiments with a range of different forms of radiation (X-rays, light, neutrons etc.) to be compared easily. The varieties of radiation typically used in biological scattering experiments are shown in Table 13.1, which includes the wavelength of the radiation and the length scales in the sample that can be typically

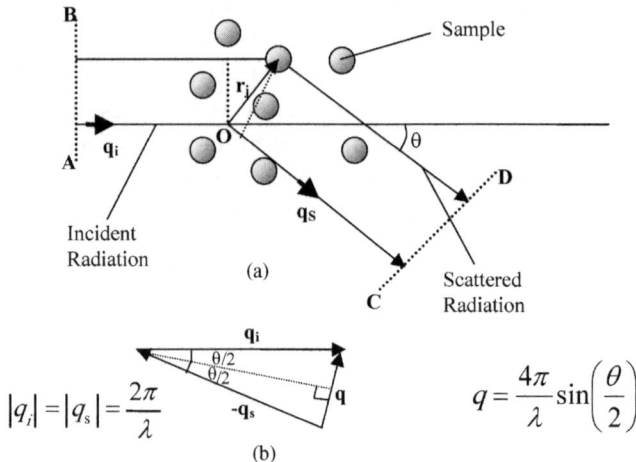

Figure 13.1 (a) The elastic scattering processes involved with a wide range of radiation (X-rays, light, neutrons and electrons) can all be understood using the same schematic diagram. (b) The momentum transfer can be calculated from the wavelength and scattering angle using a simple geometrical construction

Table 13.1 Comparison of scattering techniques applied to biological samples

Technique	Typical wavelength (nm)	q values (nm^{-1})	Real space distances (nm)	Sample contrast	Comments
Small angle X-ray scattering	0.15	0.009–6.284	1–700	Electron density	Beam damage can be a problem
Small angle neutron scattering	0.4	0.0031–6.284	1–200	Scattering length density	Samples can be labelled with deuterium
Light scattering	450	0.0003–0.126	50–2000	Refractive index	Multiple scattering is a problem at high sample concentrations (>2%)
Wide angle X-ray scattering	0.15	6.284–62.84	0.1–1	Electron density	Beam damage can be a problem
Wide angle neutron scattering	0.4	6.284–62.84	0.1–1	Scattering length density	Useful to measure hydrogen bonds
Electron scattering	0.0037	0.006284–6.284	1–1000	Electron density	Very thin sections are required due to strong Coulombic interaction with samples

probed. Specialised detectors and optics are required for each different form of radiation.

There are a number of novel techniques that utilise the coherence of electromagnetic radiation and in particular that of X-rays. Two promising new coherent methods are *X-ray diffraction imaging* and *quasi-elastic x-ray scattering*. Images of completely aperiodic magnese stained bacteria have now been reconstructed using coherent X-ray diffraction with 100 nm resolution and quasi-elastic X-ray scattering offers dynamic measurements from soft matter systems with unprecedented sensitivity to length scale.

A further modern advance with X-ray and neutron scattering is the introduction of effective *focusing techniques*. Focusing of X-rays is routinely made to submicron levels at third generation synchrotron sources and micron sized beams can be made with laboratory based microfocus sources. Such beams can be rastered across heterogeneous biological materials and the molecular structure probed as a function of position on the sample (Figure 13.2). New focusing devices have now been made to create micron sized beams that include capillaries, fresnel lens, mirrors and even simple compound lens (e.g. lenticular holes in a block of aluminium). Such scanning X-ray microdiffraction techniques

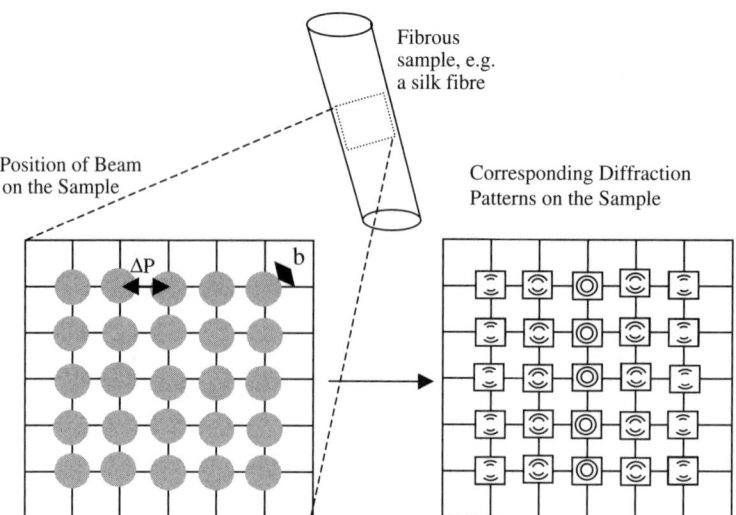

Figure 13.2 Scanning X-ray microdiffraction across a semi-crystalline anisotropic material
(b is the diameter of the X-ray beam. Two dimensional diffraction images are recorded at micron spaced steps (ΔP) across the sample and provide detailed information on the molecular structure of the fibres (see Figure 4.23).)

are helping to revolutionise the field of fibrous carbohydrates and proteins.

The *contrast* that is measured during a scattering process can be varied in both neutron and X-ray scattering experiments. With neutrons isotopic substitution can be used to label biomolecules through the replacement of hydrogen atoms with chemically equivalent deuterium atoms. This labelling scheme is particularly attractive when electronically light atoms need to be located in a crystalline structure, e.g. the elucidation of the structure of hydrogen bonds. With X-rays the wavelength of the radiation can be matched to the absorption edge of a heavy atom that exists in a biological structure, and the contrast can be varied to elucidate both the crystalline and solution state structures with much improved resolution, e.g. through the method of anomalous small-angle X-ray diffraction.

13.2 DYNAMIC SCATTERING TECHNIQUES

Once the structure of a biological sample is well understood, quantifying the dynamics of the components of the material poses some important questions. A challenge is to examine the dynamics of the material without perturbing the sample morphology. With soft biological materials the dynamics can be studied quantitatively with scattering methods by observing the time decay of stimulated emission (*fluorescence techniques*) or by measuring the change in energy of the scattered particles (*quasi elastically* or *inelastically scattered*).

Fluorescence intensity correlation spectroscopy is a modern example of a scattering technique that has been adapted to single molecule experiments. Fluorescent probes are added to the biological molecules whose dynamics are of interest (there is a vast range of possible ways to do this and large commercial catalogues exist of the available fluorescent probe molecules) and made to fluoresce using a tightly focused laser beam under an optical microscope (Figure 13.3). The intensity of the fluorescently emitted radiation ($I_f(t)$) is proportional to the concentration of fluorescent molecules in the scattering volume ($c(r,t)$), the fluorescent yield (Q) and the intensity of the incident laser beam (I):

$$I_f(t) = Q\varepsilon \int I(r)c(r,t)d^3r \qquad (13.3)$$

where ε is the extinction coefficient of the molecular species and the integral over d^3r is over the complete scattering volume. The dynamic

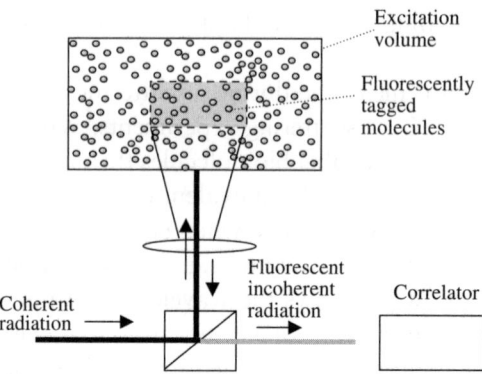

Figure 13.3 Schematic diagram of a fluorescence intensity correlation spectroscopy experiment (FICS)
(Incident coherent light excites fluorescence in a small volume of the sample. Fluctuations in the emitted fluorescence are measured with a correlator and can be related to the dynamics of biomolecules to which the fluorescent tags are attached.)

information is contained in the fluctuations of the emitted fluorescent intensity with time, which are directly related to the fluctuations of the concentration of the fluorescent molecules ($\langle \delta c(r,t) \delta c(r',0) \rangle$):

$$\langle \delta I_f(t) \delta I_f(0) \rangle = (\varepsilon Q)^2 \int\int I(r)I(r') \langle \delta c(r,t) \delta c(r',0) \rangle d^3r d^3r' \qquad (13.4)$$

where $\langle \rangle$ denotes a time average and δ is a small fluctuation in a quantity. For *translational diffusion* of the fluorescent probes the intensity autocorrelation function has a simple exponential form. Thus, an exponential can be fitted to the autocorrelated fluorescent signal from a fluorescently tagged biomolecule, which gives the characteristic time constant (τ) to diffuse out of the scattering volume (e.g. a cube of side b), and hence the diffusion coefficient (D) of the molecules ($D = 6b^2\tau$). Fluctuations in the intensity of the fluorescent light emitted as particles move across the irradiated volume can thus be related to the diffusion coefficients of the fluorescent species. Unfortunately, the information that relates to the momentum transfer (equation (13.2)) is lost in this inelastic scattering method and, for larger scattering volumes of biological materials, data from quasi-elastic scattering are much more rich in information.

Fluorescence depolarisation experiments are a further powerful tool. A pulsed laser excites fluorescent probes attached to biological molecules, whose motion can be detected by the change in polarisation of the reemitted photon. If the molecule reorientates a considerable amount

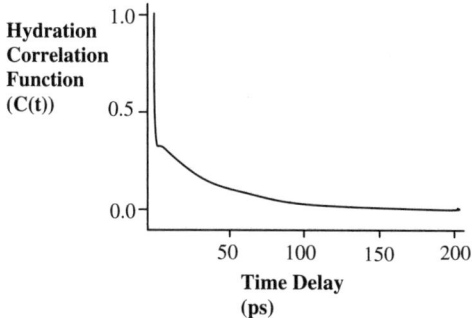

Figure 13.4 Ultra fast dynamics were probed using fluorescence depolarisation pulsed laser experiments with the protein subtilisin Carlsberg
[*Reprinted with permission from S.K. Pal, J. Peon, B. Bagchi and A.H. Zewail, J. Phys.Chem. B, 106, 12376–12395, Copyright (2002) American Chemical Society*]

over the picosecond time scale of fluorescent emission, the polarisation state of the emitted photon is changed. The utility of fluorescence depolarisation stems from the fact that the technique can probe dynamics in the ultra fast picosecond time regime due to the availability of intense ultra-fast pulsed laser sources (Figure 13.4). Furthermore, the high yield of fluorescent reemission makes the experimental measurement of correlation functions over ultra-fast time scales and with single molecules feasible.

Quasi-elastic scattering experiments cover a wide range of techniques that monitor small energy changes in scattered radiation due to the motion of a sample, i.e. a Doppler shift of the energy of the scattered particle. Typically, it is the normalised *intermediate scattering function* ($F(q,t)$) that is measured in a quasi-elastic scattering experiment (Figure 13.5). The intermediate scattering function is a useful general

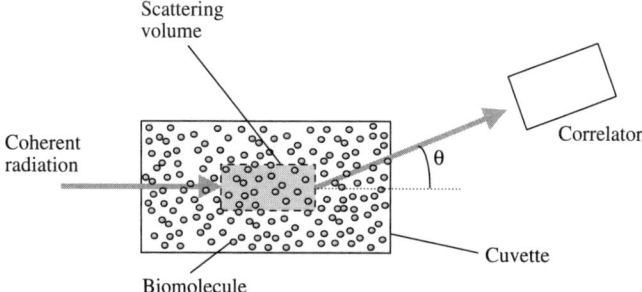

Figure 13.5 Schematic arrangement of a dynamic light scattering experiment (Coherent light is Doppler shifted by the motion of the biomolecules in the illuminated scattering volume and this shift is subsequently detected using a correlator.)

tool, which is amenable to accurate quantitative theoretical analysis. For the scattering of coherent light (called equivalently quasi-elastic light scattering, photon correlation spectroscopy or dynamic light scattering) the quantity examined experimentally ($g_1(\tau)$, the field correlation function) relates to the correlation function of the scattered electric field. In terms of the scattered electric field (E) at time (τ) the field correlation function measured at a certain angle is defined as:

$$g_1(\tau) = \frac{\langle E^*(0)E(\tau)\rangle}{\langle I\rangle} \quad (13.5)$$

where $\langle\,\rangle$ is the ensemble averaged quantity over the array of scatterers, and I is the intensity measured on the detector. The electric field strength (E) scattered by the collection of diffusing particles in the sample is given by:

$$E = \sum_{j=1}^{N} A_j e^{iq\cdot r_j} E_0 e^{-i\omega_0 t} \quad (13.6)$$

where E_0 is the magnitude of the incident wave, A_j is the scattered wave amplitude of the jth particle, r_j is the position of the jth particle, q is the momentum transfer, t is the time and ω_0 is the frequency of the incident radiation. The sum from $j = 1, \ldots, N$ is over all the particles that scatter radiation in the sample. The form of the electric field can be substituted in equation (13.5) and the correlation function can therefore be written:

$$g_1(\tau) = \langle e^{iq\cdot[r_l(\tau)-r_l(0)]}\rangle e^{-i\omega_0\tau} \quad (13.7)$$

Fick's second law of diffusion (equation (5.19)) in three dimensions is:

$$\frac{\partial c(r,t)}{\partial t} = D\nabla^2 c(r,t) \quad (13.8)$$

where $c(r,t)$ is the concentration of molecules in the scattering volume and D is the translational diffusion coefficient. Let $P(O/r,t)$ be the conditional probability that a particle can be found in volume element d^3r at time t. For low particle concentrations the conditional probability $P(O/r,t)$ also obeys the diffusion equation:

$$\frac{\partial P(O|r,t)}{\partial t} = D\nabla^2 P(O|r,t) \quad (13.9)$$

A Fourier transform can be taken of either side of (13.9) and allows the equation to be solved:

$$\int_0^\infty e^{i\underline{q}\cdot\underline{r}} \frac{\partial P(O|r,t)}{\partial t} d^3r = D \int_0^\infty e^{i\underline{q}\cdot\underline{r}} \nabla^2 P(O|r,t) d^3r \qquad (13.10)$$

A general property of Fourier transforms is that the Fourier transform of a nth order differential is equal to $(-iq)^n$ times the Fourier transform of the argument of the differential:

$$\int_{-\infty}^\infty e^{iqy} \frac{\partial^n}{\partial y^n} P(y) dy = (-iq)^n \int_{-\infty}^\infty e^{iqy} P(y) dy \qquad (13.11)$$

The *intermediate scattering function* ($F(q,t)$) is equal to the Fourier transform of the probability distribution ($P(O|r,t)$):

$$F_s(\underline{q},t) = \int_0^\infty P(O|r,t) e^{i\underline{q}\cdot\underline{r}} d^3r \qquad (13.12)$$

This definition and the Fourier transform identity (equation (13.11)) allows equation (13.10) to be simplified in terms of the intermediate scattering function:

$$\frac{\partial F_s(\underline{q},t)}{\partial t} = -Dq^2 F_s(\underline{q},t) \qquad (13.13)$$

This is a simple variables separable differential equation which has the solution:

$$F_s(\underline{q},t) = F_s(\underline{q},0) e^{-Dq^2 t} \qquad (13.14)$$

where the initial condition is given by $F_s(q,0) = 1$. From equation (13.7) the field correlation function can therefore be written as:

$$g_1(\tau) = F_s(q,t) e^{-i\omega_0\tau} = e^{-Dq^2\tau} e^{-i\omega_0\tau} \qquad (13.15)$$

The intensity correlation function ($g_2(\tau)$) is measured by time correlation of the signal on a photomultiplier tube due to the scattered radiation and is related to the electric field correlation function ($g_1(\tau)$) by:

$$g_2(\tau) = 1 + |g_1(\tau)|^2 \qquad (13.16)$$

By substitution of equation (13.15) in (13.16) it is seen that diffusional process introduces a $\exp(-Dq^2\tau)$ term in the intensity correlation function, $g_2(\tau)$. A typical correlation function for a fibrous protein that experiences translational diffusion is shown in Figure 13.6.

A qualitative understanding of the form of a correlation function can be achieved. At short times the particles have insufficient time to move anywhere, they do not dephase the scattered light (equivalently the scattered speckle pattern does not move), and the correlation is nearly perfect $g_1(\tau) \sim 1$. At longer times the motion of the particles decorrelates the phase of the scattered radiation and reduces the value of $g_1(\tau)$. Eventually at very long times the correlation function reduces to zero due to a complete random phasing of the scattered photons.

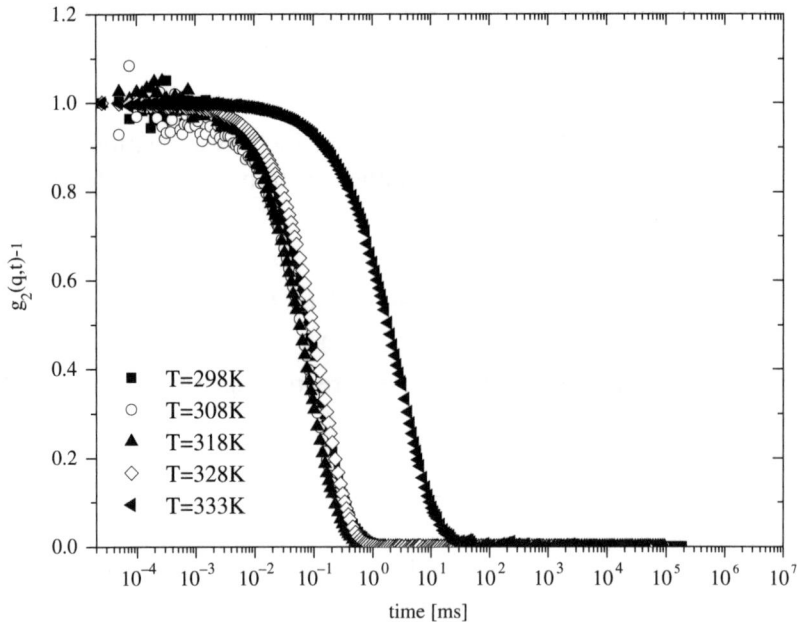

Figure 13.6 Intermediate scattering (intensity correlation, $g_2(q,t)$) functions from a quasi-elastic light scattering experiment on titin, a giant protein from skeletal muscle (The dynamics slow down as the protein unfolds with increasing temperature (T) due to the increase in protein length [*Reprinted with permission from E. Di Cola, T.A. Waigh, J. Trinick, et al., Biophysical Journal, 88, 4095–4106, Copyright (2005) Biophysical Society*])

A wide range of other radiation can be used to probe dynamic processes in mesoscopic biological materials with quasi-elastic scattering experiments that includes X-rays (time scales 10^{-7}–1000 s)) and neutrons (time scales 0.1–100 ns). X-ray quasi-elastic techniques are similar in conception to dynamic light scattering, whereas neutron spin echo measurements use the spin of scattered neutrons to clock the dynamics of the scattering process.

13.3 OSMOTIC PRESSURE

The phenomena associated with osmotic pressure are important in a series of biological processes: how cell metabolism is regulated (animal cells walls are ruptured if the external osmotic pressure is too high or low), how intermolecular forces are mediated by solvent molecules and the molecular crowding of the intracellular environment. Consider an idealised experiment with a semi-permeable membrane that separates two polymer solutions (Figure 13.7). This apparatus is an example of a membrane osmometer, a device for the measurement of osmotic pressure. From standard thermodynamic theory the partial differential of the Gibbs free energy ($G = F - TS + PV$) with respect to the pressure (P) is equal to the volume (V) at constant temperature (T):

$$\left(\frac{\partial G}{\partial P}\right)_T = V \qquad (13.17)$$

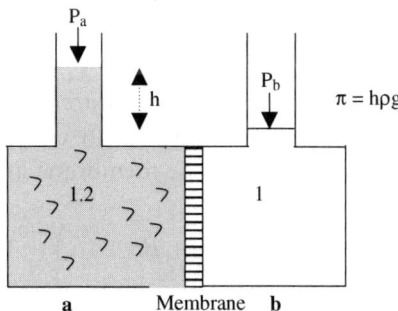

Figure 13.7 A membrane osmometer can be used to measure the difference in osmotic pressure between two solutions, a and b, which have P_a and P_b external applied pressures respectively. The difference in fluid heights (the capillary rise h) provides the osmotic pressure ($\pi = h\rho g$), where ρ is the fluid density and g is the acceleration due to gravity

The chemical potential (μ, the Gibbs free energy per particle) with respect to one of the components in the solution (μ_1) is:

$$\left(\frac{\partial \mu_1}{\partial P}\right)_T = \overline{V_1} \tag{13.18}$$

where $\overline{V_1}$ is the partial molar volume of the particles. Consider the chemical potentials on either side of the membrane in sections a and b (Figure 13.7). *Compartment b* only contains solvent molecules and the chemical potential is that of the solvent (μ_1^0):

$$\mu_1^b = \mu_1^0 \tag{13.19}$$

In *compartment a* the chemical potential has two additional effects, that due to the solute and that due to the reservoir used to measure the pressure:

$$\mu_1^a = \mu_1^0 - (solute\_effect) + (pressure\_effect) \tag{13.20}$$

The effect of the solute on the chemical potential is given by a van der Waals type expansion in the component concentration:

$$\mu_1^a = \mu_1^0 - RTV_1^0\left(\frac{c}{M} + Bc^3 + ..\right) + \int_{P_0}^{P_0+\pi} \overline{V_1} dP \tag{13.21}$$

where c is the concentration of the species too large to permeate the membrane (the solute), M is the molecular weight of the solute and B is the second virial coefficient of the solute. $V_1^0 = \overline{V_1}$ is the molecular volume of solvent at one atmosphere pressure, which allows the last term in equation (13.21) to be evaluated. In thermal equilibrium the chemical potentials on each side of the membrane are equal ($\mu_1^b = \mu_1^a$):

$$\mu_1^0 = \mu_1^0 - RTV_1^0\left(\frac{c}{M} + Bc + ...\right) + V_1^0\pi \tag{13.22}$$

This expression can be solved for the osmotic pressure (π) of the solution:

$$\pi = RT\left(\frac{c}{M} + Bc^2 + ...\right) \tag{13.23}$$

OSMOTIC PRESSURE

For a dilute solution the second virial coefficient is very small ($B = 0$), so to a good approximation:

$$\pi = \frac{RTc}{M} = nkT \qquad (13.24)$$

The osmotic pressure of a solution of non-interacting particles is directly proportional to the number of solute molecules ($n = c/N_A M$, where N_A is Avogadro's number) and, if the concentration is known, an accurate determination of the molecular weight of the particles can be made. An example of the osmotic pressure of a protein solution is shown in Figure 13.8. With the simple osmometer shown in Figure 13.6, the osmotic pressure (π) is given by:

$$\pi = hg\rho \qquad (13.25)$$

where h is the difference in fluid height, g is the acceleration due to gravity and ρ is the fluid density. Thus a simple measurement of the height of the fluid in the capillary leads to a direct calculation of the solution's osmotic pressure.

The effects of osmotic pressure are extremely important in determining the physical state and morphology of a biological material. For example, a charged polyelectrolyte gel placed in a dilute solution expands many times in volume due to the pressure exerted by the counterions associated with the polyelectrolyte chains. Polyelectrolyte gels can exhibit significant elasticity at very low volume fractions, e.g. 'wobbly solid' gelatine gels are 2–3%

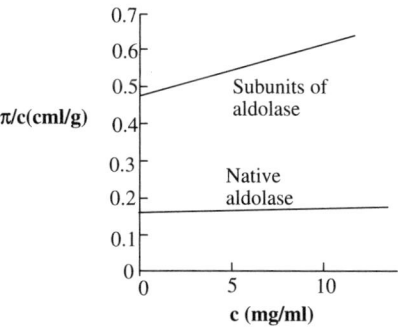

Figure 13.8 The osmotic pressure of a solution of globular proteins (A number of aldolase subunits assemble into the native structure which causes a decrease in the osmotic pressure of the solution. The osmotic pressure per gram of protein (π/c) is plotted as a function of protein concentration [*Reprinted with permission from F.J. Castellino and O.R. Baker, Biochemistry, 7, 2207–2217, Copyright (1968) American Chemical Society*])

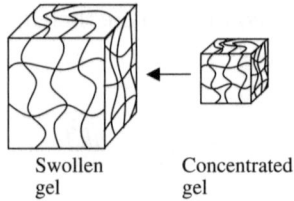

Figure 13.9 The swelling of polymer gels is intimately related to the osmotic pressure of the network and the swelling is resisted by the elasticity of the polymer chains

polymer and 97–98% water. Simple macroscopic measurements of gel sizes and concentrations thus provide an important tool to understand their molecular structuring (Figure 13.9).

There are a number of other methods typically used to measure osmotic pressure. These include the *vapour pressure osmometer* (it examines the depression of the boiling point by the osmotic effect) and *optical tweezers* (a direct measurement of the piconewton pressures on colloidal particles is made).

13.4 FORCE MEASUREMENT

There is a wide range of techniques for the measurement of mesoscopic forces; the general principles were briefly examined in Chapter 2. Some of the more modern developments in the field of force measurement include *atomic force microscopy* (AFM), *glass fibers*, *surface force apparatus* (SFA) and *magnetic/optical tweezers*. The range of forces that can typically be measured with each of the techniques is compared in Figure 13.10. In general terms AFM and SFA offer the largest forces, and magnetic/optical tweezers offer the greatest sensitivity.

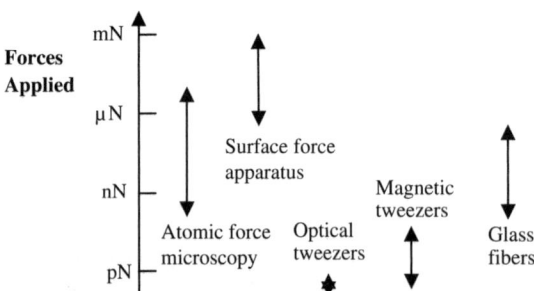

Figure 13.10 The range of forces that can typically be applied to biomolecular systems with a range of modern force apparatus

FORCE MEASUREMENT

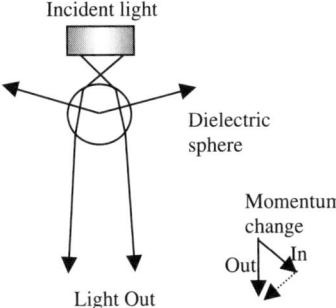

Figure 13.11 A focused laser beam can be used to provide an optical trap for a dielectic sphere. The momentum transfer due to the change in direction of the refracted beam induces a force on the sphere

Optical and *magnetic tweezers* are both similar in application, a feedback mechanism is used to clamp the position of a colloidal particle in three dimensions under an optical microscope, and this force is subsequently used to manipulate biological molecules. However, the physical processes which control the mode of action of the two types of tweezer are radically different. Optical tweezers focus laser light to trap a dielectric particle using the pressure of photons in the incident laser beam (Figure 13.11). Magnetic tweezers use the magnetic force on superparamagnetic or ferromagnetic beads exerted by gradients in an applied magnetic field (Figure 13.12).

For *optical tweezers*, the electric dipole (p) induced in a trapped particle is given by:

$$\underline{p} = \underline{\alpha}.\underline{E} \qquad (13.26)$$

where $\underline{\alpha}$ is the polarisability of the irradiated material and $\underline{E}$ is the electric field of the incident laser. The induction of the optical dipole moment in

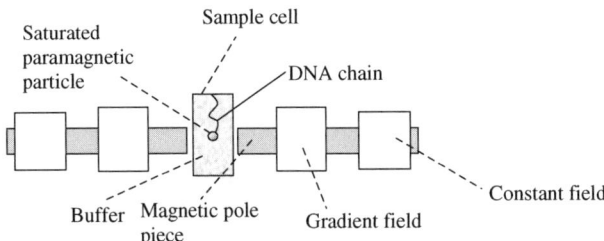

Figure 13.12 Two pole piece magnetic tweezers for single molecule extension experiments. Multiple pole piece tweezers can provide particle manipulation in three dimensions

the trapped particle by the laser beam provides a force (F_{light}), proportional to the laplacian of the electric field, or equivalently the gradient of the intensity of the incident light ($\underline{\nabla} I$):

$$F_{light} = \alpha \underline{\nabla}^2 E = \alpha \underline{\nabla} I \qquad (13.27)$$

A single well focused laser beam can trap dielectric particles (~ 0.5 μm) in three dimensions using the pN forces induced from the optical dipolar force. For a specific laser, microscope and optical set up, the tweezer force (F_{light}) is related to the incident laser power by the equation:

$$F_{light} = \frac{Q n_m P}{c} \qquad (13.28)$$

where Q is the efficiency of the trap, n_m is the index of refraction of the particle, c is the speed of light and P is the incident laser power. Double optical traps can be used to extend single molecules that are attached at either end to colloidal probes. The correct choice of the laser in a trapping experiment is important to reduce the damage on fragile biological molecules. Often, with delicate biological materials, infra red lasers are used to minimise this damage.

In contrast to optical tweezers, with *magnetic tweezers* the potential energy (U) of a magnetic dipole (m) placed in a magnetic field (B) is given by the scalar product:

$$U = -\underline{m}.\underline{B} \qquad (13.29)$$

Thus a free permanent magnetic dipole experiences a torque as it minimises its energy through the alignment of the dipole with the applied magnetic field. The magnetic forces experienced by a probe particle depend sensitively on its type of magnetism. The colloidal probes used in magnetic tweezer experiments are typically either ferromagnetic or superparamagnetic. The corresponding magnetic force (F_{mag}) is the gradient of the potential (∇U):

$$F_{mag} = -\underline{\nabla}(\underline{m}.\underline{B}) \qquad (13.30)$$

For superparamagnetic spheres the magnetisation is approximately equal to the saturated value (M_{max}), $m \cong M_{max}$, and the magnetic field gradient occurs parallel to the x-axis, so:

$$F_{mag} \approx M_{max} V \frac{dB}{dx} \qquad (13.31)$$

FORCE MEASUREMENT

where V is the particle volume. With superparamagnetic particles the application of a magnetic field gradient can provide forces on the order of 100 pN. The torque on a large ferromagnetic particle (4 µm) can be quite considerable ($\sim$ 1000 pNµm) and magnetic tweezer cytometry has found applications for determining the elasticity of cells adhered to magnetic beads. Hysteresis effects in the magnetism curves of the probe particles and pole pieces pose a number of technical challenges for accurate quantitative analysis of magnetic forces, particularly with ferromagnetic beads.

The hydrodynamic drag force (F_{drag}) experienced by a trapped particle moved (velocity, v) through its surrounding solvent with optical or magnetic tweezers is given by Stoke's law:

$$F_{drag} = 6\pi\eta a v \tag{13.32}$$

This equation provides a method for calibrating both optical and magnetic traps. The trapped colloidal probe is held at rest with respect to the laboratory and the solvent is given a constant velocity using a flow cell. The critical velocity at which the trapped bead becomes dislodged is measured, which allows the force applied to the tweezers to be calculated using equation (13.32).

A more accurate method for calibrating traps uses an analysis of the thermal fluctuations of the trapped particle and is based on Langevin's equation for the particle's motion:

$$m\frac{dv}{dt} = F_{thermal}(t) - \gamma v - \kappa x \tag{13.33}$$

This is just Newton's second law, a balance of the inertial force (mdv/dt), the thermal force $F_{thermal}(t)$, the drag force γv and the elastic trap force (the effective lateral trap spring constant is κ and x is the particle displacement). To solve equation (13.33), the thermal force ($F_{thermal}$) is assumed to be completely random over time (t); mathematically this is equivalent to:

$$\langle F_{thermal}(t)F_{thermal}(t-\tau)\rangle = 2kT\gamma\delta(\tau) \tag{13.34}$$

where $\delta(\tau)$ is the dirac delta function, kT is the thermal energy and γ is the frictional coefficient. In the low Reynolds number regime the inertial term can be neglected ($mdv/dt = 0$); this greatly simplifies equation (13.33) and is typically the case in most tweezer experiments at low frequencies. The Laplace transform of the Langevin equation (13.33) can be taken to provide an expression of the power spectrum ($S_x(\omega)$, Section 5.2) of the fluctuations of the bead displacement:

$$S_x(\omega) = \frac{kT}{\pi^2 \gamma (\omega_c^2 - \omega^2)} \qquad (13.35)$$

where kT is the thermal energy, γ is the drag coefficient, ω_c is the cornering frequency and ω is the frequency. The power spectrum of bead fluctuations can be easily determined experimentally using a numerical fast fourier transform of the bead square displacement as a function of time (Figure 13.13). Equation (13.35) then allows the cornering frequency (ω_c) to be calculated and the spring constant (κ) of the trap can be subsequently found using the equation:

$$\omega_c = \frac{\kappa}{2\pi\gamma} \qquad (13.36)$$

The technique of *atomic force microscopy* (AFM) allows the force between the tip of a cantilever (with a small radius of curvature) and virtually any kind of surface to be measured (Figure 13.14). The AFM technique also has the significant advantage that the tip can be used to form an image of the surface. In a typical AFM experiment a small pyramid shaped tip is mounted on a cantilever which acts as a spring, with a spring constant $\sim 0.1\,\text{Nm}^{-1}$. The cantilever is arranged on a piezo electric driver that moves the tip in the vertical direction whilst the

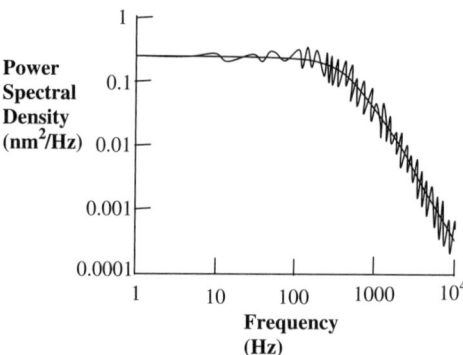

Figure 13.13 Power spectral density ($< r^2(f) > = S_x(\omega)$) of the position of a particle trapped in the laser of an optical tweezer set up as a function of frequency (f) [Reprinted with permission from K. Svoboda and S.M. Block, Ann. Rev. Biophys. Biomol. Struct. 23, 247–285, Copyright (1994) Annual Reviews]

FORCE MEASUREMENT

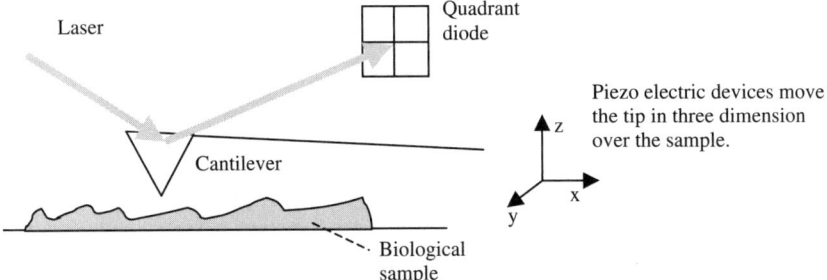

Figure 13.14 Schematic diagram of an Atomic Force Microscopy (AFM) experiment to study the surface of a biological material
(The laser reflects off the back of the cantilever and is then detected by a quadrant diode. A piezo electric device moves the tip over the surface to produce a three dimensional map of the surface topography.)

resulting displacement is measured by reflecting a laser beam from the back of the cantilever onto a split photodiode. A range of feedback methods is used to control the position of the cantilever, for example, that holds the cantilever at a constant force. The detailed construction of an AFM is shown in the Figure 13.15.

AFM allows much larger forces to be applied to a sample than with optical/magnetic tweezers and imaging is also possible (Figure 13.16).

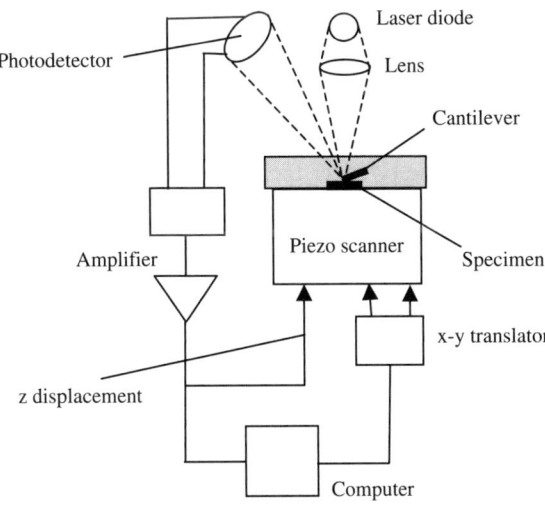

Figure 13.15 Detailed schematic diagram of an AFM apparatus. The feedback scheme can be used to hold the cantilever at a constant force on the sample.
[Ref.: J. Yang, L.K. Tamm, A.P. Somlyo and Z. Shao, Journal of Microscopy, 1993, 171, 183–198]

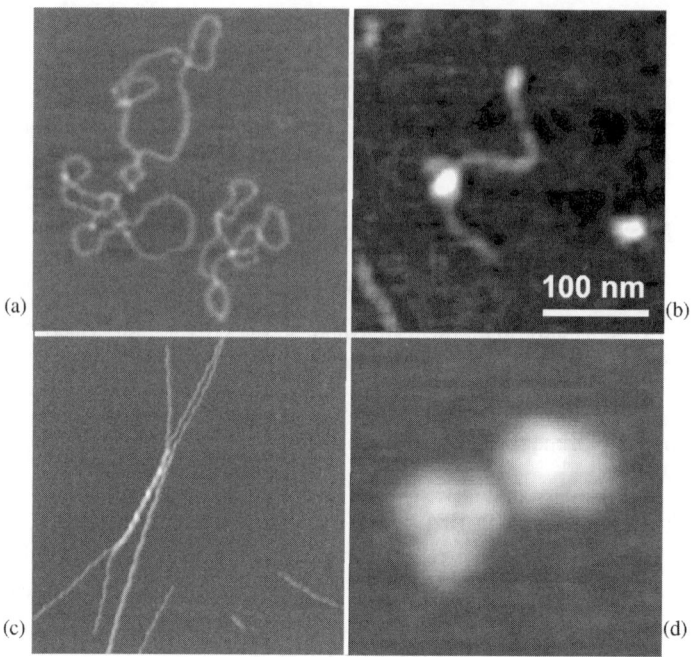

Figure 13.16 Atomic force microscopy image of circular DNA (a, b) and amyloid fibrils (c, d)
[*Ref.: Neil Thomson, University of Leeds, 2005*]

However, it is much more difficult to model the viscoelastic response of materials close to surfaces with AFM due to the effects of lubrication hydrodynamics, and the cantilever geometry causes the sensitivity to be reduced compared to the tweezer techniques (important for single molecule applications). Soft surfaces can be perturbed (indented) during the process of image collection due to the contact with the cantilever and sensitive feedback systems have been implemented to reduce this damage (the so called 'non contact mode'). Generally, the magnitude of the cantilever displacement in response to a force (F) at the surface is given by:

$$F = K_{cantilever}\Delta x \qquad (13.37)$$

where Δx is the displacement of the tip and $K_{cantilever}$ is the spring constant of the cantilever.

A typical value for $K_{cantilever}$ is $10^{-3}\,\text{Nm}^{-1}$ and a typical tip radius is 3×10^{-8} m.

FORCE MEASUREMENT

Surface force apparatus (SFA) examine the forces between surfaces on macroscopic dimensions. The technique involves the measurement of the distance of separation as a function of the applied force of crossed cylinders coated with molecularly cleaved mica sheets (Figure 13.17). The separation between the surfaces is measured interferometrically to a precision of 0.1 nm and the surfaces are driven together with piezo electric transducers with a resolution of 10^{-8} N. Much of the most accurate fundamental information on mesoscopic forces has been established using SFA.

The *split photodiode detector* is a critical piece of technology for a series of force probe techniques that include AFM, glass fibres and optical tweezers. The detector allows fast accurate measurement of light intensities. It can provide sub-nanometer resolution of probe positions on the time scale of 100 μs–100 s through comparison of the scattered light intensity projected onto the two sections of the split photodetector.

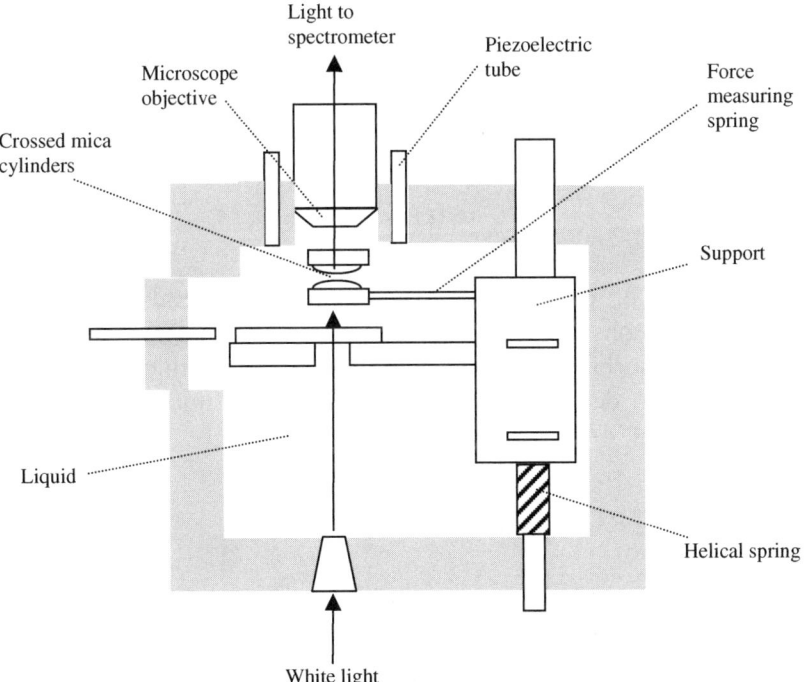

Figure 13.17 A schematic diagram of the arrangement of a surface force apparatus to measure mesoscopic forces
(The distance between the two mica cylinders is measured using an interferometric technique and the force is measured with a finely calibrated spring [*Ref.: J.N. Israelachvili, Chemtracts-Analy. Phys. Chem, 1989, 1, 1–12*])

The mean photocurrent ($\langle i \rangle$) at time t measured by a section of the split detector is:

$$\langle i \rangle = n \int_0^\infty g(t)dt = nze \quad (13.38)$$

where $g(t)$ is the photo current detected at a time (t) given by:

$$g(t) = \frac{ze}{\tau_0} \exp\left(-\frac{t}{\tau_0}\right) \quad (13.39)$$

where z is the total number of charges displaced upon absorption of a photon on the detector, e is the electronic charge, τ_0 is the time constant of the detector and n is the total number of photons collected. The position of a probe (e.g. the cantilever with an AFM or the colloidal probe with optical tweezers) measured using a split photodiode is found by comparing the difference in current signals (Δi) between the two photodiodes. The displacement noise on the determination of the probe position (with standard deviation $\sigma_x(f)$, where f is the frequency) quantifies the accuracy of the split diode in a particular geometry and can be calculated as:

$$\sigma_x^2(f) = \frac{d^2}{2qn} \quad (13.40)$$

The resolution of a split diode experiment thus depends on the total number of photons collected by the detector (n), the efficiency of the detector for absorbing photons (q) and the spatial width of the detector (d). It does not depend on the electronic charge (e), the instrument amplification (z), or the magnification.

13.5 ELECTROPHORESIS

Electrophoresis is a cheap, powerful tool for the analysis and separation of charged biological molecules such as proteins and nucleic acids. Electrophoresis can be used to measure the size of biopolymer molecules and also to deduce the chemical sequence of the chains.

The force experienced by a particle (F) in an electric field (E) is given by Coulomb's law:

$$F = ZeE \quad (13.41)$$

where Z is the number of charges on the particle and e is the electronic charge. The mobility of a charged particle in an electric field is proportional to the ratio of the net charge on the particle (which provides the Coulombic force) to its frictional coefficient (f). Electrophoresis can be used to obtain information about either the relative charge or the relative size of charged molecules. For steady state electrophoretic motion the frictional force (the frictional coefficient (f) multiplied by the velocity (v), fv) is balanced by the force due to the electric field. The electrophoretic mobility (U) with colloids is defined as:

$$U = \frac{v}{E} = \frac{Ze}{f} \qquad (13.42)$$

Combined with Stokes law for the frictional force, this equation becomes:

$$U = \frac{Ze}{6\pi\eta R} \qquad (13.43)$$

Thus the mobility of the colloids measured in an electrophoresis experiment can be related to the charge fraction (Z) and the radius of the particles (R).

Conceptually the simplest method of measuring the mobility of a colloidal particle in an electric field is by using *moving boundary* (free) *electrophoresis* (Figure 13.18). Particle velocities are measured directly with an optical microscope as they move in the electric field. However, this technique suffers from artefacts such as convection and multicomponent interactions. It is possible to circumvent these problems using gels and ion exchange papers, and these are the electrophoresis methods that are predominently used today.

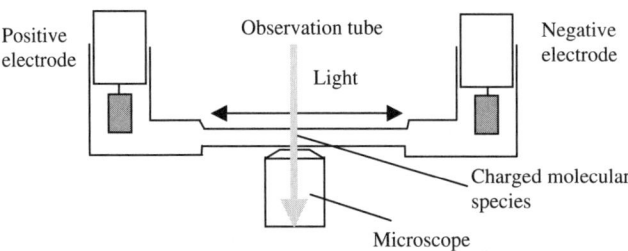

Figure 13.18 A moving boundary apparatus for the examination of free solution electrophoresis. The motion of the colloids that experience electrophoresis is measured with an optical microscope

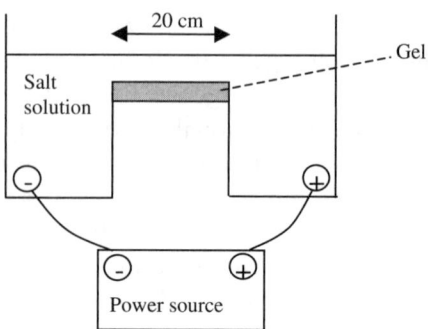

Figure 13.19 Arrangement of a gel used in a simple electrophoresis experiment to measure the mobility of polyions loaded on the gel, e.g. DNA

The charge on a protein depends on the pH of the buffer (Section 1.1). Electrophoretic motion can be studied as a function of pH to calculate the isoelectric point. The isoelectric point is the pH at which the average net charge on a macromolecule is zero. Isoelectric focusing (the pH is adjusted until there is no particle motility) can provide quantitative molecular information on charged macromolecules using simple table top apparatus.

The standard method of examining DNA chains is with *gel electrophoresis* (Figure 13.19). The use of the gel removes the problems with convection inherent in free boundary electrophoresis. The gel is placed across a constant applied voltage in a salt solution and the DNA chains are loaded onto the gel near the negative electrode. The gel is 'run' for a fixed period and the mobility of the chains (the time taken to travel a certain distance) can be simply related to their size. Surprisingly, detailed information on the complex topological nature of the gel is not required for quantitative predictions to be made on the molecular weight of DNA chains as they move across the gel.

There are two common ways to locate DNA on a gel to measure the distance it has travelled. Ethidium bromide can be used to label the chains and fluoresces strongly under ultra violet light. Alternatively, it is possible to incorporate radiative phosphorus atoms that will darken a photographic film into DNA at one of its ends.

An example is provided by electrophoresis with super coiled DNA molecules. Gel electrophoresis is a relatively easy method to separate closed super coiled DNA from the relaxed (cut) molecules. There is a large increase in the mobility of the super coiled DNA due to its compact form and it therefore experiences a reduced frictional coefficient (f) compared with the extended relaxed form.

For detailed sequencing of DNA chains restriction enzymes are used. These enzymes cut the DNA chains whenever they find the GAATTC sequence. If there are n such sequences there are $n + 1$ bands that occur. Other specific enzyme/DNA reactions allow individual DNA molecules to be cut in different places and the resultant information can be combined to sequence chains of up to 400 base pairs.

Isoelectric focusing is also possible with gel electrophoresis and can be used as an effective separation technique if the bands that contain the required charged molecules are cut out of the gel.

The theory of *reptation* is used to explain the ability of gel electrophoresis to separate DNA chains of different lengths (Figure 13.20, Section 8.5). The components of the electric force perpendicular to the axis of the tube are cancelled by the tube reaction force and the longitudinal components induce an electrophoretic motion of the chain along the tube (forced reptation).

In the limit of *very strong* electric fields the front end of the DNA chain moves forward and creates new parts of the tube (Figure 13.21(a)). The stretching force is proportional to the number of monomers (N), since the total electric charge on the chain is also proportional to N. Furthermore, the coefficient of friction (μ) for the whole chain is proportional to N (as for reptation, $\mu \propto N$). Thus the speed of motion (v) in a strong electric field is independent of N ($v = f/\mu$). The method of strong field electrophoresis is therefore not useful for the separation of DNA fragments.

However, *weak* electric field gel electrophoresis (Figure 13.21(b)) is much more successful, since the DNA molecules remain Gaussian coils.

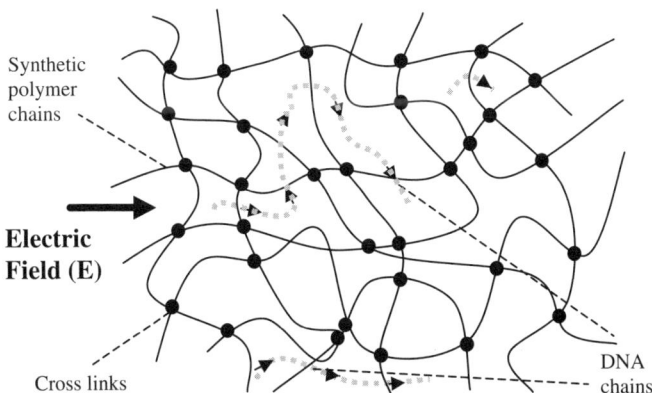

Figure 13.20 The electrophoresis of DNA fragments across a cross-linked gel is driven by an electric field. Smaller chain fragments migrate more quickly. A process of driven reptation occurs

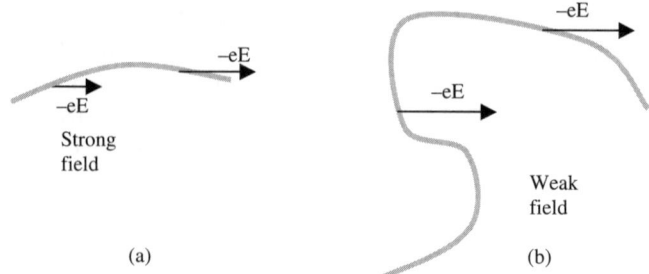

Figure 13.21 Schematic diagram indicating the difference in conformation of charged macromolecules during electrophoresis in a gel ((a) In a *strong electric field* the chains are completely stretched and electrophoresis is not a sensitive measure of the mobility; (b) in a *weak field* the chains adopt a Gaussian conformation and mobility measurements are much more successful. eE is the electrostatic force on a section of a chain.)

The force that the electric field exerts on the DNA molecules is proportional to the displacement of the chain parallel to the electric field ($N^{-\frac{1}{2}}$), since only the motion of segments of the chain parallel to the electric field are not restricted by the cross-links of the gel. The speed of reptation (v_r) is:

$$v_r = \frac{f}{\mu} \sim \frac{N^{-\frac{1}{2}}}{N} \sim N^{-\frac{3}{2}} \tag{13.44}$$

where N is the number of monomers in the chain. The speed of the centre of mass motion (v) is a factor of $N^{\frac{1}{2}}$ slower than the speed of reptation and the speed of centre of mass reptation is therefore inversely proportional to the chain length ($v \sim 1/N$) in a weak field. Weak field electrophoresis thus provides a practical method for the separation of DNA chains. A more accurate calculation of the velocity of the DNA fragments in a weak field gives:

$$\underline{v} = \frac{q}{3\eta}\left(\frac{1}{N} + w\left(\frac{EqL}{kT}\right)^2\right)\underline{E} \tag{13.45}$$

where $\underline{E}$ is the electric field vector, q is the charge per unit length of the DNA, N is the number of Kuhn segments in the DNA chain, L is the length of a Kuhn segment, η is the viscosity of the medium and w is a constant of the order of unity. The technique is, therefore, not very sensitive at separating long DNA chains (Figure 13.22), which is a big

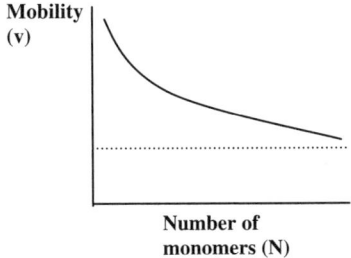

Figure 13.22 Dependence of the mobility of DNA fragments (v) on the number of monomers in a chain (N) during a gel electrophoresis experiment. The mobility becomes a less sensitive function of N as the chains increase in length

problem if micron long pieces of genomic DNA require sequencing. A trick to increase the sensitivity of electrophoresis for the separation of long DNA chains is to periodically switch off or rotate by 90°, the applied external field at the typical time for the renewal of the reptation tube ($\tau^* \sim N^3$, Section 8.5). In this case, electrophoretic motion only occurs for chains of the length (N) defined by the periodicity of the rotating electric field and the method can be used to select the longer chains. This technique of *pulsed electrophoresis* works very well.

The *polymerase chain reaction* (PCR) is a biochemical technique for amplifying short (10 000 base pairs) stretches of nucleic acid. It is often used in conjunction with electrophoresis methods in the process of genetic fingerprinting. Thus a single DNA molecule can be amplified to provide sufficient quantities of DNA to be sequenced using electrophoresis by means of the PCR technique.

SDS electrophoresis can be used to calculate the molecular weights of proteins. SDS is a surfactant that is an effective protein denaturant. It binds to all proteins qualitatively to the same degree and causes them to adopt extended conformations. The apparent electrophoretic mobility ($u(c)$) of a protein denatured with SDS at a particular gel concentration (c) is phenomenologically given by:

$$\ln u(c) = -k_x c + \ln u(0) \tag{13.46}$$

where k_x depends on the extent of cross-linking of the gel and $u(0)$ is a constant for a particular protein. The mobility ($u(0)$) is related to the molecular weight (M) of the protein through a simple relationship:

$$u(0) = b - a \log M \tag{13.47}$$

where b and a are standard constants. The molecular weight of a denatured protein can therefore be calculated from the measurement of its mobility on a gel at a series of different gel concentrations.

There have been a number of modern developments in electrophoretic techniques. Problems with convection in free boundary electrophoresis can be reduced by using a very fine bored capillary in *capillary electrophoresis* (Figure 13.23). This is a useful microanalytical separation technique. The importance of convection in a fluidic system is described by a dimensionless group, the Rayleigh number (Ra), which is defined as:

$$Ra = \frac{R^4 g}{\eta \alpha} \left(\frac{\Delta \rho}{\Delta r} \right) \qquad (13.48)$$

where R is the radius of the channel, g is gravity, η is the viscosity, α is the thermal diffusivity of the medium and ($\Delta \rho / \Delta r$) is the density change per unit radial distance caused by heating. For small Rayleigh number ($Ra < 1$) convection is suppressed in an electrophoresis tube and this corresponds to a small capillary bore (R). Typically, capillary diameters for electrophoresis experiments are in the order of a few microns to provide low Rayleigh number dynamics for the charged molecules examined.

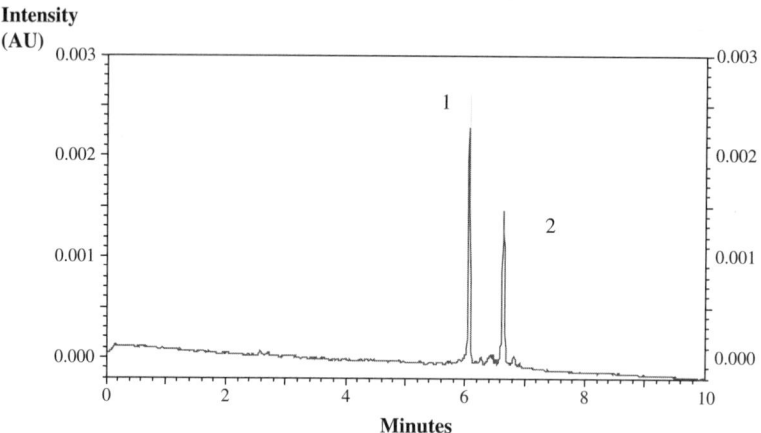

Figure 13.23 Typical data from a capillary electrophoresis experiment that shows the separation of two samples of short double stranded DNA
((1) is AAATTATATTAT/ATAATATAATTT and (2) is GGGCCGCGCCGC/GCGGCGCGGCCC. Sample (1) travels down the column faster than sample (2) [Ref.: I.I. Hamden, G.G. Skellern, R.D. Waigh, *Journal of Chromatography, A, 806 (1), 165–168, 1998*])

SEDIMENTATION

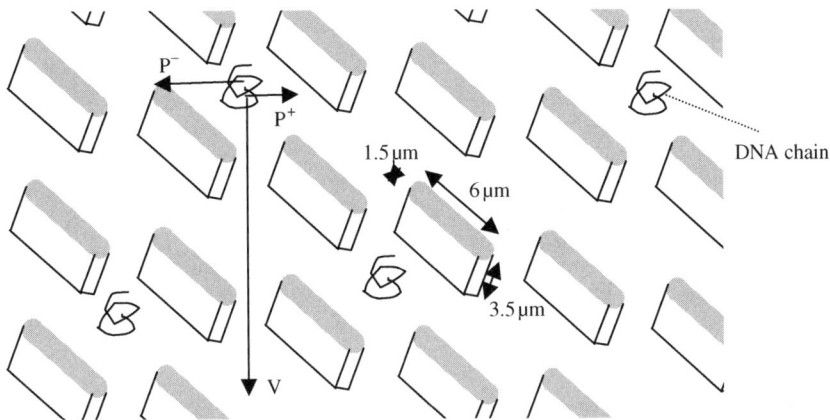

Figure 13.24 Etched microarrays used for electrophoresis experiments (An electric field moves the DNA molecules vertically downwards with a velocity v. Due to the anisotropic nature of the obstacle orientation there is a larger probability for the molecules to change channels to the right (P^+) than to the left (P^-), and this probability is a function of the size of the chains [*Ref.: C.F.Chou, R.H.Austin, O.Bakajin et al., Electrophoresis, 2000, 21, 81–90*]

Etched obstacle arrays on silicon chips can also be used for electrophoresis (Figure 13.24). Silicon microarrays offer a number of advantages over standard gel techniques: smaller samples can be explored and the microstructure of the etched silicon can be better defined than with gels and, consequently, so too can the microfluidics.

13.6 SEDIMENTATION

Sedimentation is a key separation technique used to extract the particular biomolecule of interest from the complex soup of species found in the cell. Separation by sedimentation is a standard first step in a molecular biophysics experiment. An external force acting on a mixture of suspended particles is used to separate them by means of their varying buoyancies with respect to the background solvent. In an analytical ultracentrifuge the radial acceleration provides the external force and causes the molecules to be separated as a function of both their density and shape (Figure 13.25).

From simple Newtonian mechanics the radial force (F) on a suspended particle that is rotating in an ultracentrifuge is given by:

$$F = m^* \omega^2 r \tag{13.49}$$

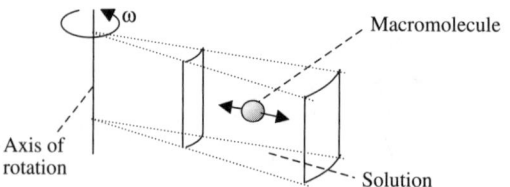

Figure 13.25 Schematic diagram of a sedimentation experiment (The sample is rotated about an axis at an angular velocity of ω and macromolecules adjust their position with respect to the solution due to their relative densities.)

where m^* is the effective mass, ω is the angular velocity and r is the radial distance of the particle from its centre of motion. From Archmides principle, the mass of a particle (m) suspended in a solvent needs to be corrected by the density of the surrounding solvent and the effective mass of the particle (m^*) in the solvent is given by:

$$m^* = m(1 - \bar{v}\rho_0) \qquad (13.50)$$

where $\bar{v}$ is the partial specific volume of the molecule and ρ_0 is the solvent density. The velocity at which the particles move due to the centripetal force is given by equation (13.49) divided by the frictional resistance:

$$v = \frac{dr}{dt} = m^*\omega^2 \frac{r}{f\eta_0} \qquad (13.51)$$

where f is a frictional coefficient and η_0 is the solution viscosity. The variation of sedimentation velocity (v) with particle size and density forms the basis of a method to separate particles using sedimentation. When a centrifugal field is applied to a solution of molecules a moving boundary is formed between the solvent and the solute. This boundary travels down the sample cell with a velocity determined by the sedimentation velocity of the macromolecules. Concentration gradients can be accurately measured using ultra violet absorption (Figure 13.26) and therefore the sedimentation velocities can be calculated.

The velocity of sedimentation (dr_b/dt) is equal to the rate of motion of the boundary:

$$\frac{dr_b}{dt} = r_b\omega^2 s \qquad (13.52)$$

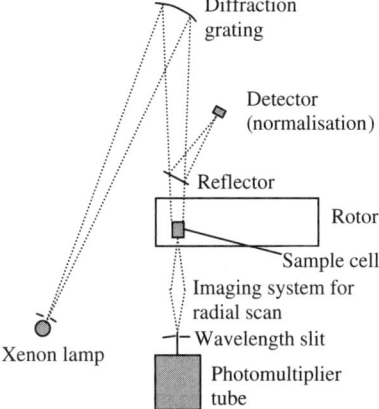

Figure 13.26 Experimental arrangement of a modern analytic centrifuge. UV absorption with a Xenon lamp is used to measure particle concentrations

where s is the sedimentation coefficient, equal to the velocity of sedimentation divided by the centrifugal strength $(\omega^2 r_b)$ at the radius at which the boundary occurs (r_b). Integration of equation (13.52) provides an expression for the position of the boundary as a function of time:

$$\ln\left[\frac{r_b(t)}{r_b(t_0)}\right] = \omega^2 s(t - t_0) \quad (13.53)$$

where t_0 is a reference time at which the boundary is found at $r_b(t_0)$. Diffusion broadens the boundary as it progresses down the column (Figure 13.27) and the rate of motion allows the sedimentation coefficient to be calculated from equation (13.53). The sedimentation coefficient depends on the size, shape and degree of hydration of a macromolecule. Fortunately, for globular proteins there is a well defined relationship between the sedimentation coefficient and the molecular weight due to their spherical geometry.

It is also possible to make focusing measurements with sedimentation experiments if the particles are suspended in a dense salt, e.g. a solution of caesium chloride (CsCl) or caesium sulfate (CsSO$_4$). Particles collect together in a narrow band at the point of matching buoyancy. The calculation of the sedimentation profile during a centrifugation experiment is an elegant illustration of the predictive power of equilibrium statistical mechanics. The work required $(E(r) - E(r_0))$ to lift a particle

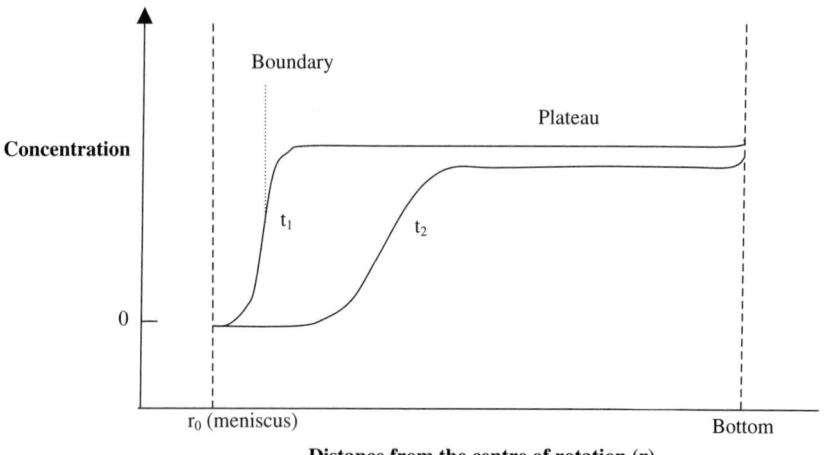

Figure 13.27 Schematic diagram showing the progress of a moving boundary sedimentation experiment at two different time steps (t_1 and t_2). The boundary moves towards the centre of rotation (the cell bottom) as a function of time

from a radius r_0 to a radius r in a centrifugal force field is equal to the work done against the centrifugal force:

$$E(r) - E(r_0) = -\int_{r_0}^{r} m'r\omega^2 dr = m'\omega^2 \frac{(r_0^2 - r^2)}{2} \quad (13.54)$$

where m' is the effective mass of the particles adjusted for the solvent density. In thermal equilibrium the range of concentrations ($c(r)$) as a function of the radius is given by a simple Boltzmann distribution ($e^{-E'/kT}$) and therefore:

$$\frac{c(r)}{c(r_0)} = \exp\left(-m'\omega^2 \frac{[r_0^2 - r^2]}{2kT}\right) \quad (13.55)$$

The density near the radius of a particular band (r_b) can be expressed as a Taylor expansion:

$$\rho(r) = \rho(r_b) + \left.\frac{\partial \rho}{\partial r}\right|_{r=r_b}(r - r_b) + \ldots \quad (13.56)$$

Substitution of the density expansion in equation (13.50) allows equation (13.55) to be expressed as:

$$\frac{c(r)}{c(r_b)} = \exp\left[-mr_b\omega^2 \bar{v}\rho'(r_b)\frac{(r-r_b)^2}{2kT}\right] \quad (13.57)$$

where $\rho'(r_b)$ is the density gradient defined by:

$$\rho'(r_b) = \left.\frac{\partial \rho}{\partial r}\right|_{r=r_b} \quad (13.58)$$

and m is the true mass of the particles. The concentration profile at radius (r_b) at which the band of particles occurs is therefore a Gaussian distribution with standard deviation (σ_r) and is given by:

$$\sigma_r = \frac{kT}{[mr_b\omega^2 \bar{v}\rho'(r_b)]^{\frac{1}{2}}} \quad (13.59)$$

The band of particles is narrow and well focused for particles of large mass (m), in high centrifugal fields (large $r_b\omega^2$) and in steep density gradients (large $\rho'(r_b)$). Sedimentation focusing is thus another extremely useful technique for particle separation.

13.7 RHEOLOGY

All real materials demonstrate behaviour intermediate between the idealised cases of solids and liquids (Chapter 12). Rheology is the study of this phenomenon of *viscoelasticity* and rheometers are instruments for measuring the rheology of materials.

There are two broad categories of techniques for measuring the viscoelasticity of a material. Firstly, there are *bulk methods* where the response of a macroscopic amount of a material to an externally applied stress or strain is recorded. These bulk methods have traditionally been used to examine the viscoelasticity of biological samples. Secondly, there is the measurement of the viscoelasticity of a sample as a function of length scale using *microrheology techniques*. Here probes are typically injected into the system of interest; the probes can be passive (e.g. marker colloids) or active (e.g. magnetic colloids). The motion of the probes is recorded with a video camera or measured with light scattering and the

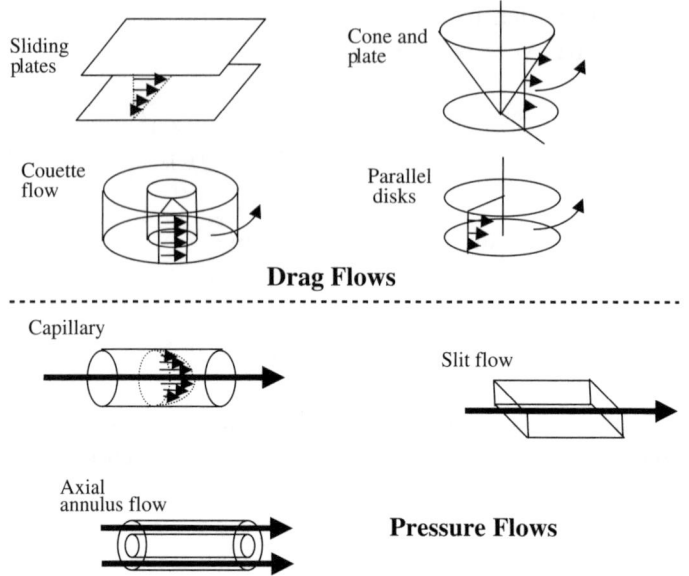

Figure 13.28 Schematic diagram of the flow geometries commonly encountered in rheological experiments
(The geometries are separated between *drag flows* where the surfaces move relative to one another, and *pressure flows* in which the flow rate is determined by the pressure drop across the pipe [*Adapted with permission from C.W.Macosko, Rheology, Principles, Measurements and Applications, Copyright (1994) Wiley-VCH*])

resultant fluctuation spectrum of the particle displacements is related to the viscoelasticity of the material in which they are embedded.

In *bulk rheology* experiments a series of different geometries can be used and each tends to have different advantages in terms of the mechanism of sample loading, the time window that can be explored and the sensitivity of the measurements (Figure 13.28). Different geometries for rheometers require different corrections to analyse the dependence of the stress on the strain. How the geometry grips the sample is also important and, combined with the type of force or displacement transducers, this determines the sensitivity of the measurements. Bulk rheometers measure the large scale viscoelastic properties of assemblies of biological molecules. Rheometers can function in linear (Chapter 12) or non-linear modes. Non-linear rheology corresponds to large deformations and deformation rates, in which both Deborah and Peclet numbers are appreciable.

In *drag flow rheometry* the velocity or displacement of a moving surface is measured simultaneously with the force on another surface

RHEOLOGY

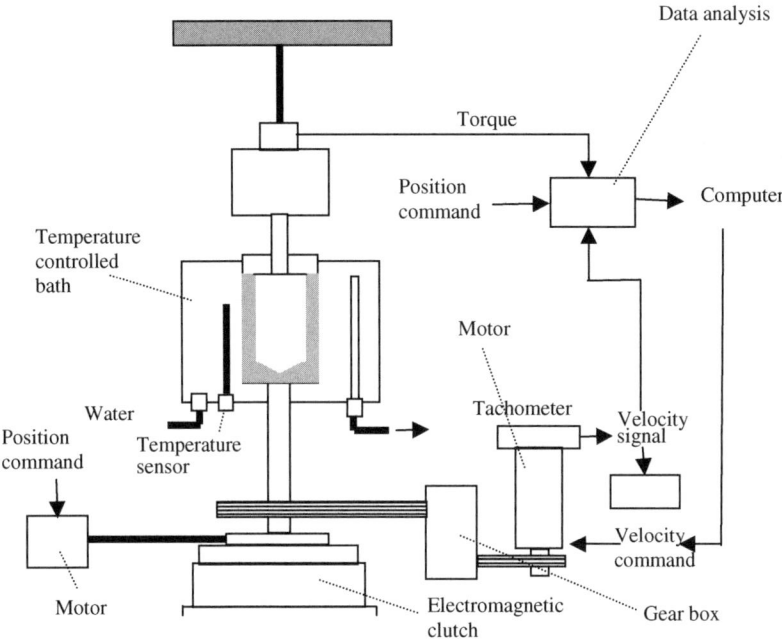

Figure 13.29 Schematic diagram of a modern shear rheometer
[*Ref.*: L. Bohlin, in *Progress in Trends in Rheology, II*, Eds H. Gieseku and M.F. Hibberd, Steinkopf, 1988, 151]

that moves in response to its motion. Couette's original concentric cylinder drag flow rheometer was a controlled strain apparatus. The angular velocity of the outer cup was fixed and the torque on the inner cylinder was measured from the deflection of a suspended wire. The measured variable in a controlled strain rheometer is the torque. Couette measured the twist in a torsion bar whereas modern electronic rheometers use a linear variable differential transducer to do the same job (Figure 13.29).

More sophisticated modern rotary rheometers measure normal stresses (the stresses normal to the direction of shear). It is an experimental challenge to find steady state behaviour with normal stresses, as they are easily disturbed by fluctuations in the temperature and axis of rotation. Rheometers thus need to be machined to high precision. Often commercial rheometers are mechanically accurate to within 2 µm over the 25 mm cup diameter. The control of the torque and the subsequent measurement of the angular motion in a controlled stress rheometer is also a standard technique in rotational rheometry. Furthermore, it is important to control the temperature, pressure and humidity to make accurate rheological measurements with biological specimens.

The most commonly measured linear viscoelastic material function is the complex shear modulus, $G^*(\omega)$ (Section 12.2). There are three standard techniques used to measure G^*: in the *shear wave propagation* method the time for a pulsed deformation to travel through a sample is measured; the sample can be made to oscillate at its *resonant frequency*, and the response at this single frequency observed; and the *forced response* to a sinuisoidal oscillation in stress/strain can be measured in terms of the resultant strain/stress (Figure 13.30). Forced resonance devices are better suited to low elasticity materials such as polymer solutions and soft biomaterials.

There are two basic design types of *pressure driven rheometers*. One features control of the pressure and measures the flow rate (e.g. capillary

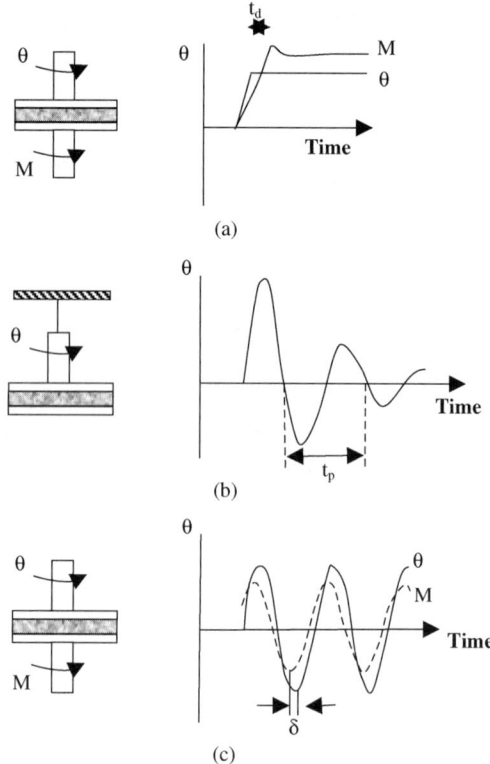

Figure 13.30 Methods for the measurement of the shear modulus (a) wave speed, (b) resonance and (c) forced oscillations [*Adapted with permission from C.W. Macosoko, Rheology, Principles, Measurements and Appications, Copyright (1994) Wiley-VCH*])

rheometers) and the other uses a controlled flow rate and measures the pressure drop. Such capillary type geometries have direct analogues in biological circulatory systems (e.g. blood flow) which motivate the analysis.

Microrheology has experienced a number of important recent developments. Often biological samples, which are homogeneous on the macro scale, are inhomogeneous on the micron scale (e.g. inside living cells); a range of microrheology techniques have been developed to measure this behaviour. The range of frequencies and moduli that can typically be accessed using the different microrheology techniques are shown in Figure 13.31. In particular, the measurable frequency range can be increased by many orders of magnitude using microrheology when compared with standard bulk rheology methods.

Particle tracking microrheology is, practically, the simplest microrheology technique to implement. It requires a video camera, optical microscope, oil immersion objective and digital recording apparatus. The fluctuation–dissipation theory is used to relate the fluctuations in the displacements of tracer particles embedded in a material to its viscoelastic response. The fluctuation spectrum of the mean square displacements of colloidal particles embedded in a viscoelastic material is calculated as a function of time (Figure 13.32). For simple viscous liquids one would expect (Section 5.1) a linear dependence of the mean square displacement in two dimensions of the embedded probes ($\langle r^2 \rangle$) under the microscope on time (t):

$$\langle r^2 \rangle = 4Dt \qquad (13.60)$$

A viscoelastic component is introduced as a sub-linear diffusive process at short times modifying equation (13.60) ($\langle r^2 \rangle \sim t^\alpha, \alpha < 1$) and includes the information on the viscoelasticity of the material. The linear viscoelastic shear moduli (G^*) of the material can subsequently be calculated from the mean square displacement using the Generalised Stokes–Einstein equation (compare with equation (5.10)):

$$G(s) = \frac{kT}{\pi a^2 s \langle r^2(s) \rangle} \qquad (13.61)$$

where $r^2(s)$ is the Laplace transform of $\langle r^2(t) \rangle$, $G(s)$ is the Laplace transform of the relaxation modulus ($G(t)$) and a is radius of the probe particle. The Laplace frequency (s) has been introduced to provide a compact solution of the fluctuation dissipation theorem. $G'(\omega)$ and

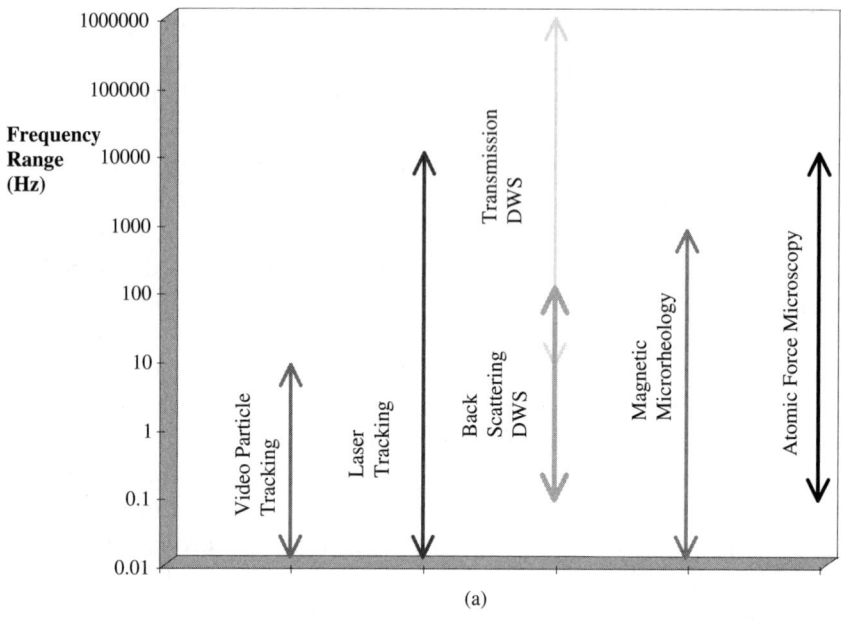

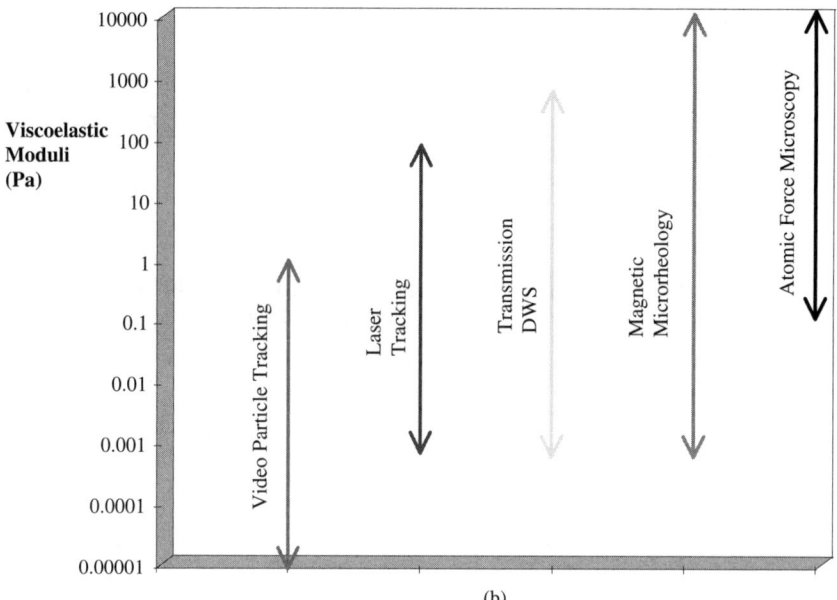

Figure 13.31 Comparision of the range of frequencies (a) and moduli (b) that can typically be measured using different microrheology techniques
[*Reprinted with permission from T.A. Waigh, Reports on Progress in Physics, 68, 685–742, Copyright (2005) IOP Publishing*]

RHEOLOGY

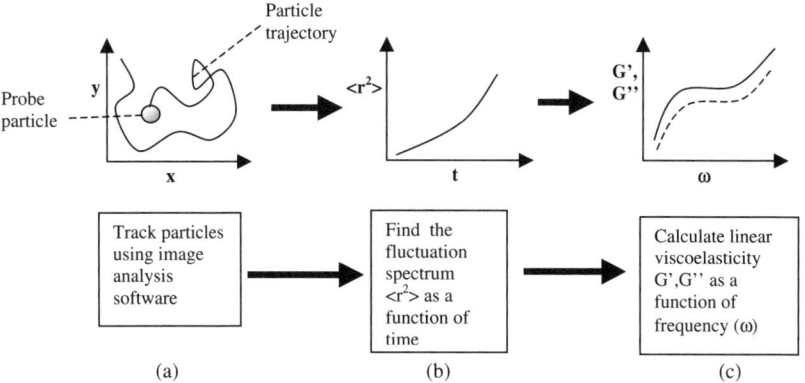

Figure 13.32 The strategy used in passive particle tracking micorheology experiments
((a) The trajectory of a fluctuating colloidal sphere is recorded, (b) the mean square displacement ($\langle r^2(t)\rangle$) fluctuations are calculated and (c) the shear moduli (G', G'') are found using the generalised Stokes–Einstein equation.)

$G''(\omega)$ can be determined mathematically by Fourier transform of the relaxation modulus as a function of time ($G(t)$). It is also possible to determine the linear viscoelastic spectrum by the analysis of the mean square fluctuations in angular displacement of a probe as a function of time, and again a generalised Stokes–Einstein equation for rotational motion is used to calculate the complex shear modulus.

Laser deflection techniques allow the high frequency viscoelastic behaviour of materials to be probed. Back focal plane interferometry is a particularly sensitive laser deflection method for measuring the small fluctuations (nanometres) of probe spheres that occur at high frequencies (Figure 13.33). Multiply-scattered laser light from colloidal spheres also can be used to yield the fluctuation spectrum of colloidal spheres embedded in biological specimens through the technique of *diffusing wave spectroscopy* (DWS, Figure 13.34). The intensity correlation function is measured as the autocorrelation of the scattered intensity (Section 13.2) and is used to construct the mean square displacement ($\langle r^2(t)\rangle$) of the probe spheres. The viscoelastic moduli can then be calculated in a similar manner to the particle tracking technique. DWS microrheology is useful for ultra high frequency viscoelastic measurements, since the process of multiple-scattering amplifies the sensitivity of the measurements to small particle displacements (Å) and allows particle motions to be detected at high frequencies (MHz). Single scattering photon correlation spectroscopy techniques can also be used at lower

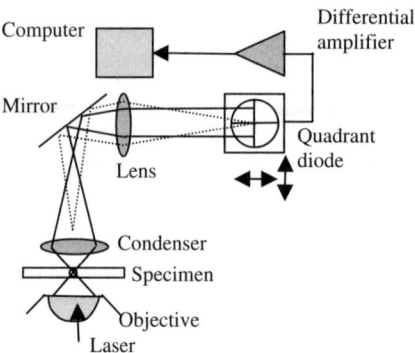

Figure 13.33 The fluctuations of the displacement of a single bead embedded in a biological specimen can be followed with scattered laser light projected onto a quadrant diode. This provides a sensitive laser scattering microrheology technique

colloidal concentrations and provide information on particle motion at slightly lower frequencies. DWS microrheology methods enable a very wide range of frequencies for the linear rheology of solution state biological materials to be accessed (Figure 13.35). Both transmission and back scattering geometries are possible for DWS experiments (Figure 13.34). Optical tweezers also find many applications in microrheology studies and are particularly well suited for measuring the elasticity of membranes due to their low moduli.

Further reduction in sample volumes for *nano and pico rheology* are possible, but data analysis often becomes more difficult and can reduce the sensitivity of the methods. Examples of submicrolitre rheometers that

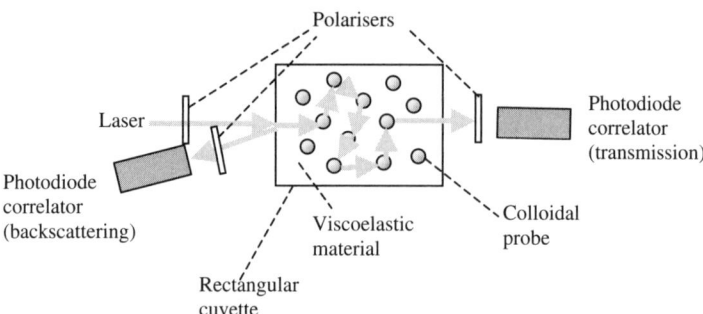

Figure 13.34 Schematic diagram of a diffusing wave spectroscopy experiment (Coherent laser light is multiply-scattered from a dense suspension of colloidal particles. Analysis of the resultant correlation functions can provide the high frequency visoelasticity of a biological specimen.)

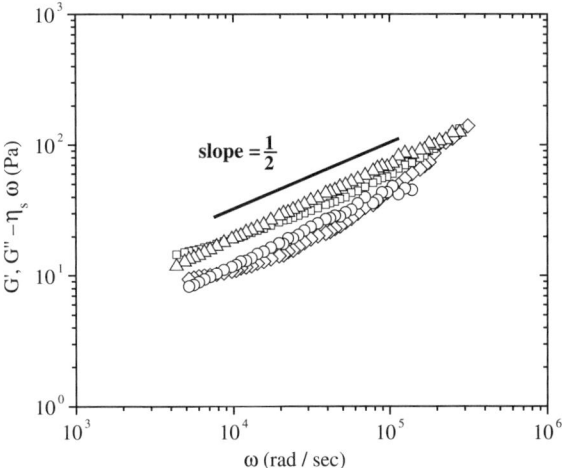

Figure 13.35 High frequency linear viscoelasticity from aggrecan solutions (The slope of $G' \sim G'' \sim \omega^{\frac{1}{2}}$ indicates that Rouse modes are present in these flexible polyelectrolyte solutions at high concentrations. Two separate concentrations are shown 2 mg/ml and 12 mg/ml [*Ref.: A.P.Papagiannopoulos T.A.Waigh, T.Hardingham and M.Heinrich, Biomacromolecules, 2006, 7, 2162–2172*])

are currently being investigated are backscattered DWS with optical fibres (sensitive to picolitre volumes), fluorescent correlation spectroscopy (sensitive to picolitre volumes) and oscillatory AFM (sensitive to nanolitre volumes).

Other micromechanical techniques specialise solely in the measurement of the elasticity of biological systems and neglect the behaviour of the viscosity. These include micropipette aspiration, steady state deformation using AFM and the use of internal markers to drive or record the deformation of the cytoplasm, e.g. magnetic beads or fluorescent markers.

13.8 TRIBOLOGY

A range of tribometers have been developed to quantify frictional behaviour at surfaces. In a typical modern device adapted for the measurement of solid–solid friction mediated by a thin viscoelastic film, forces are obtained in a direct manner through the measurement of the deflection of a spring with nanometre resolution (Figure 13.36). The stiffness of the employed bending beam is known exactly. The instrument is calibrated both in the normal and tangential direction. The force

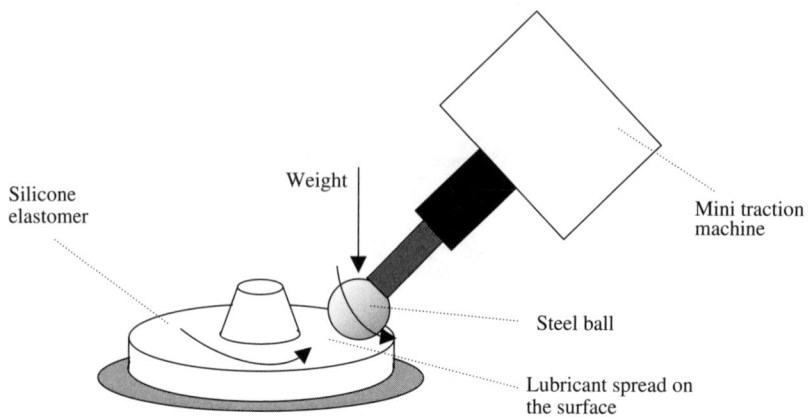

Figure 13.36 Modern ball on plate tribometer used to measure the friction coefficient of thin viscoelastic films as a function of shear rate (Stribeck curves) [Ref.: J. de Vicente, J.R. Stokes and H.A. Spikes, Tribology International, 2005, 38, 515–526]

measurement has a resolution of nN in the range nN to mN in both directions. The springs for force measurement are often made of photo-structurable glass and the spherical ball probes are made of silicon or steel with a well defined diameter. Interferometers can be used to measure the deflection of the spring. With biological specimens challenges are presented by their non-planarity and the requirement for hydrated environments, e.g. the cartilage in articulated joints. AFMs are sometimes used to measure frictional forces, since they are not confined to planar specimens. However, quantitative measurements of frictional coefficients with AFM continue to be challenging, since it is difficult to infer both the normal and frictional forces simultaneously using light scattered from the back of a cantilever. Drag flow rheometers can also be adapted to provide high precision frictional measurements (e.g. a plate–plate rheometer with a section of material attached to either plate), but the utility of the technique is dependent on the geometry of the specimens matching that of the cell.

13.9 SOLID PROPERTIES

Solid materials with high elasticity and minimal flow behaviour require a separate set of techniques for their measurement, since extremely large forces must be applied to provide significant sample displacements

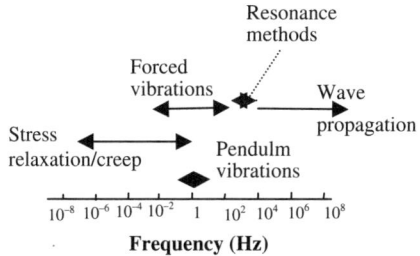

Figure 13.37 Methods for the measurement of the linear viscoelasticity of solid materials over different frequency ranges
[*Ref.: Becker, Mater. Plast. Elast., 1969, 35, 1387*]

(Figure 13.37). Dynamic mechanical testing apparatus (DMTA) are compressional analogues of oscillatory shear rheology that are often used on solid biomaterials. DMTA can provide the complex Young's modulus (E^*) of a material in compression or extension as a function of frequency.

Highly anisotropic biomaterials provide a challenge for the experimentalist, since a large number of parameters need to be measured to fully characterise the stress response of the oriented material (Section 11.1). The relative orientation of the applied stress and resultant strain needs to be carefully monitored. Other material properties are also important for the mechanical properties of biomaterials such as how samples buckle under compressive stress (measured using three point Euler buckling apparatus) and indentation tests for fracture mechanics.

FURTHER READING

J.L. Viovy, T. Duke and F. Caron, *The Physics of DNA Electrophoresis*, Contemporary Physics, 1992, 33, 1, 25–40. Useful introduction to the physics of electrophoresis.

C.R. Cantor and P.R. Schimmel, *Biophysical chemistry part II, Techniques for the Study of Biological Structure and Function*, W.H. Freeman, 1980. Old fashioned, but excellent account of a wide range of biophysical experiments.

T.A. Waigh, *Microrheology of Complex Fluids*, Reports of Progress in Physics, 68, 2005, 685–742. Detailed introduction to the field of microrheology.

I.N. Serdyuk, N.R. Zaccai, J. Zaccai, *Methods in Molecular Biophysics*, CUP, 2007. Expansive modern treatment of biophysical techniques.

TUTORIAL QUESTIONS

13.1) How would you calibrate the force of a single optical laser trap? The power spectral density of an optical trap is shown in the figure below. What is the trap stiffness for a spherical particle of radius 1 μm in water ($\eta = 0.001\,\mathrm{Pas}$)? Could you use the same method of calibration for a magnetic trap?

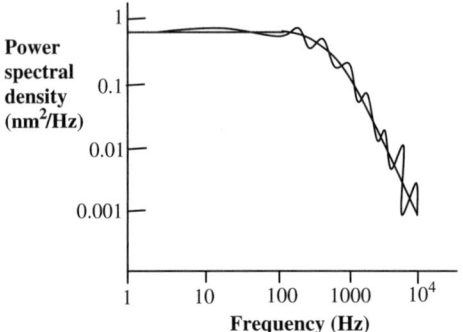

13.2) The mean square displacements (MSD) of the probe particles in a video particle tracking experiment embedded in two different fluids are shown in the figure below. Which of the fluids (A or B) could be viscoelastic over the time scale probed? What would the effect of a static error in the measurement of the particle positions be on the resultant mean square displacements?

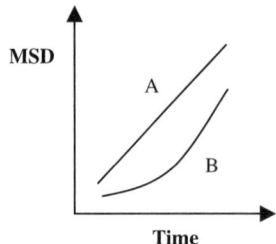

13.3) Estimate the velocity of a DNA chain that contains one million base pairs in a polyacrylamide gel if the electric field is $2\,\mathrm{Vcm^{-1}}$. By what factor would this velocity increase if DNA fragments of one tenth the size were chosen for the electrophoresis experiment? Assume the DNA chain is in the B form with

TUTORIAL QUESTIONS

1.7 Å spacing between phosphate groups, there are no charge condensation effects, the solution viscosity is 0.002 Pas, and there is 300 monomers in a Kuhn segment.

13.4) What is the power spectral density of a colloidal sphere diffusing in a purely viscous material? How would the power spectral density change if a colloidal sphere was placed in a viscoelastic fluid that has a power law mean square displacement i.e. $\langle r^2(t) \rangle \sim t^\alpha$?

14
Motors

A current challenge for the nanotechnology industry is how to transport chemical cargoes at the molecular scale in order to construct new materials, remove waste products and catalyse reactions. Nature has already evolved a wide range of efficient nanomotors that are used in a vast number of biological processes. Cells actively change their shape and move with respect to their environment, e.g. the contraction of muscle cells in the arm, movement of macrophages to capture and remove hostile cells, division of cells during mitosis and the rotation of flagella to propel bacteria. As a common theme chemical energy derived from the hydrolysis of ATP (or GTP with microtubules) or stored in a proton gradient is transformed into mechanical work to drive the cell motility. There are currently thought to be five separate mechanisms for molecular motility that occur naturally: *self-assembling motors*, *linear stepper motors*, *rotatory motors*, *extrusion nozzles* and *prestressed springs* (Figure 14.1).

Adenosine triphosphate (ATP) is the central currency in energy transduction in biological systems and it is useful to examine the chemical reaction of the molecules in more detail. The dissociation of ATP into ADP and a free phosphate ion liberates a reasonable amount of energy and is used to power a wide range of biochemical reactions:

$$ATP \leftrightarrow ADP + P_i \qquad (14.1)$$

where ATP signifies a range of species with different degrees of ionisation, e.g. $MgATP^{2-}$, ATP^{4-} etc, P_i is the free phosphate ion

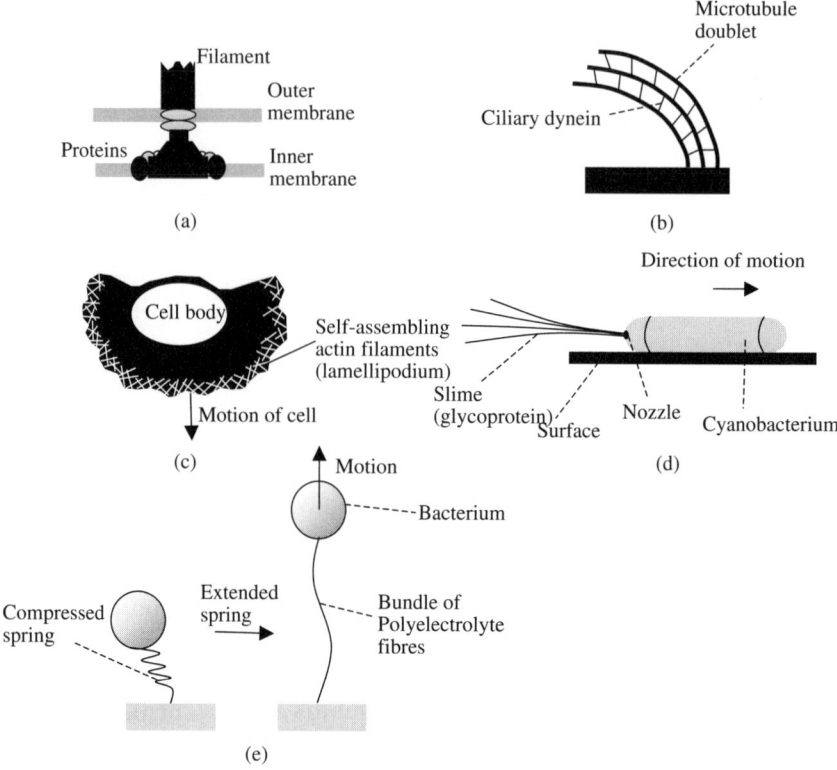

Figure 14.1 Examples of the five separate categories of biological motors ((a) rotatory motors – bacterial locomotion, (b) linear stepper motors – cilia, (c) self-assembling motors – actin filaments in lamellipodium, (d) extrusion nozzles–cyanobacterium, and (e) prestressed springs – bacterial locomotion.)

and ADP is adenosine diphosphate. The equilibrium constant (K) for energy transduction from ATP has the same units as concentration (moles):

$$K = \frac{c_{ADP} c_{Pi}}{c_{ATP}} = 4.9 \times 10^5 \, \text{M} \qquad (14.2)$$

where c_{ADP}, c_{Pi} and c_{ATP} are the concentrations of ADP, Pi and ATP respectively. The value of the equilibrium constant depends on a number of factors, including the free magnesium concentration, the pH and the ionic strength. The value given for the equilibrium constant in equation (14.2) corresponds to the standard conditions found in the cytoplasm of the vertebrate cell. The amount of energy liberated (ΔG) by the ATP reaction can be calculated with:

$$\Delta G = \Delta G_0 - kT \ln \frac{c_{ATP}}{c_{ADP}c_{Pi}} \quad (14.3)$$

where ΔG is the standard free energy and $\Delta G_0 = -54 \times 10^{-21}$ J. The free energy of the hydrolysis reaction depends on both the standard free energy and the concentrations of ATP, ADP and Pi. Motor protein enzymes can thus liberate energy from ATP to drive conformational (mechanical) changes in their low Reynolds number aqueous environments that give rise to motility.

The standard speed for many biological processes driven by simple molecular motors is on the order of 1 µm/s. The growth of actin filaments occurs at rates of $10^{-2} - 1\, \mu m\, s^{-1}$ and is dependant on the concentration of the actin filaments. Actin based cell crawling is in the range $10^{-2} - 1\, \mu m\, s^{-1}$, and this involves the growth and disassembly of actin filaments at the leading edge of a lamellipodium (Figure 14.1(c)). Myosin interacts with actin and leads to motility with a range of rates, $10^{-2} - 1$ µm/s. Striated muscle parallelises the myosin/actin interactions and provides much larger forces than available from individual molecules, but with a similar time response to that of the individual molecules. Microtubule growth and shrinkage is on the order of $0.1 - 0.6\, \mu m\, s^{-1}$, which is similar to the rate of motion of self-assembling actin. Fast and slow axonal transport occurs at rates in the range $10^{-3} - 10^{-1}\, \mu m\, s^{-1}$ as the motor proteins kinesin and dynenin walk towards the plus and minus ends of microtubules.

14.1 SELF-ASSEMBLING MOTILITY – POLYMERISATION OF ACTIN AND TUBULIN

The polymerisation of actin and tubulin are examples of one dimensional aggregating self-assembly (Section 6.4). An additional complication in this process is that chemical energy is used to drive the self-assembly process and this permits non-equilibrium dynamic structures to evolve. In the simplest models of this behaviour the rate of addition of subunits is found to be proportional to the concentration of free monomers in solution (c_m) and there is a constant of proportionality for the addition of monomers (k_{on}). The number of monomers captured per unit time

(dn/dt) is proportional to the number of monomers available for capture:

$$\frac{dn}{dt} = k_{on} c_m \qquad (14.4)$$

In contrast, it is found that the release rate does not depend upon the free monomer concentration. k_{off} is a constant for the subtraction of monomers and is independent of the monomer concentration:

$$\frac{dn}{dt} = -k_{off} \qquad (14.5)$$

The total elongation rate of the filament is the sum of the processes for addition (equation (14.4)) and release (equation (14.5)) of the monomers provided a nucleation site for filament growth is available:

$$\frac{dn}{dt} = k_{on} c_m - k_{off} \qquad (14.6)$$

The critical concentration (c_{mcrit}) for self-assembly occurs when the elongation rate (dn/dt) vanishes, i.e. placing dn/dt equal to zero in equation (14.6) gives:

$$c_{mcrit} = \frac{k_{off}}{k_{on}} \qquad (14.7)$$

A graphical solution of equation (14.6) for this process of one dimensional aggregating self-assembly is shown in Figure 14.2. Above the

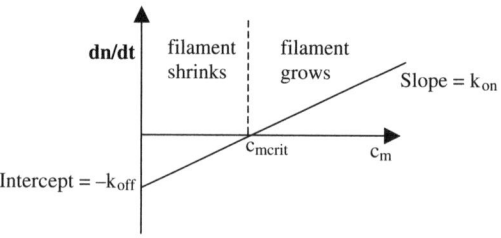

Figure 14.2 The rate of polymerisation of actin filaments (dn/dt) as a function of monomer concentration (c_m)
(c_{mcrit} is the critical monomer concentration for self-assembly. Below c_{mcrit} the filaments shrink and above c_{mcrit} they grow. The gradient of the figure gives the association rate constant (k_{on}), and the dn/dt intercept gives the dissociation rate constant ($-k_{off}$))

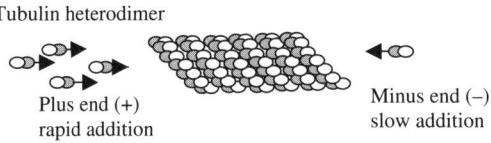

Plus end (+)
rapid addition

Minus end (−)
slow addition

Figure 14.3 Actin and tublin self-assembly is anisotropic, due to the anisotropy of the constituent subunits. Fast addition occurs at the positive end (+) and slow addition on the negative end (−)

critical monomer concentration the fibres expand, whereas below this concentration they shrink.

Similar processes of self-assembly are observed experimentally for both actin and tubulin filaments. In principle it is easy to extract the rate constants for addition and subtraction of the monomer subunits (k_{on} and k_{off}) from an in vitro experiment by plotting the elongation rate as a function of the monomer concentration (as in Figure 14.2). There are, however, some additional complications with real self-assembling biological motors. Subunits are often asymmetrical and assemble side-by-side with a preferred orientation that gives rise to orientated filaments (Figure 14.3). The two ends of the polymer are not chemically equivalent. The faster growing end is referred to as the plus end (+), and the slow growing end is labelled with a minus sign (−). Thus experimentally the two ends (+ and −) of the self-assembling filaments need to be considered separately to extract the two sets of rate constants required to describe the separate processes of addition and subtraction. It is found that the rate constants depend on both the solvent and salt concentration, so the aqueous environment that surrounds the filaments needs to be carefully controlled.

This situation of anisotropic self-assembly can be analysed through an extension of the Oosawa model described by equation (14.6) (Figures 14.4 and 14.5). Since the two ends of the filament are not equivalent, two

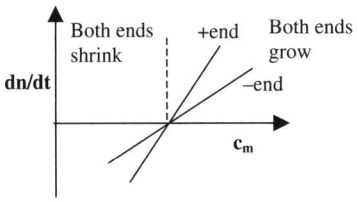

Figure 14.4 Model for the dynamics of actin self-assembly that considers the different rate constants for both ends of the anisotropic filament
(The rate of assembly (dn/dt) is shown as a function of the monomer concentration (c_m). In the case illustrated $c_{m+crit} = c_{m-crit}$.)

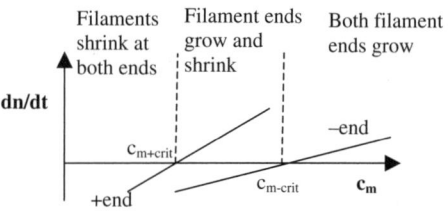

Figure 14.5 Model for the dynamics of actin self-assembly in which $c_{m+crit} \neq c_{m-crit}$ during the assembly of the anisotropic filaments. The rate of assembly (dn/dt) is shown as a function of the monomer concentration (c_m)

equations are needed for the rate of elongation of the two ends (+ and −):

$$\frac{dn^+}{dt} = k_{on}^+ c_m - k_{off}^+ \tag{14.8}$$

$$\frac{dn^-}{dt} = k_{on}^- c_m - k_{off}^- \tag{14.9}$$

Each of the equations has a separate critical monomer concentration for the process of self-assembly:

$$c_{m+crit} = \frac{k_{off}^+}{k_{on}^+} \tag{14.10}$$

$$c_{m-crit} = \frac{k_{off}^-}{k_{on}^-} \tag{14.11}$$

In the special case that the critical concentration of both ends are equal ($c_{m+crit} = c_{m-crit}$) both ends grow or shrink simultaneously, although the rates of assembly may be different. For steady state conditions ('treadmilling') the rate of growth and shrinkage of the two ends must be equal. This can be expressed mathematically as:

$$\frac{dn^+}{dt} = -\frac{dn^-}{dt} \tag{14.12}$$

And there is therefore a single critical concentration (c_{tm}) for this process of treadmilling self-assembly:

$$c_{tm} = \frac{(k_{off}^+ + k_{off}^-)}{k_{on}^+ + k_{on}^-} \tag{14.13}$$

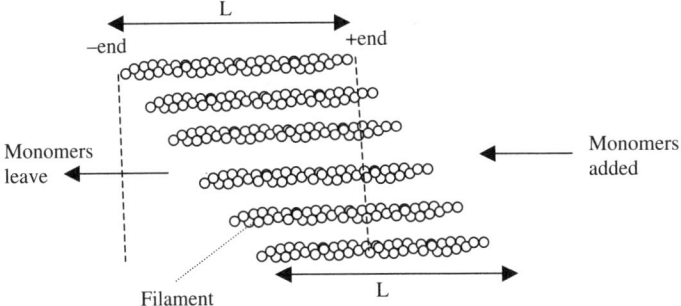

Figure 14.6 The treadmilling process involved in actin self-assembly (Monomers leave from the negative end and are added to the positive end. The filament length (L) is conserved during the process, as the centre of mass of the filament moves to the right.)

The process of treadmilling is schematically shown in Figure 14.6, the length of the filament is invariant during the process, but its centre of mass is displaced. Typical values for the rate constants and critical concentrations that occur during the self-assembly of actin and microtubules are given in Table 14.1. Treadmilling is often the dominant process in vivo, since it is highly efficient in the reuse of subunits.

The interaction and pattern formation in active self-assembling motor protein networks can be very complex. An example of a dynamic morphology created during cell division is shown in Figure 14.7. Here an animal cell is shown in the final stages of cell division (cytokinesis) with an actin-myosin ring contracting to pinch off the two divided cells. Also shown in the figure are the remains of the mitotic spindle formed from microtubules that drive the movement of the dividing chromosome during the initial stages of cell division.

Table 14.1 Rate constants for actin and microtubule self-assembly [Ref.: T.D. Pollard, *J. Cell. Biol.*, 1986, 103, 2747–2754]

Monomer in solution	k_{on}^{+} $(\mu Ms)^{-1}$	k_{off}^{+} s^{-1}	k_{on}^{-} $(\mu Ms)^{-1}$	k_{off}^{-} s^{-1}	c_{m+crit} μM	c_{m-crit} μM
Actin						
ATP-actin	11.6	1.4	1.3	0.8	0.12	0.6
ADP-actin	3.8	7.2	0.16	0.27	1.9	1.7
Microtubules						
Growing (GTP)	8.9	44	4.3	23	4.9	5.3
Rapid disassembly	0	733	0	915	n/a	n/a

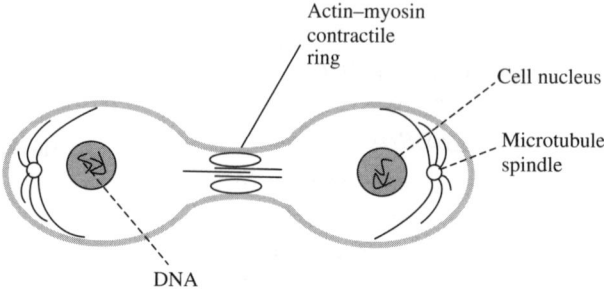

Figure 14.7 Motor proteins involved during the cytokinesis of cell division (The actin–myosin ring pinches off the cell in the final stages of replication. The microtubule spindle is used in a prior step in the process of chromosomal division.)

14.2 PARALLELISED LINEAR STEPPER MOTORS – STRIATED MUSCLE

The basic constituents of striated muscle are actin and myosin which are arranged in a parallel array (Figure 14.8). These motors are perhaps the most important for human health, since heart disease provides the largest contribution to annual human mortality rates and heart muscle is striated.

A scheme for the chemomechanical transduction process of ATP to provide motility in striated muscle is provided by the *rotating crossbridge model* (Figure 14.9). This involves two key ideas: the myosin motor cycles between attached and detached states, and the motor undergoes a conformational change (working stroke) that moves the load bearing region of the motor in a specific direction along the filament. The rotating cross-bridge model incorporates the Lymn–Taylor scheme, which describes chemically how nucleotides regulate the attachment and detachment of myosin from the filament, the swinging lever arm hypothesis, which provides a mechanism for amplifying small structural

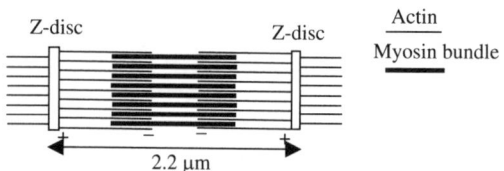

Figure 14.8 The arrangement of actin and myosin that are parallelised into arrays in striated muscle. The distance between the Z-discs decreases during muscular contraction as the myosin molecules walk along the actin filaments

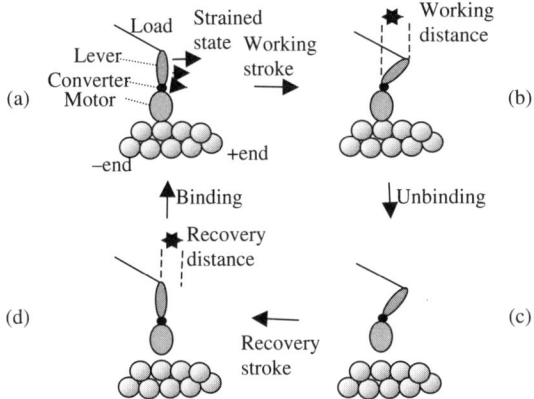

Figure 14.9 The rotating cross-bridge model for myosin–actin association consists of four distinct steps
((a) The myosin attachs to the actin filament, (b) the myosin molecule does work as it stresses the binding site, (c) the myosin unbinds from the actin filament and (d) the stress in the myosin molecule is dissipated as it moves one step along the actin filament.)

changes around the nucleotide-binding pocket into the much larger conformational changes of the cross-bridge, and the powerstroke model, which accounts for how the motor generates force through the use of an elastic element within the cross-bridge that is strained during the power stroke.

During the action of striated muscle there is a sequence of transitions between different chemical states of the myosin molecules; ATP binding, hydrolysis and ADP release. These transitions alter the association between the motor domain and the filament, which leads to the alternating between attached and detached states.

There are three distances required to understand the inch worm motion of myosin along actin filaments. The *working distance* (δ) is the distance a cross-bridge moves during the attached phase of its hydrolysis cycle. The *distance per ATP* (Δ), is the distance that each motor domain moves during the time it takes to complete a cycle, which is also equal to the speed of movement divided by the ATPase rate per head. The *path distance* is the distance between consecutive myosin binding sites (or stepping stones) (Figure 14.10) along the actin fibre.

A series of single molecule techniques have been used to measure the force and characteristic distances used by myosins associated with single fibres of actin. Force transducers that are typically used are the cantilevered glass rod, atomic force microscope (AFM) and dual trap optical tweezers (Section 13.4). A particularly elegant variety of experiment uses

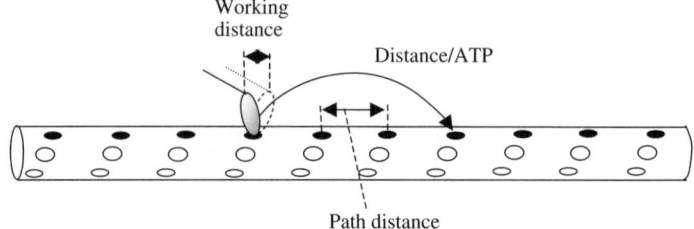

Figure 14.10 Three distances associated with the inch worm motion of myosin molecules along actin
(The working distance is the length moved by a myosin molecule in each cycle of the rotating cross-bridge model, the path distance is the lateral distance between binding sites and the distance/ATP is the length moved by a myosin molecule that uses one molecule of ATP [*Reprinted with permission from J. Howard, Mechanics of Motor Proteins and the Cytoskeleton, Copyright (2001) Sinauer Associates*])

an actin filament attached at either end to two optically trapped spheres, and the filament is allowed to interact with single myosin II molecules (Figure 14.11). Individual working strokes of the myosin molecules are resolvable with this method. Single molecule fluorescence is another powerful technique for following the pathway of single motor protein motility and, similarly, microrheology techniques can resolve the changes in viscoelasticity due to the motion of molecular motors which provides important dynamic information.

It is useful to consider the exact nature of the molecular steps of myosin II travelling along the actin filament. The myosin has five structural configurations during its interaction with actin in muscular motion (Figure 14.12) (compare with Figure 14.9). Initially there is tight binding of the myosin head to the actin filament, called the rigor position (Figure 4.12(b), as in rigor mortis where the additional cross-links account for the rigidity of dead muscle). Next, the myosin filament is released on capturing ATP (Figure 14.12(c)), which provides the energy for the force

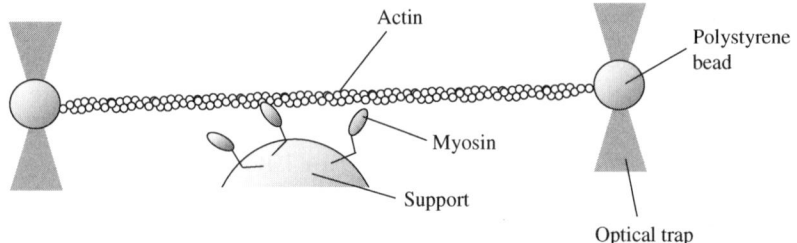

Figure 14.11 Double trap optical tweezers can be used to measure the step size of myosin II interacting with actin filaments. The actin filament is attached at either end to optically trapped polystyrene beads

PARALLELISED LINEAR STEPPER MOTORS – STRIATED MUSCLE

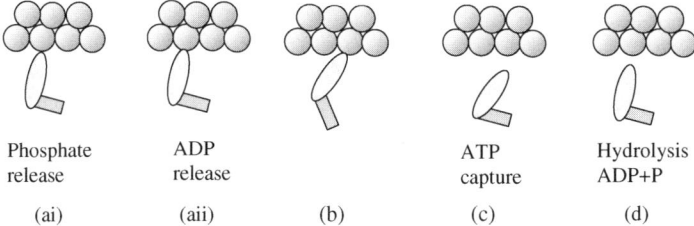

Figure 14.12 Chemical steps in the cyclic attachment of myosin to actin filaments that corresponds to the rotating cross-bridge model (Figure 14.9)

on the actin fibre. There is then a configurational change of the cocked position during hydrolysis of ATP (Figure 14.12(d)). Subsequently, there is weak binding of the head to the myosin filament in a new position and, finally, a phosphate group is released (Figure 14.12(a)).

The speed and processivity of the cross-bridge motion can be understood using the concept of the duty ratio, which is the fraction of time each motor domain spends attached to its filament. There is a cyclic process (Figure 14.13) in which the motor repeatedly binds to and unbinds from the filament. During each cross-bridge cycle, a motor domain spends an average time attached to the filament (τ_{on}) when it makes its working stroke, and an average time detached from the filament (τ_{off}) when it makes its recovery stroke. The duty ratio (r) is the fraction of time that each head spends in its attached phase:

$$r = \frac{\tau_{on}}{\tau_{on} + \tau_{off}} \qquad (14.14)$$

The minimum number of heads (N_{min}) that associate with a filament and are required for continuous movement (N_{min}) is thus related to the duty ratio:

$$r \approx \frac{1}{N_{min}} \qquad (14.19)$$

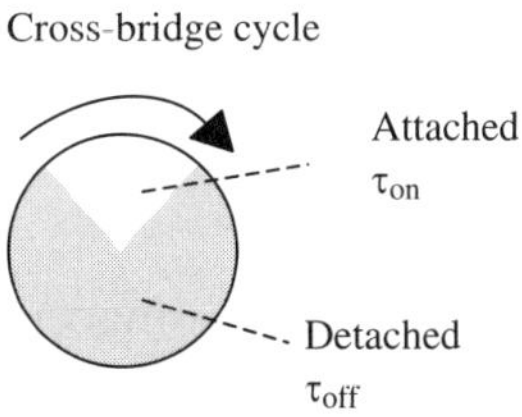

Figure 14.13 Cross-bridge cycle with myosin binding to actin filaments (τ_{on} is the attached time and τ_{off} is the detached time. The cycle rotates through alternating periods of attachment and detachment.)

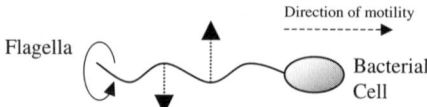

Figure 14.14 Helical flagellar filaments provide bacteria with motility in its low Reynold's number environment

14.3 ROTATORY MOTORS

Following on from the discussion of Poisson motility processes in Section 5.3, the molecular biophysics of the rotatory flagellar motor for the propulsion of bacteria is considered (Figure 14.14). A curved segment separates the motor from the main length of the filament. The filament is bent away perpendicularly from the surface of the membrane for several nanometres. This filament executes a helical motion as it is rotated by the motor and acts like a propeller, which provides a source of motility for the bacteria.

A series of proteins form the flagellum and each has a specific function; the bushings seal the cell membrane, the circular stator is attached to the cell and the rotor attached to the flagellum (Figure 14.1(a)). The flagellar propellor is not run directly by ATP. Instead, protons run down a pH gradient across the membrane and produce an electric potential. Sodium ions can also fulfil the same function in marine bacteria. As bacteria move through a solution their flagella can rotate at 100 revolutions per second, which is comparable to the rate at which an automobile engine (30 Hz) functions. The flagellar motor works equally well in both clockwise and counterclockwise modes. These bacterial motors are relatively complicated devices and over twenty separate protein components are required to provide motility. The evolutionary history of such a finely orchestrated engine is a fascinating story. The rate of rotatory motion is found to be proportional to the potential difference across the motor under physiological conditions.

14.4 RATCHET MODELS

An interesting, but inefficient (and thus inaccurate) model of molecular motility is provided by the thermal ratchet. The model indicates how directed motility of a muscle protein can be derived from rectified Brownian motion (Figure 14.15), i.e. a constant bias on the probability of motion is superposed on the thermal fluctuations of displacement of a

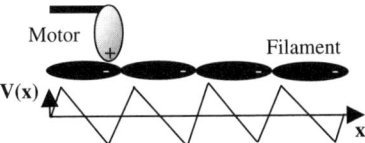

Figure 14.15 The interaction of motor proteins (e.g. myosin) with actin can be modelled with a single one dimensional potential. The monomers of the biofilament have a dipolar charge distribution and the myosin motors experience a saw tooth interaction potential ($V(x)$) as they interact with the filament

particle in a particular direction. Widely differing processes of motility, such as the self-assembly of actin and the action of rotatory motors, can be described in terms of rectified Brownian motion and ratchet models have therefore been used to analyse the motion of these systems.

The thermal ratchet is a simple means for producing motion in a low Reynold's number environment. It uses a spatially asymmetric potential that oscillates with time (Figure 14.16). The probability distribution of motor proteins ($P(x)$) evolves in the standard manner due to thermal diffusive motion when the potential is switched off (Section 5.1). The asymmetry of the oscillating potential, when superposed on the thermal fluctuating force, causes a net motion of the proteins in a given direction. The net probability of directed motion (P_{net}) is the difference between the probability to move right (P_R) and that to move left (P_L):

$$P_{net} = P_R - P_L \qquad (14.16)$$

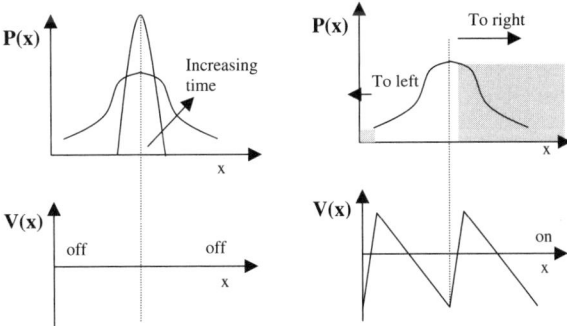

Figure 14.16 An oscillating saw tooth potential ($V(x)$) can be used to move a series of particles (probability density $P(x)$). Within this realisation of the ratchet model particles are moved to the right by the asymmetry of the saw tooth potential
[Ref.: J. Prost, J.F. Chauwin, L. Peliti and A. Ajdari, Physical Review Letters, 1994, 72, 16, 2652–2655 and R.D. Astumian and M. Bier, Physical Review Letters, 1994, 72, 11, 1766–1769]

A simple mathematical form for the probability distribution of the motor proteins results from the action of a saw tooth potential. If the proteins do not diffuse a sufficient distance there is no net flux:

$$P_{net} = 0 \qquad 0 < \sigma < \alpha\lambda \qquad (14.17)$$

where λ is the wavelength of the sawtooth potential, σ is a measure of the spread of the particle distribution and $\alpha\lambda$ is the peak-to-trough separation of the potential. If the probability distribution created by the thermal motion is sufficiently broad, a net flux occurs:

$$P_{net} = (1 - \alpha\lambda/\sigma)^2/2 \qquad \alpha\lambda < \sigma < (1-\alpha)\lambda \qquad (14.18)$$

The probability distribution initially broadens when the potential is not applied due to thermal diffusion of the motor proteins. When the potential is switched on again there is a higher probability of the particles being drawn to the right than to the left due to the asymmetric nature of the potential. This allows the dipolar nature of the motion of the monomers along a biofilament to be modelled.

The major problem with such a simple Brownian ratchet model is its efficiency. A thermal ratchet can take the hydrolysis of up to 10 ATP molecules for one step ($P_{net} = 0.1$) of the ratchet. In real biological systems the efficiency is typically five times better than that found for the model (Figure 14.8). More sophisticated extensions of such models have recently been proposed that aim to resolve this shortfall.

Ratchet models have also been applied to rotatory motors (Figure 14.1(a)). An elastic link is invoked between the stator unit and the cell wall that rectifies the angular thermal fluctuations in a certain sense (anticlockwise or clockwise) that provides a mechanism for rotational motion.

14.5 OTHER SYSTEMS

Other less common mechanisms for biological motility have been discovered. *Extrusion nozzles* are present in the myxobacteria, cyanobacteria and flexibacteria. Slow uniform gliding motion is achieved for these organisms by a continuous secretion of a glycoprotein slime (Figure 14.1(d)).

Supramolecular springs store conformational energy in chemical bonds that act as latches for the release of the energy, which provides a one shot mechanism of motility (Figure 14.1(e)). The specific power of such motors can be very high. One example is the scruin–actin system in which scruin captures actin in a slightly overtwisted state. Calcium dependent changes in the scruin are then used to release the conformational energy of the actin and provide a force for motility.

FURTHER READING

D. Boal, *Mechanics of the Cell*, Cambridge University Press, 2002. Contains a useful section on active molecular networks.

J. Howard, *Mechanics of Motor Proteins and the Cytoskeleton*, Sinauer, 2001. Very good introductory text on motor proteins.

TUTORIAL QUESTION

14.1) From Table 14.1 check that the relationship between the critical monomer concentration and the dissociation constants holds (equations (14.10) and (14.11)). What do you predict are the critical concentrations for treadmilling of actin and microtubules?

15
Structural Biomaterials

A wide range of biomaterials that are optimally matched to their structural roles have evolved naturally. Examples include cartilage in synovial joints, spider silk for web building, resilin in the hinges of dragon fly wings, mollusc glue for adhesion and cancellous bone in the skeletons of a range of animals. These examples are chosen to illustrate the rich variety of physical phenomena involved and the exquisite nature of the design principles that evolution has used in solving structural biomaterial problems.

15.1 CARTILAGE – TOUGH SHOCK ABSORBER IN HUMAN JOINTS

Normal healthy human joints have friction coefficients (μ) in the range 0.001-0.03, which is lower than that found with the materials that coat non-stick frying pans ($\mu \approx 0.01$ for teflon on teflon). These values are also remarkably low when compared with hydrodynamically-lubricated bearings that are constructed in efficient mechanical engines, such as those in cars. However, hydrodynamic lubrication is not in effect (it occurs in car engines, aircraft turbines etc. at high speeds), since the bone surfaces in synovial joints never move relative to one another at more than a few cms^{-1} (Section 7.6). Synovial joints function in the boundary lubrication regime (Figure 7.20).

A schematic diagram of an articulated joint is shown in Figure 15.1. It consists of three main mechanical components: *bone* (a living mineral

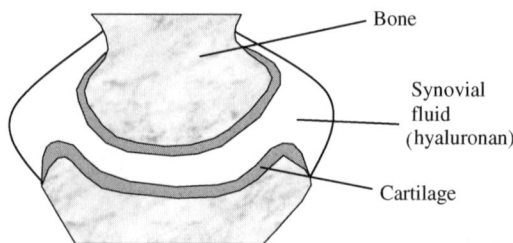

Figure 15.1 Schematic diagram of an articulated joint that shows the two sections of bone, the synovial fluid and the two cartilage shock absorbers

foam composite), a *viscoelastic fluid* (a semi-dilute solution predominantly composed of the polyelectrolyte hyaluronic acid and water) and *cartilage* (an elastic protein/proteoglycan composite).

Cartilage acts as a shock absorber in a series of applications throughout the body, including articulated joints. It is a living tissue and specialised cells (chondrocytes) contained within the tissue play a role in repairing damage from wear of the moving surfaces and protect it against bacterial attack (Figure 15.2).

In human knee joints the average pore size of the cartilage between collagen fibres is approximately 60 Å. The surface of cartilage has ripples (amplitude 3 µm and wavelength 40 µm) superimposed on a microroughness (amplitude 0.3 µm and wavelength 0.5 µm). The synovial fluid contained between the sections of cartilage is a non-Newtonian liquid having the property of shear thinning; its viscosity decreases almost linearly with shear rate. This rheological behaviour is typical of non-associating polyelectrolyte solutions (Section 12.3).

Cartilage presents one of the biggest challenges in tissue engineering, the creation of replacement materials to treat arthritic conditions. The sections

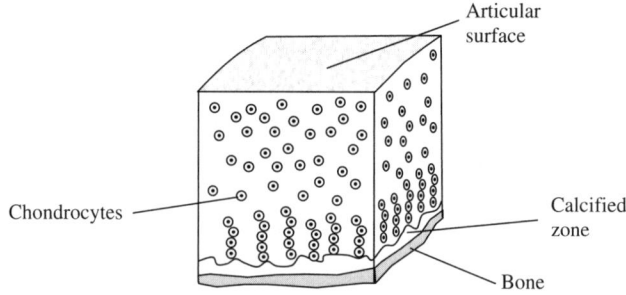

Figure 15.2 Cartilage is a living shock absorber, chondrocytes are arranged throughout its structure. The condrocytes replenish the extracellular material that constitutes the cartilage

of bone in articulated joints can be successfully replaced with synthetic materials such as polyethylene, and new hyaluronic acid can be injected in to the knee cavity to replace damaged boundary lubricants. However, osteoarthritis involves the break down of cartilage on the bones surface, as cracks form in the material due to a number of wear mechanisms, and currently no effective replacement exists for these low friction shock absorbers. The lifetime of polyethylene replacement joints is seriously compromised by the high friction wear mechanism that results from the absence of a cartilage covering and they often need to be replaced after ten to fifteen years. New alternatives for replacement materials are required and this necessitates an improved physical understanding of synovial joints.

The collagen in cartilage exists with an anisotropic distribution of fibre orientation (Figure 15.3) and thus it has anisotropic mechanical properties. The shear modulus of the material is higher perpendicular to the chain orientation than in the parallel direction (the load bearing direction).

In general, the function of articular cartilage in articulated joints is to increase the area of load distribution and to provide a smooth wear resistant surface optimised for low friction. Biomechanically articular cartilage can be viewed as a two phase (solid-fluid) material; the collagen/proteoglycan solid matrix (25% contribution to the wet weight) is surrounded by freely moving interstitial fluid (75% by wet weight). The important biomechanical properties of articular cartilage are the resistance of the solid matrix to deformation and the frictional resistance to the flow of the interstitial fluid through the porous permeable solid matrix. Articular cartilage has the ability to provide joints with a self-lubrication behaviour that operates under normal physiological conditions. Pressure on the surface of the cartilage forces water through the porous matrix, moving out through the surface, which provides a lubricating fluid film on the surface of the cartilage (Figure 15.4). Damage to

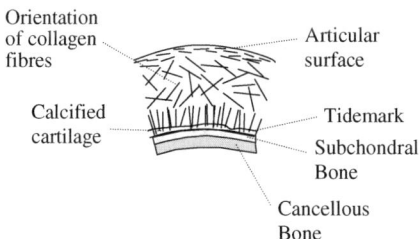

Figure 15.3 Cartilage has an anisotropic fibrous structure that leads to anisotropic mechanical properties. Collagen fibres are attached perpendicularly to the surface of the bone and the arrangement shifts to a parallel alignment at the surface of the articular cartilage

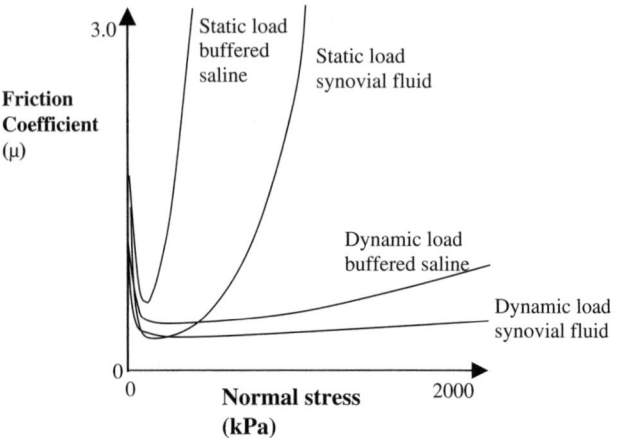

Figure 15.4 The friction coefficient as a function of normal stress between two cartilage surfaces
(Synovial fluid is seen to cause an important reduction in the friction coefficients of both static and dynamic tests compared with the saline control [Ref.: L.L. Malcom, 1976, University of California San Diego])

articular cartilage can disrupt the normal load carrying ability of the tissue and the lubrication process that operates in the joint. Insufficient boundary lubrication is thought to be a primary factor in the development of osteoarthritis, which causes acute damage to the cartilageneous surfaces and extreme pain for the sufferers.

Cartilage is a charged cross-linked elastomeric composite material and can be compared with resilin and elastin, two uncharged bioelastomers examined in Section 15.3. Although cartilage is reasonably elastic, it is not resilient; energy dissipation is maximised. The cross-links forming the elastic matrix are provided by the collagen in cartilage and the dissipative properties are provided by giant polyelectrolyte combs (the aggrecan, Figure 15.5).

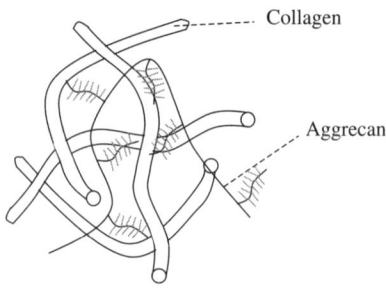

Figure 15.5 Cartilage is a composite mixture of collagen and aggrecan molecules. The collagen molecules provide strength and elasticity to the network whereas aggrecan is used to dissipate energy

The physical phenomena that contribute to the behaviour of cartilage are very rich. Principle questions include the repulsive forces between the two charged cartilaginous plates, the friction coefficient of the polymeric cartilage gels, the modulus of the cartilage gels with their rigid molecular nematic inserts (collagen) and the time effects observed with longitudinal stress relaxation after the cartilage has been mechanically loaded. A simple model for the extremely low friction coefficients found in articular cartilage is examined first. The model still requires development and is only at a qualitative level of understanding. However, it does demonstrate the bottom up approach of molecular biophysics to explain some sophisticated material properties.

First of all consider *the forces between two charged plates*. The Poisson–Boltzmann equation can be used for the potential (ψ) due to the surface charges at a perpendicular distance (r) from the plates (Section 2.3):

$$\nabla^2 \psi = -\left(\frac{e\rho_0}{\varepsilon}\right) e^{-e\psi(r)/kT} \qquad (15.1)$$

where ρ_0 is the ion density profile at the point of zero potential ($\psi = 0$), i.e. the mid-point between the planar plates. e is the electronic charge, ε is the dielectric constant, and kT is the thermal energy. The charge density as a function of the distance from a single charged surface is shown schematically in Figure 15.6. For a charged homopolymer gel carrying one charge on each monomer unit, the solution for the surface charge density (σ, units of electrons per m^3) is:

$$\sigma = (1000\, cN_A)^{\frac{2}{3}} = \left(\frac{10^6 N_A}{qM_w}\right)^{\frac{2}{3}} \qquad (15.2)$$

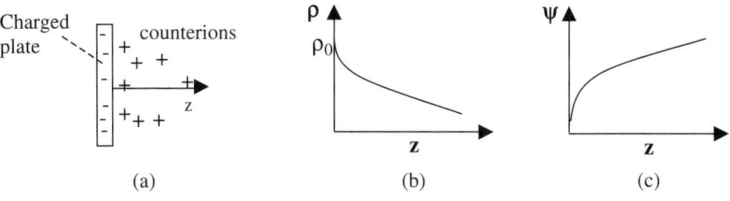

Figure 15.6 Schematic diagram of the charge density near a charged plate ((a) the arrangement of the counterions, (b) the counterion density (ρ) and (c) the potential (ψ) near the surface as a function of the perpendicular distance from the plate (z))

where N_A is Avogadro's number, c is the molar polymer concentration, M_w is the molecular weight and q is the degree of swelling of the gel (q = swollen sample volume/dry sample volume). For positively charged counterions in one dimension the Poisson–Boltzmann equation (15.1) can be written:

$$\frac{d^2\psi}{dz^2} = -\frac{e\rho_0}{\varepsilon}e^{-e\psi/kT} \tag{15.3}$$

where z is the distance from the plates. The Poisson–Boltzmann equation is subject to two boundary conditions on the cartilage gel surfaces and on the symmetric plane between the two sections of cartilage. From the Poisson equation of electrostatics the boundary conditions can be written mathematically. On the gel surface the gradient of the potential equals the surface charge density:

$$\left(\frac{d\psi}{dz}\right)_{z=\pm l} = \mp\frac{e\sigma}{\varepsilon} \tag{15.4}$$

where l is the solvent layer thickness on the gels' surface. On the symmetric plane the gradient of the potential is zero:

$$\left(\frac{d\psi}{dz}\right)_{z=0} = 0 \tag{15.5}$$

Electrical neutrality requires that the surface charge density (σ) is equal to the total charge of the oppositely charged counterions associated with the polyelectrolyte chains:

$$\sigma = \rho_0 \int_0^l e^{-e\psi/kT} dz \tag{15.6}$$

Solution of the Poisson–Boltzmann equation subject to the three requirements of equations (15.4), (15.5) and (15.6) gives:

$$\sigma = \sqrt{\frac{2\rho_0}{l_b}} \tan\left(l\sqrt{\frac{\rho_0 l_b}{2}}\right) \tag{15.7}$$

Here $l_b = e^2/\varepsilon kT$ is the Bjerrum length (a constant). The repulsive osmotic pressure (π) between two charged surfaces is determined by the ion charge distribution (density ρ_0) at the symmetry plane from the contact value theorem (Section 2.4):

$$\pi = \rho_0 kT \quad (15.8)$$

where kT is the thermal energy. In the equilibrium state with a constant pressure on the cartilage, the osmotic pressure (π), predominantly due to the counterions in a charged gel (Section 9.4), is counter balanced by the applied pressure (P), e.g. the weight of a person's upper body distributed across the area of their knees:

$$P = \pi \quad (15.9)$$

The solvent layer thickness ($2l$) that remains between the two sections of cartilaginous gels can then be calculated. Equation (15.7) is rearranged in terms of the solvent layer thickness and the results of equations (15.8) and (15.9) are used:

$$2l = 2\sqrt{\frac{2kT}{Pl_b}} \tan^{-1}\left(\sigma\sqrt{\frac{kTl_b}{2P}}\right) \quad (15.10)$$

It is concluded that highly charged surfaces are able to sustain more pressure (at fixed pressure the equilibrium distance is larger) than the equivalent neutral surface (Figure 15.7). Cartilage has a relatively rigid cross-linked network, so swelling of the chains by the osmotic pressure at equilibrium is neglected in this model.

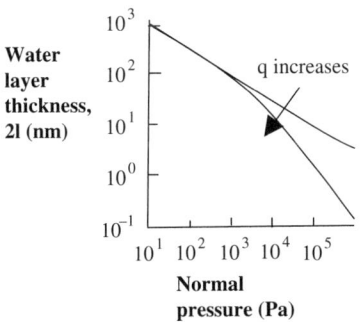

Figure 15.7 Thickness of water between two charged plates as a function of the normal pressure. As the swelling of the gels (q) increases they move closer together due to the corresponding decrease in surface charge density
[Reprinted with permission from J. Gong, Y. Iwasaki, Y. Osada, et al., J. Phys. Chem. B, 103, 6001–6006, Copyright (1999) American Chemical Society]

The *friction coefficient* of a polymer gel can also be motivated using a scaling approach (Section 8.2). Amonton's law for the friction of 'conventional' materials states that the frictional coefficient is independent of load (Section 7.6). The frictional force (F) of many solids on solids is therefore related to the normal force (W) by the universal law:

$$F = \mu W \quad (15.11)$$

Where μ is the frictional coefficient. Experimentally in the case of solids with repulsive interfacial interactions (e.g. like-charged polyelectrolyte gels) it is found that the velocity dependence of the friction is strongly dependent on the normal compressive strain. The smaller the strain, the weaker the velocity dependence of the friction. Indeed, with gels it is found that the frictional force is no longer linearly proportional to the normal force and equation (15.11) needs to be replaced by:

$$F \propto AP^\alpha \quad (15.12)$$

where P is the average normal pressure, equal to the weight (W) divided by the contact area (A) and α is a constant in the range 0 to 1. Furthermore, when two pieces of negatively charged gels are allowed to slide past each other the frictional force is found to be proportional to a power law of the velocity (v) with a constant exponent (β):

$$F \propto v^\beta \quad (15.13)$$

where β depends on the normal compressive strain. This power law dependence on the velocity indicates that not only hydrodynamic lubrication, but also the viscoelasticity of the polymer networks play an important role in the resultant frictional properties.

A qualitative *molecular theory* for polymer friction at the surface of a neutral gel can be motivated. For an uncharged semi-dilute polymer solution the mesh size (ξ) (equation (8.37)) depends only on the polymer concentration (c):

$$\xi \sim ac^{-\frac{3}{4}} \quad (15.14)$$

where a is the effective monomer length and ξ can be considered a measure of the size of the pores in the polymer mesh. It is found experimentally that polymer gels have the same scaling with regard to their correlation length (ξ) and osmotic pressure (π) as the equivalent semi-dilute solution prepared at the same polymer concentration (this is called the *c** *theorem*). The change in interfacial energy $(A - A_0)$ between a polymer gel and a solid surface is:

$$A - A_0 \approx \pi_0 \xi \qquad (15.15)$$

where π_0 is the osmotic pressure of the bulk solution, A is the interfacial energy between the solid and the gel and A_0 is the interfacial energy between the substrate and the pure solvent. From polymer scaling theory the osmotic pressure of the polymer gel is known to be related to its correlation length:

$$\pi_0 \approx T\xi^{-3} \qquad (15.16)$$

where T is the temperature. The work done by the solid surface to repel the polymer from the surface a distance ξ_g against the osmotic pressure is equal to the increase in surface free energy:

$$P\xi_g \approx A - A_0 \approx \pi_0 \xi \qquad (15.17)$$

where P is the average normal pressure. When no surface adsorption of the polymer occurs the frictional force is due to the viscous flow of the solvent at the interface. Viscous solvent flow obeys Newton's second law and hydrodynamic lubrication theory can be applied between two particles separated by a solvent layer to obtain the frictional force (f) using non-slip boundary conditions:

$$f = \frac{\eta v}{\xi_g + D} \qquad (15.18)$$

where ξ_g is the thickness of the solvent layer, D is the thickness of the polymer film (the thickness that is sheared), η is the viscosity of the solvent and v is the relative velocity of the surfaces. A combination of these results can be used to show that the frictional coefficient of a *neutral gel* depends on the temperature (T), Young's modulus of the gel (E) and applied pressure (P):

$$f = \frac{\eta v P}{E^{\frac{2}{3}} T^{\frac{1}{3}}} \qquad (15.19)$$

A similar calculation for a *charged gel surface* gives:

$$f = \frac{\eta v}{2(D + \sqrt{k_{gel}})} \quad (15.20)$$

where k_{gel} is the hydraulic permeability of the gel (see equation (15.23)). The frictional coefficient (μ) can be shown to depend on the pressure (P) in a non-linear manner with charged gels:

$$\mu \sim P^{-\frac{3}{5}} \quad (15.21)$$

This is in agreement with experiment (Figure 15.8).

Simply put, the higher the charge on the two like-charged polyelectrolyte gels in a frictional experiment (assuming constant swelling), the lower the resultant frictional coefficient (Figure 15.9). However, charge effects are only one important factor in determining the frictional properties. Practically the situation is often more complicated in synovial joints. Elastohydrodynamic fluid films of both the sliding and squeeze type could play an important role in lubricating synovial joints. There is thus a mixed method of reducing friction in cartilage, with contributions from both the repulsion of charged polymers at the surface and the water exuded from inside the cartilage. Microcontacts (asperities) between the surfaces would also be expected to play a role in determining frictional coefficients in regions where the double layer forces are insufficient to

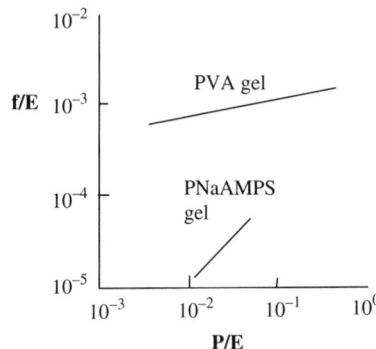

Figure 15.8 Experimental dependence of the frictional force (f) on the normal pressure (P) for a flexible neutral polymer (PVA) gel and a flexible polyelectrolyte (PNaAMPs) gel. The frictional force and pressure are renormalised by the Young's modulus for comparison. The frictional force is much lower for the charged gels in water
[*Reprinted with permission from J. Gong and Y. Osada, J. Chem. Phys., 109, 8062, Copyright (1998) American Institute of Physics*]

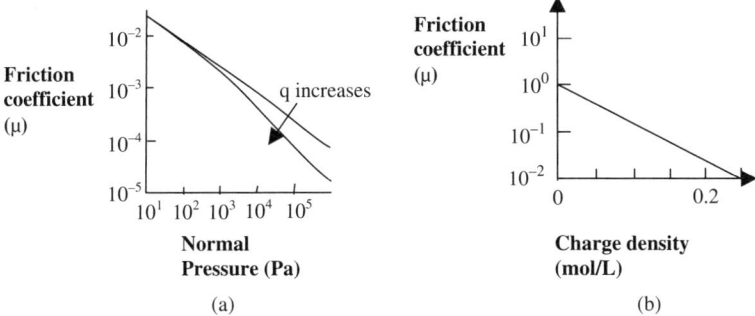

Figure 15.9 (a) Dependence of the frictional coefficient on the normal pressure (q is the swelling of the gel). The frictional coefficient decreases as the gel swelling increases. (b) Dependence of the frictional coefficient on the charge density for a polyelectrolyte gel. The friction coefficient of the charged gel decreases with increasing charge density of the gel
[*Reprinted with permission from J. Gong, Y. Iwasaki, Y. Osada, et al., J. Phys. Chem. B, 103, 6001–6006, Copyright (1999) American Chemical Society*]

withstand the local pressure increases (Figure 15.10). Highly charged biomolecules occur in a series of motility mechanisms (e.g. actin and myosin in striated muscle) and it is expected that double layer repulsive forces contribute to a reduction in friction in these processes.

The time dependence of the *relaxation modulus* of cartilage can also be considered. When cartilage is sheared or longitudinally compressed/extended the material properties become time dependent due to the motion of fluid through the pores of the gel. Fluid motion through porous materials occurs in a range of biological situations, e.g. when blood plasma moves through blood clots. The problem has been solved by D'arcy in the case of a Newtonian fluid moving through an ideally porous material. Cartilage is a composite material (Figure 15.5); a rigid collagen scaffold combined with a dissipative proteoglycan matrix. The modulus of the composite is 10^5 times that of the concentrated

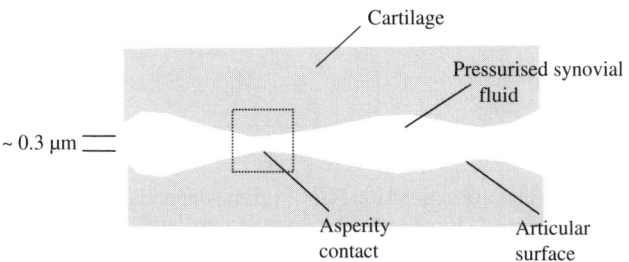

Figure 15.10 Microscopic contacts (asperities) could provide an important contribution to the frictional properties of cartilage

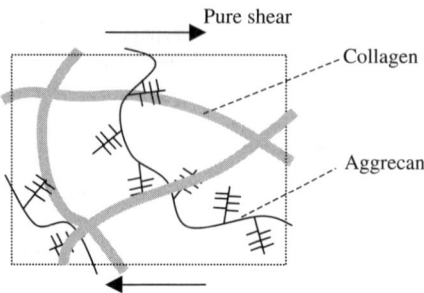

Figure 15.11 Schematic diagram of collagen and proteoglycans (aggrecan) in cartilage that experience a shear deformation

proteoglycan solutions that can be extracted from it, which implies the proteoglycans do not contribute to the shear stiffness of articular cartilage (Figure 15.11). The two major contributions to the shear stiffness are thus the cross-linked anisotropic collagen molecules and the flow of fluid through the network.

The stress relaxation curve as a function of time after a step change in shear strain is shown in Figure 15.12. Thus shearing the cartilage in the knee (for example picking up a heavy object) has a long time effect on the elasticity of the material. It can be shown that the relaxation time for a cross-linked gel is proportional to the mutual diffusion coefficient of the polymer gel, and using *D'Arcy's law* (equation (15.23)) it is possible to show that the slowest relaxation time (τ_1) in a polymer gel is given by:

$$\tau_1 = \frac{\delta \delta_{eq}}{\pi^2 E k} \quad (15.22)$$

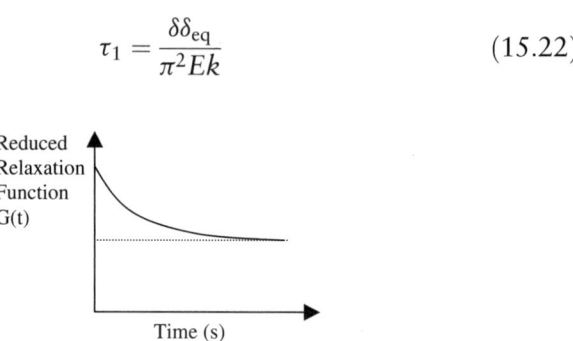

Figure 15.12 The relaxation function ($G(t)$) is time dependent with a typical relaxation time of the order of 30 minutes for human cartilage. This relaxation time is thought to be due to the dynamics of the water that moves through the pores of the gel [Ref.: V.C. Mow, C.S. Proctor and M.A. Kelly, Biomechanics of articular cartilage in 'Basic Biomechanics of Musculoskeletal System', Eds M. Nordin and V.H. Frenkel, 1989, Lippincott, Williams and Wilkins]

where k is the hydraulic permeability, δ is the compressed sample thickness, δ_{eq} free swelling sample thickness and E is Young's modulus. The relaxation time is thus inversely proportional to the ease with which water can move through the pores of the gel (k).

D'Arcy's law
The average fluid velocity (U) through a porous system is linearly related to the pressure gradient ($\partial P/\partial x$):

$$U = -k\frac{\partial P}{\partial x} \tag{15.23}$$

where k is the hydraulic permeability and x is the position in the system.

In tension the mechanical properties of cartilage are strongly *anisotropic*. Cartilage is stiffer and stronger in the standard direction of load, i.e. perpendicular to the surface. It exhibits viscoelastic behaviour in tension, which is attributed to both the internal friction associated with polymer motion and the flow of the interstitial fluid. A typical equilibrium tensile stress/strain curve for articular cartilage is shown in Figure 15.13. For small amounts of strain the collagen molecules are

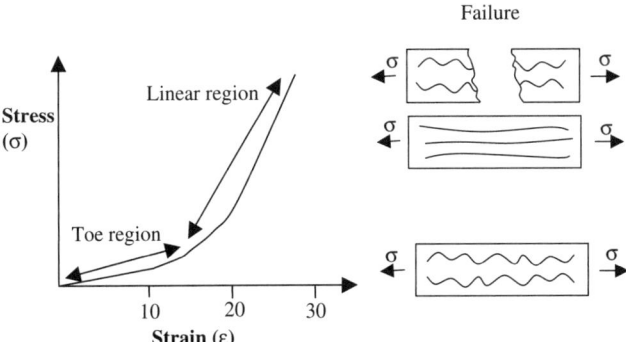

Figure 15.13 Typical stress/strain curve for the tensile properties of cartilage (The schematic behaviour of the collagen fibres under extension is also shown on the right. The toe region corresponds to the hookean elasticity of the collagen fibres. The linear region is due to straightening of the fibres which causes the Young's modulus to be increased. At high stresses/strains the cartilage fails and the collagen network is broken.)

slowly extended and reoriented in the toe region. Further extension straightens the collagen molecules, and finally the molecules break and fail at high strains.

15.2 SPIDER SILK

Spider silk is a classic example of a nanostructured polymer composite. Evolution has designed the silk to provide the spider with a structural material that is both super tough and strong. The silk can be rapidly produced by the spider to an external stimulus (e.g. to escape a predator) and is fabricated inside the spinneret (Figure 15.14). The silk protein is produced in a nematic liquid crystalline state inside the spider and is extruded into an orientated solid polymer with remarkable structural properties. The orientation of polymers is directly related to their tensile strength, as observed with synthetic analogues such as Kevlar, which is used in bullet proof jackets (a synthetic liquid crystalline polymer). The extreme toughness of spider silk can be observed in the large area contained under the stress/strain curve (Figure 15.15). This toughness is a key feature of the silk. It is five times greater than that of Kevlar. The spider can produce a wide range of silk materials (up to eight) whose mechanical properties are optimised for their different roles. The different mechanical properties found for two different types of spider silk are shown in Figure 15.16. Dragline silk is optimised for the maximum stress before fracture and catching silk has a high strain before fracture. The range of mechanical properties offered by spider silks encompasses both rubber-like and extremely rigid behaviour. Viscous silk is used in the glue covered spiral of the orb web, and rigid silk is used as a safety line when the spider moves around.

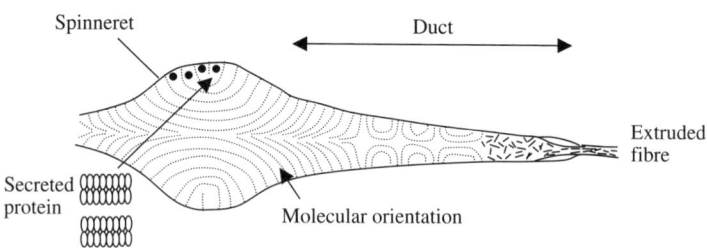

Figure 15.14 Schematic diagram of the structure of spider silk inside the spinneret. The proteins adopt a nematic liquid crystalline phase as they are extruded through the spinneret and solidify to form spider silk
[*Reprinted with permission from D.P. Knight and F. Vollrath, Phil. Trans. R. Soc. Lond. B, 357, 155–163, Copyright (2002) Royal Society of London*]

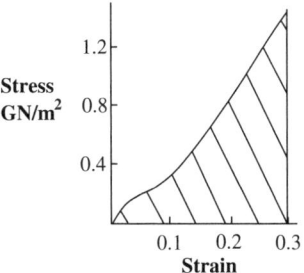

Figure 15.15 Stress/strain curve from a fibre of spider silk. The high toughness of the spider silk is indicated by the large area under the stress/strain curve
[*Reprinted with permission from Gosline, DeMont and Denny, Endeavour, 10, 1, 37–43, Copyright (1986) Elsevier*]

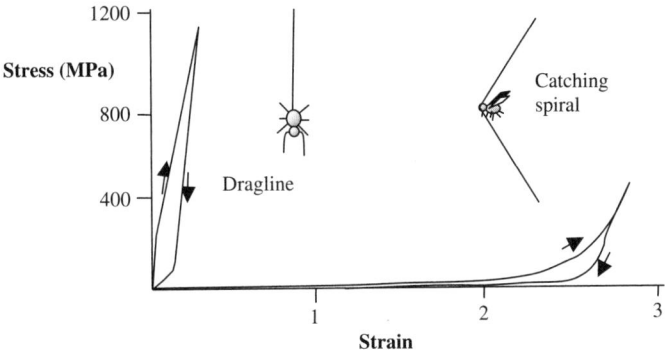

Figure 15.16 Comparison of the mechanical properties of dragline and catching spiral silk
(Dragline silk has a high stiffness, but is relatively inextensible. The opposite is true with the catching spiral silk. [*Reprinted with permission from J. Gosline, M. Lillie, E. Carrington et al., Phil. Trans. R. Soc. Lond. B, 357, 121–132, Copyright (2002) Royal Society of London*])

15.3 ELASTIN AND RESILIN

The key function of elastin and resilin is to provide a low stiffness, highly extensible, efficient elastic energy storage mechanism in animals. Elastin is a major component of arteries and allows them to adjust to pressure differences in blood flow. Elastin is frequently used as a shock absorbing material, e.g. in the necks of cows to cushion the motions of their heads. Resilin is also used for elastic energy storage in a series of roles in different animals such as the jumping mechanism in fleas and the hinges in the wings of dragon flies. In such mechanical roles a predominantly elastic response is required over a wide range of

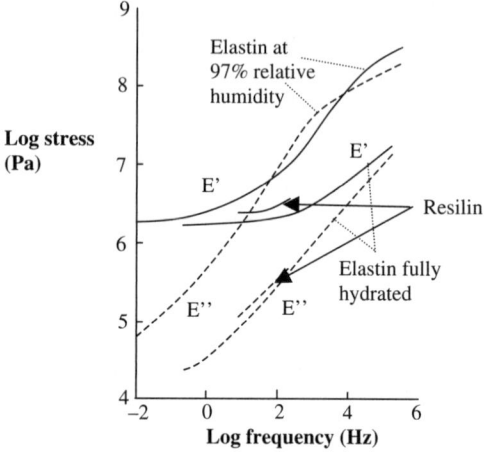

Figure 15.17 Complex Young's modulus for elastin and resilin as a function of frequency
(Both samples are predominantly elastic with $E' > E''$. Addition of water to elastin causes a decrease in its moduli. [*Reprinted with permission from J. Gosline, M. Lillie, E. Carrington et al., Phil. Trans. R. Soc. Lond. B, 357, 121–132, Copyright (2002) Royal Society of London*])

frequencies and both elastin and resilin are extremely well optimised elastomers. Structurally this implies flexible proteins strands are held between cross-links to provide entropic elasticity in these rubbery proteins (Section 8.3).

The stress–strain curves for elastin and resilin are shown in Figure 15.17. The *resilence* (R) is the fraction of work that is stored in a mechanically stressed system and can be calculated as:

$$R = e^{-2\pi\delta} \qquad (15.24)$$

where δ is the damping factor equal to the ratio, $\delta = E'/E''$, where E' is the storage Young's modulus and E'' is the dissipative Young's modulus. Both resilin and elastin are extremely resilient materials according to this measure.

A number of structural proteins such as resilin and spider silk can now be expressed using recombinant DNA technology and these materials could have a range of biomedical applications, e.g. replacement arteries. The synthetic processing of the genetically expressed proteins presents a number of challenges to provide well defined material properties, and this is currently a bottle neck that restricts the application of the technology.

BONE

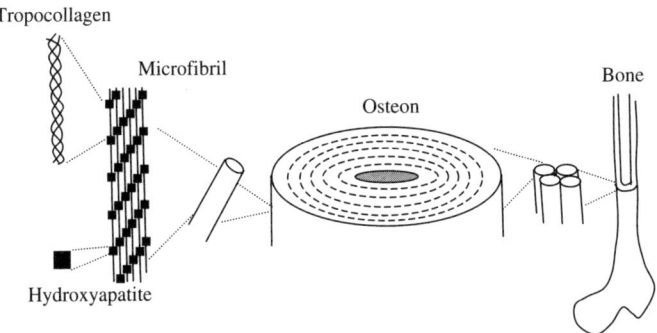

Figure 15.18 Schematic diagram of the morphology of bone. Tropocollagen forms a composite fibrous material with hydroxyapatite, which is incorporated into osteons and then into bone
[*Reprinted with permission from J. Vincent, Structural Biomaterials, Copyright (1990) Princeton University Press*]

15.4 BONE

Bone is a protein/inorganic crystalline composite material (Figure 15.18). Compact bone is similar in structure to nacre and is discussed in Section 15.6. Cancellous bone has a more porous structure. It is a cellular solid and is well optimised for strength and weight. At small strains the linear elastic response of isotropic cancellous bone is due to the elastic bending of the cell walls (Figure 15.19). At higher strains the cell walls fail by elastic buckling. This buckling plateau continues until the cell

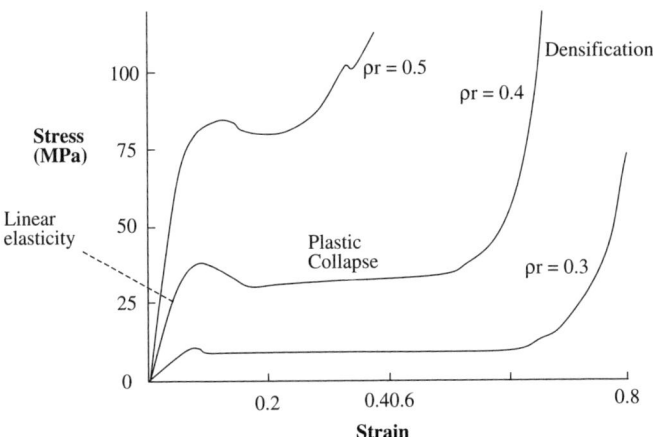

Figure 15.19 Compressive stress-strain curves for cancellous bone at a series of relative densities (ρ_r) of the open-celled foams
(Compare with Figure 11.9 [*Reprinted with permission from L.J. Gibson and M.F. Ashby, Cellular Solids, Copyright (1997) Cambridge University Press*])

walls meet and touch, which causes a large increase in the stress the material experiences as the strain increases. The modulus of the material is very sensitive to its degree of hydration, since the plasticisation of the adhesive proteins attached to the hydroxyapatite crystallites radically alters their mechanical properties.

15.5 ADHESIVE PROTEINS

Surface coatings of proteins play a crucial role in a number of biological scenarios. The proteins that attach molluscs onto rocks are important in the ship building industry, since molluscs adhere equally well to the hulls of boats as to the rocks (Figure 15.20). Nature has produced a well

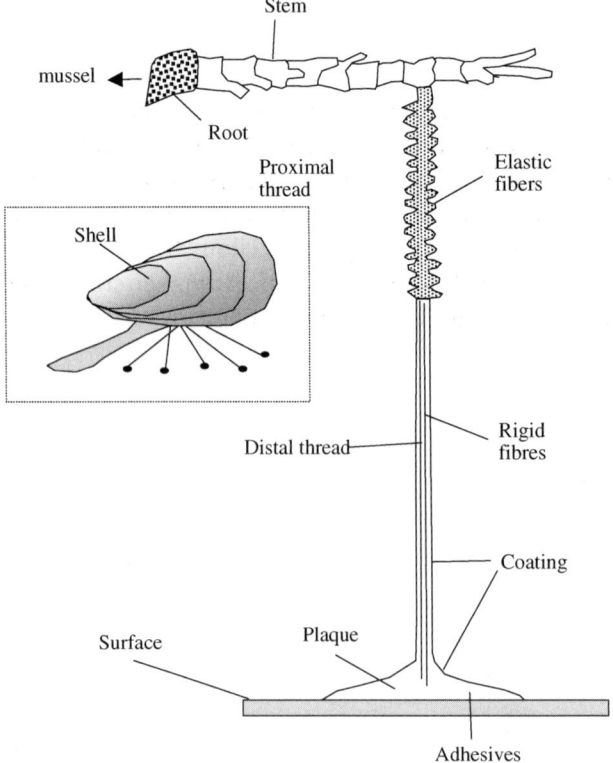

Figure 15.20 Schematic diagram of the process of adhesion of a mussel onto a surface, e.g. a rock or the hull of a boat
(Adhesive proteins displace water on the surface plaques and are attached to rigid fibres (distal threads) which connect the mussel securely to the surface [*Reprinted with permission from S.W. Taylor and J.H. Waite in Protein Based Materials, Eds K. McGrath and D. Kaplan, Copyright (1997) Birkhauser Boston Inc.*])

optimised adhesive that acts in a harsh hydrated environment. Often the primary function of the protein is to act as a sealant so the muscular foot of the organism can hold itself on to the rock with suction. Moluscs appear to adhere preferentially to high energy surfaces and the adhesive proteins are optimised to displace water at these surfaces. Glues are also thought to occur with the feet of starfish, but with most organisms that attach to surfaces mixed adhesive mechanisms are in effect. Van der Waals forces (e.g. Geckos), capillary forces (e.g. frogs) and micro hooks (e.g. plant burrs) are all thought to be important.

15.6 NACRE AND MINERAL COMPOSITES

Nacre and bone are both fibrous composite materials with hard nanocrystallites embedded in a compliant protein matrix (Figure 15.21). The stress distribution along the length of mineral crystals is assumed linear, so the maximum (σ_m) and average ($\overline{\sigma_m}$) tensile stress in the mineral component can be written as:

$$\sigma_m = \rho \tau_p \tag{15.25}$$

and:

$$\overline{\sigma_m} = \rho \tau_p / 2 \tag{15.26}$$

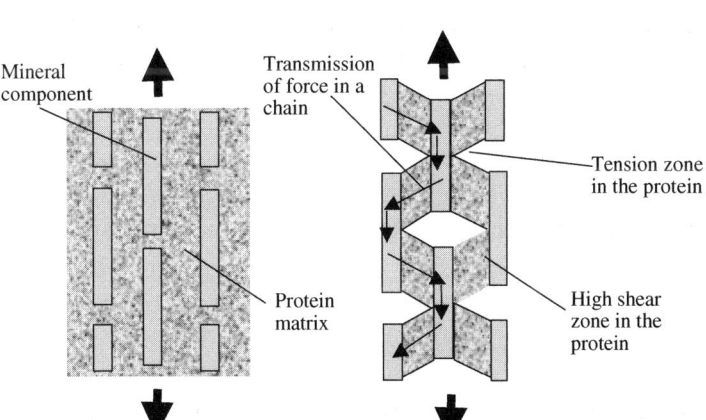

Figure 15.21 Schematic diagram of a generic biocomposite that occurs in materials such as cortical bone and nacre
((a) The mineral/protein composite is placed under stress. (b) The tensile stress is predominantly experienced by the mineral component and the protein shears under the stress. [Ref.: B. Ji and H. Gao, Journal of Mechanics and Physics of Solids, 2004, 52, 1963–1990])

where $\rho = L/h$ is the aspect ratio of mineral platelets (length L and width h) and τ_p is the shear stress of the proteins. It is assumed that the protein does not carry a tensile load and the effective tensile stress (σ) in the nanocrystalline composite is given by:

$$\sigma = \phi\overline{\sigma_m} \tag{15.27}$$

where ϕ is the volume fraction of the mineral and $\overline{\sigma_m}$ is the average stress in the mineral component. The effective strain (ε) in the composite when the mixture is stressed is the sum of that due to both the protein and mineral components:

$$\varepsilon = \frac{\Delta_m + 2\varepsilon_p h(1-\phi)/\phi}{L} \tag{15.28}$$

where h is the thickness and L is the length of the platelets. Δ_m and ε_p are the elongation of the mineral platelets and the shear strain of the proteins respectively. The elongation of the mineral platelets (Δ_m) is given by:

$$\Delta_m = \frac{\sigma_m L}{2 E_m} \tag{15.29}$$

where E_m is the Young's modulus of the mineral component. The shear strain in the protein (ε_p) is related to the shear modulus of the protein (G_p) and the shear stress (τ_p) that the protein experiences:

$$\varepsilon_p = \frac{\tau_p}{G_p} \tag{15.30}$$

The total effective Young's modulus (E) of the biocomposite is therefore:

$$\frac{1}{E} = \frac{4(1-\phi)}{G_p \phi^2 \rho^2} + \frac{1}{\phi E_m} \tag{15.31}$$

and it is concluded that reasonably high Young's moduli are possible with such biocomposites.

In addition to the rigidity, the *toughness* of biocomposites is very well optimised in nacres. Nanoscale mineral inclusions have less flaws than the macroscopic equivalents and their strength approachs that of the atomic bonds between the crystalline atoms. The viscoelasticity of the proteins that adhere to the crystallites helps the material to dissipate fracture energy, which ensures that large cracks do not occur in their mineral composites. Biocomposites achieve a high stiffness due to the large aspect ratio of their crystallite inclusions and the nanotextured staggered alignment of the nanocrystallites.

FURTHER READING

J. Vincent, *Structural Biomaterials*, Princeton, 1990. Extremely well written compact introduction to biological materials.
S. Vogel, *Comparative Biomechanics: Life's physical world*, Princeton, 2003. An exhaustive range of biomechanical examples are given.
J.P. Gong, *Surface friction of polymer gels*, Prog. Polym.Sci, 27, 2002, 3–38.
L.J. Gibson and M.F. Ashby, *Cellular Solids*, Cambridge University Press, 1997. Classic text on solid foams.
J. Gosline, M. Lillie, E. Carrington et al. *Elastic proteins: biological roles and mechanical properties*' Phil. Trans. R. Soc. Lond. B, 2002, 357, 121–132.

TUTORIAL QUESTIONS

15.1) Estimate the shear modulus of an adhesive protein in nacre if the Young's modulus of a shell is 12.4 MPa, the crystalline volume fraction is 0.95, the aspect ratio of the crystallites is 10 and the shear modulus of the crystallites is 3 GPa.

15.2) A man picks up a piano and the sections of cartilage in his knees are compressed from 1 cm to 0.95 cm. What is the relaxation time of the cartilage gels once the weight is removed if the Young's modulus of cartilage is 0.78×10^6 Pa and the hydraulic permeability is $6 \times 10^{-13} m^4 N^{-1} s^{-1}$?

16
Phase Behaviour of DNA

The in vivo behaviour of DNA presents a wide range of fascinating phenomena with respect to the molecule's structure, dynamics, and phase transitions.

16.1 CHROMATIN – NATURALLY PACKAGED DNA CHAINS

The method through which DNA is packaged into the nucleus of a cell has posed evolution an interesting problem. In a human cell the DNA is a long narrow thread 1.5 m in length and 2 nm in diameter which needs to be accommodated into a box whose volume is only a few microns cubed. The solution that nature has evolved is to have the DNA stored in a compacted form with the chains wrapped around proteins spools (histones); much like cotton is wound around bobbins in needle work. The DNA spools are then assembled into fibrous aggregates which are called chromosomes.

The first strong experimental evidence for histones was presented by Hewish and Burgoyne in 1973. They found that the majority of chromosomal DNA, when digested by a DNA cutting enzyme, formed small fragments of regular size 200, 400 and 600 base pairs (using gel electrophoresis). The explanation for this phenomenon was that the DNA binding proteins (histones) are arranged in a regular manner and only DNA between the histones could be cut by the enzymes which sets the fundamental length of the fragments.

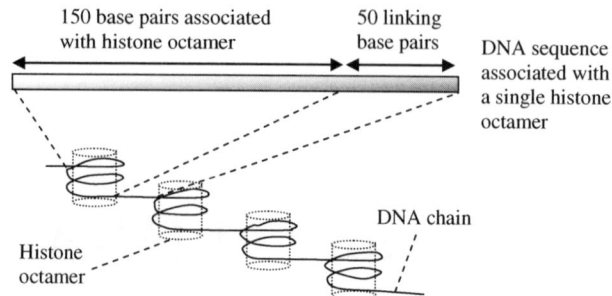

Figure 16.1 Schematic diagram of the specific association of the histone octamer with sections of DNA with well defined lengths

The method by which the nucleosomes associate with DNA chains poses many questions and the physicochemical processes driving the self-assembly of the histones on to a specific sequence of base pairs are still incompletely understood. A constant length of ~ 150 base pairs of DNA is thought to be associated with a single histone (Figure 16.1).

Wide angle X-ray and neutron diffraction experiments have examined the specific interactions at the molecular level between small fragments of DNA and histone proteins (eight histone protein subunits are required to form a single histone bobbin). Accurate molecular models have thus been made of small crystalline sections of DNA with histone octamers (Figure 16.2). Unfortunately, these techniques are not feasible with non-crystalline samples such as complete chromosomal fibres. Larger lengths of DNA chain could have markedly different conformations with the histones due to their altered elasticity, torsional resistance and counterion condensation effects. Separate experimental evidence is therefore required.

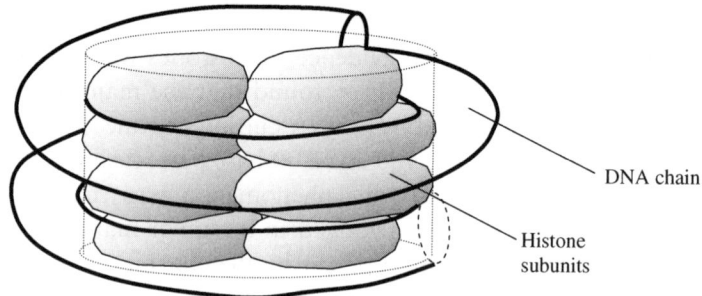

Figure 16.2 Complexation of a histone octamer (eight subunits) with a small DNA chain fragment. The structure is based on both wide angle X-ray and neutron scattering experiments on crystalline chromosomal fragments
[*Reprinted with permission from G. Arents and E.N. Moudrianakis, Proc. Nat. Acad. Sci. USA, 90, 10489–10493, Copyright (1993) National Academy of Sciences*]

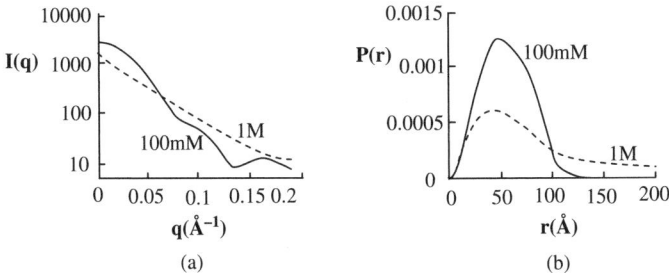

Figure 16.3 (a) Small-angle neutron scattering data (scattering intensity versus momentum transfer) from histone/DNA complexes allows (b) the solution state structure of histone/DNA complexes to be modelled with and without salt. The radial distribution function ($P(r)$) is calculated for the DNA/histone complex at two salt concentrations. A much more expanded structure is observed at high salt concentrations as the DNA chains start to unbind from the histone octamers (the electrostatic binding force decreases due to the increased screening)
[Ref.: S. Mangenot, A. Leforestier, P. Vachette et al., Biophysical Journal, 2002, 82, 345–356]

Molecular models of small-angle X-ray and neutron scattering data have extended the resolution of histone structures when combined with longer DNA chains in solution to much larger length scales (nm) (Figure 16.3) Careful modelling of the liquid state scattering data is required, but good evidence for compact and extended forms of the complexes has been found, dependent on the amount of salt in the solutions. This indicates that there is a process of electrostatic binding between the negatively charged DNA and the positively charged protein spools. However, these scattering techniques only provide general features of the chromosomal structure and little detailed information is available.

There still exist a number of questions relating to the ambient large scale structure (10 nm) of chromatin and the self-assembly of this morphology, due to the non-crystallinity and aperiodicity of the samples. Valuable information has been produced by electron microscopy, a technique that under suitable conditions can provide Angstrom level resolution of aperiodic materials. Tomographic reconstruction of transmission electron micrographs led Aron Klug and co-workers to propose the 'beads on a string model' in 1977 (Figure 16.4). This model is the result of tomographic reconstruction of a stack of transmission electron microscopy images of freeze fractured chromosomal fibres that contain a staining agent, which provides strong contrast for electron scattering. The mathematical reconstruction technique implemented to analyse the images requires careful handling to produce dependable results; the inclusion of a staining agent and the process of freeze fracture during sample preparation could both have radically affected the morphology of the chromosomal fibres. Thus

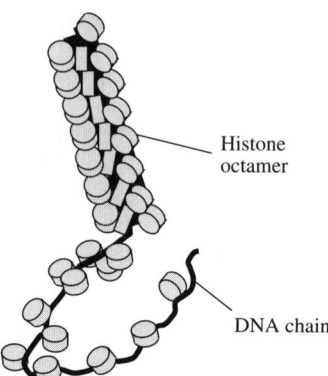

Figure 16.4 Model for chromosomal fibres based on the tomographic reconstruction of transmission electron microscopy images from freeze fractured fibres (The chromosomes consist of histone octamers assembled onto a long thread of DNA, in a beads on a string manner [*Ref.: F. Thoma, T. Koller and A. Klug, Journal of Cell Biology, 1979, 83, 403–427*])

there continues to be a degree of controversy in the field of chromosomal structural analysis with respect to the orientation of the histones along the axis of the chromosomal fibre, but all the current evidence points to the existence of tightly bundled fibres formed from beads on a string and that these fibres occur in vivo in the nucleus.

In contrast to human DNA, the length of bacterial DNA is much smaller and is not associated with histones. However, to make the chains fit inside the cell nucleus compaction is still required. Over twists are introduced into the circular DNA chains and the chains form compact plectonomic structures (Chapter 8). Circular DNA associated with topology preserving proteins are said to be 'restrained'. Nicks in 'restrained' DNA do not cause the super coil to unwind into the relaxed state and can be repaired with no loss in the degree of winding. The torsional energy stored in a restrained duplex DNA chain is therefore conserved.

16.2 DNA COMPACTION – AN EXAMPLE OF POLYELECTROLYTE COMPLEXATION

Many diseases have a genetic origin, the most important example being cancer in its many different forms. A possible strategy for treating these conditions is to replace the malfunctioning DNA in malignant cells with a benign substitute. An obstacle to this strategy of *gene therapy* is how to transfer the material to the nucleus of a cell without it being destroyed by the cell's defence mechanisms. One reasonably successful method of transferring the DNA is to combine it with an oppositely charged polyelectrolyte or a virus (the drawback is the virus can itself prove to be pathogenic) to allow transfer through the cell wall. These questions on DNA transvection, combined with more fundamental problems concerning the natural functioning of chromosomes, provide motivation for the understanding of DNA compaction with oppositely charged colloidal spheres.

The persistence length of a DNA chain is about 500 Å under standard physiological conditions. The persistence length of the uncomplexed DNA can be calculated theoretically and is thought to be a combination of the effects of counterion condensation, the electrostatic repulsion of liked charged segments and the intrinsic rigidity of the polymer backbone (the helix acts as an elastic rod). From Section 9.9 the total persistence length (l_T) of a charged polymer is given by equation (9.88). The charge density (Q) along the chain is limited by charge condensation. The distance between the charged phosphate groups along the backbone of a DNA chain is 1.7 Å, the Bjerrum length is 7 Å at 20°C and the effective charge fraction (ξ) in the Manning charge condensation model (equation (9.83)) is therefore $1.7/7 = 0.24$. The OSF (Odjik, Skolnick and Fixman) calculation given by equation (9.88) provides the correct order of magnitude for the persistence length when compared with experiment. The total persistence length can be separated into the contribution of the intrinsic persistence length ($L_P = 30$ nm) and the charge repulsion of the phosphate groups. The charged contribution to the persistence length is therefore 20 nm at a 0.1 M monovalent salt concentration.

The effect of chirality on the resultant chromosomal morphology is a further question that should be considered with twist storing polymers such as DNA. An additional term can be introduced into the free energy of semi-flexible chains (equation (8.19)) that corresponds to the propensity for torsional rotation. Theoretical studies imply that the chirality of nucleosome fibres is due to the specific histone/DNA potential and not due to the intrinsic twist/bend interaction of the DNA fibres, although chirality does provide a small contribution to the fluctuations and elasticity of the naked uncomplexed DNA chains.

Chromosomal DNA is wrapped around the cylindrical histone core on a helical path of a diameter (D) of 110 Å (Figure 16.2). This is smaller than the intrinsic persistence length of the DNA chain, so substantial elastic energy is stored upon complexation. The origin of the attraction between DNA and the histone is electrostatic but can be considered short range at physiological salt concentrations. The Debye screening length for the electrostatic interaction at physiological salt concentration is around 10 Å and is a first approximation for the length of the DNA/histone electrostatic interaction. The binding energy per unit length of the histone can be estimated through the assumption that the wrapped state represents a dynamic equilibrium in which wrapped portions of the DNA strand spend part of their time in the dissociated state. With this assumption, the binding energy (λ) of a DNA duplex to a histone is found to be 1–2 kT per 10 base pairs. Association experiments between DNA and histones find an a thermal first order phase transition as a function of the DNA histone interaction strength from a wrapped state to a dissociated state consistent with the idea of an all-or-none wrapping transition, e.g. changing the salt concentration induces a first order unwrapping transition. This implies that the powerful thermodynamics ideas that determine the behaviour of phase transitions (Chapter 3) can be applied to the complexation of DNA.

A phase diagram for the complexation between an idealised positively charged sphere and a DNA chain in aqueous solution is shown in Figure 16.5. DNA chains can be touching, have point contacts, or be wrapped onto the charged spheres. The particular phase adopted depends on the electrostatic screening length and the amount of charge on the sphere. Molecular dynamics simulations observe similar phenomena and the effect of the curvature of the spheres can be probed. A *wrapping transition* is predicted as the diameter of the positive sphere is increased using both analytic theory and Monte Carlo simulation.

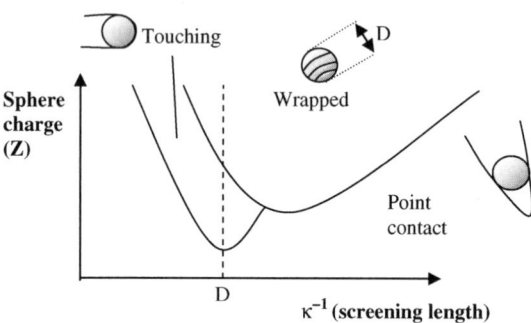

Figure 16.5 Theoretical phase diagram for the statistics of single DNA chains in contact with oppositely charged spheres
(The diagram shows the state of the DNA/sphere complexes as a function of the charge per sphere (Z) and the Debye screening length (κ^{-1}) of the solution. It is assumed that the process of binding is purely electrostatic and there is no chemical specificity. First order phase transitions are predicted between point contact, touching, and wrapped states. D is the diameter of the colloid [*Reprinted with permission from R.R. Netz, J.F. Joanny, Macromolecules, 32, 9026–9040, Copyright (1999) American Chemical Society*])

A novel counter intuitive physical phenomenon that occurs in polyelectrolyte complexation is that of overcharging. It can occur with both the complexation of colloids (e.g. histones, amine derivatised spheres etc.) with oppositely charged polymers (e.g. DNA, polystyrene sulphonate etc.) and polyelectrolytes with oppositely charged planar surfaces. More polyelectrolyte is adsorbed than required for simple charge neutralisation and the charge on the adsorbing surface is reversed. With DNA–histone complexes the charge on the DNA greatly out weighs that of the histone. This results in chromosmal complexes being strongly negatively charged. The thermodynamic origin of the effect is thought to be related to counterion condensation. The additional contribution to the electrostatic energy of the overcharged complex to the free energy of the system is compensated for by the additional entropy of the released counterions.

16.3 FACILITATED DIFFUSION

There are a range of enzymes that bind to DNA. These include enzymes that initiate transcription, RNA polymerase and endonucleases that chemically modify the DNA sequence. The rates of reaction for the DNA binding proteins are significantly faster than would be expected from calculations that assume three dimensional diffusion to reaction (Section 5.5). Indeed, with the classic example of the lac repressor the degree of association of the repressor protein for the DNA chain is underestimated by a factor of a hundred. A number of models of facilitated diffusion have been proposed to account for this shortfall. The three primary scenarios for the interaction between the DNA and the lac protein are thought to be *sliding* of the protein along the DNA, *hopping* of the protein from site to site along the DNA chain and *intersegmental transfer* of the proteins between multiple binding sites (Figure 16.6). An important clue in accounting for the increased binding rates is that the efficiency of a collision as a result of diffusion is increased when the number of dimensions in which the diffusion occurs is reduced. The protein can be compelled to execute a one dimensional random walk along a DNA chain and this greatly decreases the time taken to search the whole sequence for the correct binding site. Furthermore, the probability of a collision between the protein and the DNA is much higher when the protein is confined inside the DNA coil.

From the Stoke's–Einstein equation (5.11) the diffusion coefficient for a small globular enzyme (diameter 5 nm) in three dimensions is

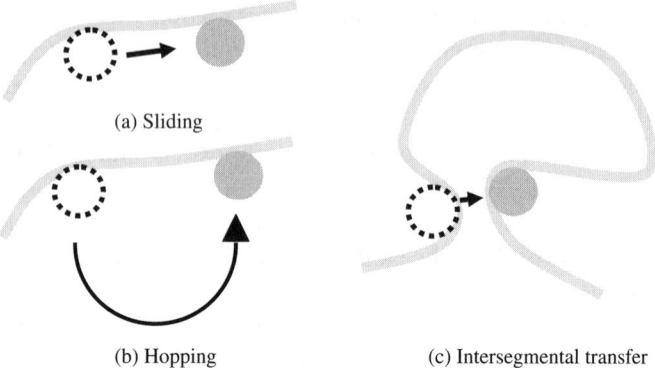

Figure 16.6 Schematic diagram of the three principle models for facilitated diffusion of a protein with respect to a DNA chain

approximately $10^8 \, \text{nm}^2\text{s}^{-1}$. It is assumed that the enzyme diffuses in three dimensions before the reaction and the DNA chain moves much more slowly that the associating protein due to its size. The association rate constant (k, equation (5.54)) is given by:

$$k = 4\pi D a \qquad (16.1)$$

where a is the size of the binding site on the protein and D is the diffusion coefficient for the protein. Typically the size of the binding site (a) is much smaller than the size of the protein, $a/d \sim 0.1$, where d is the diameter of the protein. Using equation (16.1) the value of the rate constant (k) is found to be $10^8 \, \text{Ms}^{-1}$. However experimentally the Lac repressor is found to have a rate constant greater than $1 \times 10^{10} \, \text{Ms}^{-1}$; the mechanism of facilitated diffusion is invoked to explain the shortfall.

Both sliding of the protein along the DNA and 'hopping' through three dimensions would increase the association rate in models for facilitated diffusion. The combined effect of the two processes is used for a quantitative explanation of the shortfall in association constants. The probability that a protein sliding on a DNA chain stays on the chain after N steps is:

$$(1 - P)^N = e^{N \ln(1-P)} \qquad (16.2)$$

where P is the probability of dissociation and the equality is from a simple mathematical manipulation. The process of protein dissociation is another example of a Poisson decay process (Section 5.3) where the probability is proportional to $e^{-\mu}$ and μ is the mean. The expectation value of one decay event is the probability of the Poisson process when the mean equals one ($\mu = 1$) and the number of steps over which sliding (N) can occur is therefore given by:

$$N = \frac{-1}{\ln(1-P)} \quad (16.3)$$

In the case of a very low probability of dissociation:

$$P \ll 1 \quad \ln(1-P) \approx -P \quad (16.4)$$

And, therefore:

$$N = 1/P \quad (16.5)$$

The characteristic sliding length (l_{sl}) explored by one dimensional diffusive sliding ($\langle l_{sl}^2 \rangle = 2h^2 N$, equation (5.7)) is then:

$$l_{sl} = \frac{\sqrt{2}h}{\sqrt{P}} \quad (16.6)$$

where h is the base pair step sliding length (Figure 16.7). From the definition of diffusion in one dimension (equation (5.42)) the characteristic

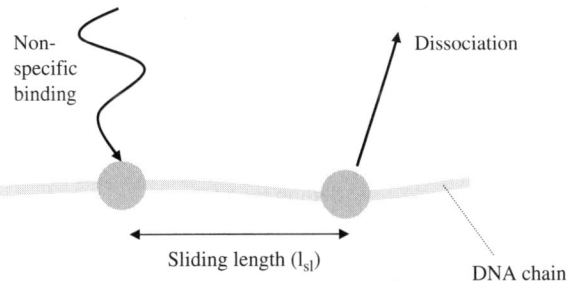

Figure 16.7 The definition of the sliding length (l_{sl}) used in the process of facilitated diffusion. The protein associates with the DNA chain by non-specific binding, it moves the sliding length along the chain and then dissociates from the chain

time (τ_{sl}) for exploring the sliding length (l_{sl}) is:

$$\tau_{sl} = \frac{l_{sl}^2}{D_1} \quad (16.7)$$

Therefore, it is concluded that very small dissociation constants (P) are required for long sliding lengths (l_{sl}, equation (16.6)), which consequently allow significant increases to occur in the reaction rates (equation (16.7)). The combined effect of both hopping and sliding motions (Figure 16.8) on the reaction rates can be included in a model and the prediction for the total reaction rate per unit length (k) is given by:

$$k = Da\left(\frac{a}{l_{sl}} + \frac{D}{D_1}aLl_{sl}c\right)^{-1} \quad (16.8)$$

where L is the total contour length of the DNA chain, a is the size of the binding site, D is the diffusion coefficient in three dimensions of the binding proteins, D_1 is the diffusion coefficient related to sliding of the proteins along the DNA chain and c is the concentration of the target DNA. The non-monotonic behaviour of the rate constant on the ionic strength predicted theoretically by such models has been observed experimentally and is a major success of the theory.

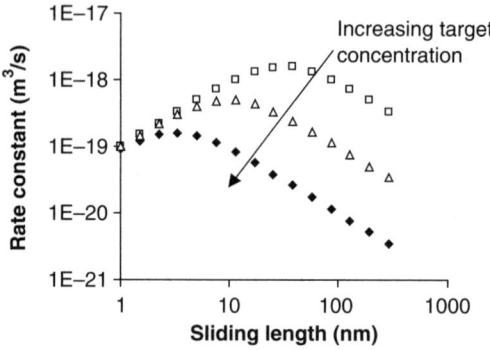

Figure 16.8 The predicted association constants of a protein (k/Da) with a DNA chain as a function of the sliding length (l_{sl})
(Three possible solutions are shown corresponding to low target concentrations and high target concentrations. The optimal sliding length depends on the target concentration. The sliding length is rescaled by the diameter of the binding site (a) in the plot [*Ref.: S.E. Halford and J.F. Marko, Nucleic Acids Research, 2004, 32, 10, 3040–3052*])

FURTHER READING

C.R. Calladine and H.R. Drew, *Understanding DNA: The Molecules and How it Works*, Academic Press, 1997. Extremely useful introductory account of the behaviour of DNA from a structural engineering point of view. Requires a minimum of mathematical ability.

S.E. Halford and J.F. Marko, *How do site specific DNA binding proteins find their targets?*, Nucleic Acid Research, 32, 10, 3040–3052. A clear explanation of the processes involved in DNA binding.

Appendix

PHYSICAL CONSTANTS

Boltzmann's constant	$k_B = 1.38 \times 10^{-23}$ JK^{-1}
Ideal gas constant	$R = N_A k_B = 8.314$ J/molK
Thermal energy at 295 K is	4.1 pNnm $= 4.1 \times 10^{-21}$ J
Electronic charge	$e = 1.6 \times 10^{-19}$ Coul
Permittivity of free space	$\varepsilon_0 = 8.9 \times 10^{-12}$ Coul^{2}N^{-1}m^2
Permittivity of water	$\varepsilon \approx 80\, \varepsilon_0$
Avogadro's number	$N_A = 6 \times 10^{23}$
Bjerrum length (room temperature)	7 Å
Debye screening length	
1:1 electrolytes e.g. NaCl	$\kappa^{-1} = \dfrac{0.304 \text{ nm}}{\sqrt{[\text{NaCl}]}}$
2:1 electrolytes e.g. CaCl$_2$	$\kappa^{-1} = \dfrac{0.176 \text{ nm}}{\sqrt{[\text{CaCl}_2]}}$
Viscosity of water	$\eta = 10^{-3}$ Pa.s
Planck's constant	$h = 6.626 \times 10^{-34}$ Js
Speed of light	$c = 2.998 \times 10^{8}$ ms^{-1}
Units of pressure	1 atmosphere $= 1.01 \times 10^5$ Pa

Applied Biophysics: A Molecular Approach for Physical Scientists   Tom A. Waigh
© 2007 John Wiley & Sons, Ltd

Appendix

Answers to Tutorial Questions

CHAPTER 1

1.1)
The length of the different forms of helix are as follows

$$\text{length}_A = \frac{4 \times 10^8}{660} \times 2.6\,\text{Å} = 158\,\mu\text{m}$$

$$\text{length}_B = \frac{4 \times 10^8}{660} \times 3.4\,\text{Å} = 206\,\mu\text{m}$$

$$\text{length}_Z = \frac{4 \times 10^8}{660} \times 3.7\,\text{Å} = 224\,\mu\text{m}$$

1.2)
The membrane has a fluid-like bilayer structure. The protein will arrange itself through the membrane, so that its hydrophobic and hydrophilic regions are correctly positioned. See Figure 6.3.

1.3)
The pH can be calculated using the definition of the equilibrium constant (K_a)

$$K_a = \frac{[H^+][Arg^-]}{[HArg]}$$

$$[H^+] \approx [Arg^-]$$

$$K_a \approx \frac{[H^+]^2}{[HArg]}$$

$$[H^+] = K_a^{\frac{1}{2}}[HArg]^{\frac{1}{2}}$$

$$pH = \tfrac{1}{2}pK_a - \tfrac{1}{2}\log[HArg] = 6.60$$

1.4)
The reader is referred to a good biochemistry textbook. Metals occur in small quantities in a wide range of biological molecules, e.g.:

- Iron in haemoglobin (oxygen transport) and ferritin (storage)
- Magnesium – hexokinases (ATP production) and chlorophyll (green plant photosynthesis)
- Calcium – prothrombin (blood clots) and troponin (muscular contraction)

CHAPTER 2

2.1)
Need to differentiate the potential with respect to the displacement to give the force:

$$F(r) = \varepsilon \left(\frac{r_0}{r}\right)^{-13}(-12)r_0 - 2\varepsilon r_0 \left(\frac{r_0}{r}\right)^{7}(-6)$$

$$F(r) = -12\varepsilon r_0 \left(\frac{r_0}{r}\right)^{-13} + 12\varepsilon r_0 \left(\frac{r_0}{r}\right)^{7}$$

The attractive component of the force is negligible when $r = r_0/2$

$$F(r_0/2) = 4.06 \times 10^{-25} \text{ N}$$

CHAPTER 2

2.2)
Formulae for the quick calculation of the Debye screening length are included in the Appendix.

For monovalent salts 0.001 M $\kappa^{-1} = \dfrac{0.304}{\sqrt{0.001}} = 9.6\,\text{nm}$

Similarly 0.01 M $\quad \kappa^{-1} = 3\,\text{nm}$
0.1 M $\quad \kappa^{-1} = 1\,\text{nm}$
1 M $\quad \kappa^{-1} = 0.3\,\text{nm}$

The concentration of ions from spontaneous dissociation in pure water is:

$$[H^+] = [OH^-] = 1 \times 10^{-7}\,\text{M}$$
$$\kappa^{-1} = 0.304/[H^+]$$

Thus the screening length of pure water is 3×10^6 nm, i.e. 3 mm, due to spontaneous dissociation! The length scale of the electrostatic interaction between molecules in water can thus be a million times bigger than their diameter.

For divalent salts 0.001 M $\kappa^{-1} = \dfrac{0.176}{\sqrt{0.001}} = 5.6\,\text{nm}$

Similarly 0.01 M $\quad \kappa^{-1} = 1.76\,\text{nm}$
0.1 M $\quad \kappa^{-1} = 0.56\,\text{nm}$
1 M $\quad \kappa^{-1} = 0.176\,\text{nm}$

A rough estimate for the equivalent salt concentration for physiological conditions is 0.1 M monovalent salt, and thus $\kappa^{-1} \approx 1$ nm.

2.3)
For a sphere the potential follows the form $\psi(r) \sim 1/r$ whereas for a cylinder $\psi(r) \sim \ln(r)$. The electrostatic potential thus decreases more quickly with distance (r) from a sphere. Close to a plane surface $\psi(r) \sim r$. The laws for the potential can all be derived from Gauss's law.

2.4)
For the steric interaction the potential takes the form:

$$w(r) \sim e^{-r/R_g}$$

for the screened electrostatic interaction the potential takes the form:

$$v(r) \sim e^{-\kappa r}$$

Assuming the prefactors are of a similar order of magnitude the steric forces become significant when $\kappa^{-1} < R_g$ and $r_{\text{sep}} < R_g$, where r_{sep} is the separation distance

CHAPTER 3

3.1)
The cooperativity of the phase transition increases with the length of the helix and thus tends to sharpen the DSC endotherm, i.e. the helix–coil transition occurs over a narrower range of temperatures.

3.2)
Hysteresis behaviour has been observed for long polymeric chains as the quality of the solvent is reduced and the chain size is increased, e.g. the size of the globular chains depends on the route by which they are globularised.

3.3)
The enthalpy change is:

$$\Delta H_m = \frac{2\gamma_{sl}T_m}{r \times \Delta T} = \frac{2 \times 1.2 \times 10^{-3} \times 323}{50 \times 10^{-9} \times 1} = 15 \text{ MJkg}$$

CHAPTER 4

4.1)
P_2 is the orientational order parameter, ψ is the lamellar order parameter and h is the helical order parameter. During heating it is possible that:

(Wet self-assembled) $P_2 > 0 \quad \psi > 0 \quad h > 0 \quad \rightarrow \quad$ (Gelatinised) $P_2 = 0 \quad \psi = 0 \quad h = 0$

(Wet self-assembled) $P_2 > 0 \quad \psi > 0 \quad h > 0 \quad \rightarrow \quad P_2 > 0 \quad \psi = 0 \quad h > 0 \quad \rightarrow$ (Gelatinised) $P_2 = 0 \quad \psi = 0 \quad h = 0$ as shown in the figure

Also the state of self-assembly can be modified through the addition of water:

(Dry unassembled) $P_2 > 0 \quad \psi = 0 \quad h > 0 \quad \rightarrow \quad$ (Wet self-assembled) $P_2 > 0 \quad \psi > 0 \quad h > 0$

Therefore there is a third possibility for heat treatment:

(Dry unassembled) $P_2 > 0 \quad \psi = 0 \quad h > 0 \rightarrow$ (Gelatinised) $P_2 = 0 \quad \psi = 0 \quad h = 0$

The process of staling can also be parameterised in a similar manner:
(Gelatinised) $P_2 = 0$ $\psi = 0$ $h = 0$ $\rightarrow$ (Stale) $P_2 = 0$ $\psi = 0$ $h > 0$

Steric constraints introduce a strong coupling between the orientation of the mesogens and the degree of mesogen helicity.

4.2)
The orientational order parameter can be calculated as:

$$\langle \cos^2 \theta \rangle = \frac{\int_0^\pi \cos^2 \theta P(\theta) d\theta}{\int_0^\pi P(\theta) d\theta}$$

$$= \frac{2}{\pi} \int_{\pi/4}^{3\pi/4} \cos^2 \theta d\theta = \frac{1}{\pi} \int_{\pi/4}^{3\pi/4} 1 + \cos 2\theta d\theta = \frac{1}{2}$$

$$P_2(\cos \theta) = \tfrac{3}{2}\cos^2 \theta - \tfrac{1}{2} = \tfrac{1}{4}$$

4.3)
The Onsager calculation for the nematic/isotropic transition gives:

$$\phi < 3.34 \frac{D}{L} = 0.167 \quad \text{i.e. 16.7\% volume fraction}$$

4.4)
The entropy of the side-chains is antagonistic to the entropy of the backbone. This increases the rigidity of the back bone chain and can induce a nematic order parameter in the backbone.

CHAPTER 5

5.1)

$$\text{Re} = \frac{2vL\rho}{\eta} = \frac{2 \times 10^{-1} \times 10^{-3} \times 1.3}{1.8 \times 10^{-5}} = 14.4$$

The Reynold's number is not small, so inertial forces could be quite considerable.

5.2)
From the definition of 1-dimensional diffusion:

$$\langle x^2 \rangle = 2Dt$$

Rearranging for the characteristic time gives:

$$t = \frac{\langle x^2 \rangle}{2D} = \frac{(2.7 \times 10^{-3})^2}{2 \times 1.35 \times 10^{-9}} = 2.7 \times 10^3 \text{ s}$$

This mechanism is far too slow.

5.3)
The rotational diffusion coefficient (D_θ) is:

$$D_\theta = \frac{kT}{8\pi\eta a^3} = \frac{4.1 \times 10^{-21}}{8 \times 3.142 \times 0.001 \times (2 \times 10^{-6})^2} = 2.04 \times 10^{-2} \text{ rad s}^{-1}$$

$$\langle \theta^2 \rangle = 2D_\theta t$$

So the characteristic time for fluctuations of 90° ($\pi/2$) is $t = 60$ seconds

In 3 dimensions:

$$\langle r^2 \rangle = 6Dt$$

$$\langle \theta^2 \rangle = 6D_\theta t_\theta$$

And:

$$(2\pi)^2 = 6D_\theta t_\theta$$

$$(2\pi a)^2 = 6Dt$$

Therefore substituting expressions for D_θ and D we have:

$$t_\theta = t^{\frac{4}{3}}$$

t_θ is the characteristic time for rotation through $2\pi a$, and t is the time for translation by $2\pi a$.

5.4)
For the motile particle the diffusion coefficient (D) is:

$$D = \frac{v^2 \tau}{3(1-\alpha)}$$

The average value of the cosine (α) is $\frac{11}{12}$, i.e. there is 23.6° between successive runs.

CHAPTER 6

6.1)

$$\alpha = \frac{4\pi R^2 \gamma}{kT} = \frac{4 \times 3.142 \times (2 \times 10^{-9})^2 \times 20 \times 10^{-3}}{4.1 \times 10^{-21}} = 245$$

$$CMC \approx e^{-\alpha/N^{\frac{1}{3}}} = e^{-245/(10000)^{\frac{1}{3}}} = 1.15 \times 10^{-5} \, M$$

6.2)
The critical concentration (cc) for self-assembly is:

$$c_c = K = e^{\Delta G_0/kT} = 0.01 \, M$$

The average degree of filament polymerisation (n_{av}) is:

$$n_{av} = \sqrt{\frac{c_t}{K}} = 10$$

The average filament length ($l_{av} = n_{av} \times monomer\ length$) is:

$$l_{av} = 10 \times 5 = 50 \, nm$$

Fairly short filaments are formed even at high monomer concentrations.

CHAPTER 7

7.1)
The contact angle can be calculated from the Young–Laplace equation:

$$\gamma_{sg} = \gamma_{sl} + \gamma_{lg}\cos\theta$$
$$18 = 73.2 + 72\cos\theta$$
$$\theta = 140°$$

The wetting coefficient

$$k = \frac{18 - 73.2}{72} = -0.77$$

The surface is unwetted which provides a useful self-cleaning mechanism for the lotus leaf.

CHAPTER 8

8.1)
The cross-linking density (v) is linearly related to the Young's modulus (E) for flexible rubbery networks:

$$E = 3kTv$$

For elastin $\quad v = \dfrac{1 \times 10^6}{3 \times 4.1 \times 10^{-21}} = 8.13 \times 10^{25}\ \text{m}^{-3}$

For collagen $\quad v = \dfrac{1 \times 10^9}{3 \times 4.1 \times 10^{-21}} = 8.13 \times 10^{28}\ \text{m}^{-3}$

The collagen chains are semi-flexible and are thus not well described by a purely flexible model for the elasticity.

8.2)

$$E_{thermal} = kT$$
$$E_{bend} = E_{thermal}$$

therefore

$$kT = \frac{kTl_p \langle \theta^2 \rangle}{2s}$$

$$\langle \theta^2 \rangle = 2.3 \times 10^{-4} \text{ rads}$$

$$\langle \theta^2 \rangle^{\frac{1}{2}} = 2.7°$$

The mean square angular displacement of the filament is therefore 2.7°.

8.3)
The length of the titin molecule trapped in the pore is:

$$R_\parallel \approx Na \left(\frac{D_b}{a} \right)^{-\frac{2}{3}} = 600 \text{ nm}$$

When stretched to 750 nm the tension blob size is smaller than the size of the pore and the conformation (and elasticity) of the chain is unaltered by the size of the pore.

8.4)
The free energy of an ideal chain is:

$$F(R) = -T \ln Z_N(R) = \text{const} + \frac{3 TR^2}{2 Nl^2}$$

If the chain now experiences an extending force (f) on both ends then:

$$f = \frac{\partial F(R)}{\partial R} = \frac{3T}{Ll} R$$

The polymer in a good solvent obeys Hooke's law.

In bad solvent there is a strong increase in the force measured by the traps (DNA $\sim$ 0.3 kT/bp). An unwinding globule coil transition is now possible and will be observed in the force/distance curves.

8.5)

The size of the DNA chain according to the worm-like chain model is:

$$\langle R^2 \rangle^{\frac{1}{2}} = \sqrt{2Ll_p} = \sqrt{(2 \times 60 \times 10^{-6} \times 450 \times 10^{-10})} = 2.32 \, \mu m$$

CHAPTER 9

9.1)

The ability of the amine group to dissociate is reduced due to the interaction between neighbouring groups along the polylysine chain (there is an energetic penalty).

9.2)

$$\xi = \frac{l_b}{b} = \frac{7}{5}$$

The effective charge fraction of the polylysine chain predicted by the Manning model is therefore $1/\xi = 5/7$.

9.3)

The total persistence length (l_p) in the OSF model is equal to the intrinsic component added to the electrostatic component.

$$l_T = l_p + \frac{l_B}{4\kappa^2 A^2}$$

The second term is the electrostatic contribution:

$$A = a/f = 1/0.5 = 2 \, nm$$
$$l_b = 0.7 \, nm$$

Thus $l_e = 0.7 \, nm$

9.4)

Critical properties of the material are:
Samples are extremely hydrophilic and swell many times their dry size (polyelectrolyte gels can be less than 1% polymer).
The materials are charged (reduces adhesion) and biocompatible.

CHAPTER 10

Assuming all the charges on the polymer chain dissociate the osmotic pressure (π) is proportional to the number of charges per unit volume (n):

$$\pi = nkT$$

The number of charged units is $n = 0.001 \times 6 \times 10^{23} = 6 \times 10^{20}/dm^3$.
Thus the osmotic pressure is 0.404 J dm^3.

The neutral polymer contribution is kT per blob or, more naively, kT per chain.

It is much smaller than the contribution of the counterions.

9.5)

The charged blob size (D) is given as:

$$D \sim a\sigma^{\frac{2}{3}} u^{-\frac{1}{3}}$$

where a is the monomer length, σ is the number of monomers between charged units (= 1 fully charged) and u is Bjerrum length.

$u = 7/3.6$ Å
a = peptide step length = 3.6 Å
$D = 2.9$ Å for fully charged blobs (fully elongated)
$D = 13.4$ Å for weakly charged blobs ($\sigma = 10$)

The chain forms a semi-flexible rod of blobs:

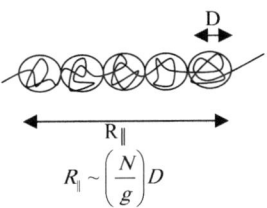

$$R_\| \sim \left(\frac{N}{g}\right) D$$

where N is the number of monomers in the chain, g is the number of monomers in a blob and D is the size of a blob.

CHAPTER 10

10.1)

The line tension is given by:

$$\lambda = \tau R^* = 2.6 \times 10^{-9} \times 0.03 = 8 \times 10^{-11} \text{ Jm}^{-1}$$

10.2)
The axial and hoop stresses are given by:

$$\sigma_{axial} = \frac{rP}{2h} = \frac{1 \times 10^5 \times 1 \times 10^{-6}}{2 \times 10^{-9}} = 0.5 \times 10^8 \text{ Pa}$$

$$\sigma_{hoop} = \frac{rP}{h} = 1 \times 10^8 \text{ Pa}$$

10.3)
For the persistence length:

$$\xi_{p1} \sim be^{10\pi}$$
$$\xi_{p2} \sim be^{20\pi}$$

The ratio of the two persistence lengths is therefore:

$$\frac{\xi_{p1}}{\xi_{p2}} = e^{10\pi}$$

For the exponent measured with X-rays:

$$\eta_{m1} \sim (BK)^{-\frac{1}{2}}$$
$$\eta_{m2} \sim (B2K)^{-\frac{1}{2}}$$

The ratio of the two exponents is therefore:

$$\eta_{m1}/\eta_{m2} = \sqrt{2}$$

10.4)

$$\frac{R_{g2}^2}{R_{g1}^2} = \frac{L_c^2}{L_c^{\frac{4}{3}}} = L_c^{\frac{2}{3}}$$

$$\frac{R_{g2}}{R_{g1}} = L_c^{\frac{1}{3}}$$

$$\frac{A_1}{A_2} = \frac{L_c^3}{L_c^2}$$

$$\frac{A_1}{A_2} = L_c$$

CHAPTER 11

11.1)
In the *parallel* arrangement the Young's modulus of the mixture (E_m) is given by:

$$E_m = E_c \phi_c + E_a(1 - \phi_c) = 50 \times 10^9 \times 0.9 + 50 \times 10^6 \times 0.1$$
$$E_m \approx 45\,\text{GPa}$$

In the *perpendicular* arrangement, E_m is given by:

$$E_m = \frac{E_c E_a}{E_a(1 - \phi) + E_c \phi} = 55\,\text{MPa}$$

Ratio of the Young's moduli is 818:1 parallel:perpendicular

11.2)
For the *unfilled foam*:

$$E_f \sim \left(\frac{t}{a}\right)^4 E$$

$E = 9\,\text{GPa}$, $a = 20\,\mu\text{m}$, $t = 1\,\mu\text{m}$

Therefore:

$$E_f \sim \left(\frac{1}{20}\right)^4 9\,\text{GPa} = 0.56\,\text{MPa}$$

For the *filled foam*:

$$E_f \sim \left(\frac{1}{20}\right)^2 9\,\text{GPa} = 22.5\,\text{MPa}$$

CHAPTER 12

12.1)

$$Shear\ Rate = 1 \times 10^{-2}/10 \times 10^{-6} = 10^3\ s^{-1}$$

$$Pe = \frac{6\pi\eta a^3 \dot{\gamma}}{kT} = 6\pi \times 0.001 \times (10^{-6})^3 \frac{10^3}{4.1 \times 10^{-21}} = 4.6 \times 10^3$$

The experiment is in the regime of high Peclet number dynamics
The shear rate could therefore be substantially affecting the microstructure.

12.2)
For the Maxwell model:

$$\eta = \tau G$$

The characteristic relaxation time is therefore:

$$\tau = 10^6\ s \sim 11.6\ days$$

12.3)
The viscosity is given by the Einstein relationship:

$$\eta = \eta_b\left(1 + \frac{5}{2}\varphi\right) = 10^{-3}\left(1 + \frac{5}{2}0.02\right) = 1.05 \times 10^{-3}\ Pas$$

CHAPTER 13

13.1)
Use the Stokes force or stochastic Langevin analysis to calibrate the apparatus.
From the figure the cornering frequency (f_c) is given by:

$$f_c \approx 500\ Hz$$

$$\gamma = 6\pi\eta r = 1.88 \times 10^{-8}$$

The trap stiffness (κ_x) is therefore:

$$\kappa_x = 2\pi f_c \gamma = 5.91 \times 10^{-5}\, \text{Nm}^{-1}$$

13.2)
A corresponds to a predominantly viscous material MSD $\sim t^1$.

B corresponds to a viscoelastic material, with a characteristic sub-diffusive behaviour for the MSDs of the probe particles $\langle \Delta r^2(t) \rangle \sim t^\alpha$, $\alpha < 1$. Static errors tend to induce a plateau in the MSDs when the value of the MSD is comparable with the square of the static displacement error. Normally that corresponds to the short time limit. Thus fluid B could be a purely viscous fluid if the static errors have not been properly accounted for.

13.3)
Using equation (13.45) the velocity of electrophoresis is:

$$v \approx \frac{q\overline{E}}{3\eta N}$$

There are 1.7 Å between phosphate groups on the DNA:

$$q = 1/1.7\, e\, \text{Å}^{-1}$$

$$N = 10^6/300$$

$$v = \frac{(1/1.7) \times 1.6 \times 10^{-19} \times 10^{10} \times 2 \times 100}{3 \times 0.002 \times 10^6/300} = 9.4 \times 10^{-9}\, \text{m/s}$$

Decreasing the size of the chains by a factor of 10, should increase the velocity by a factor of 10.

13.4)
From equation (13.35) with $\omega_c = 0$ we have the required expression for the power spectral density with no trapping force:

$$\langle \Delta r^2(\omega) \rangle = \frac{kT}{\pi \gamma \omega^2}$$

For a power law fluid, the frequency response is modified:

$$\langle \Delta r^2(\omega) \rangle \sim \frac{1}{\omega^{2-\alpha}}$$

where α is a positive constant and the MSD $\sim t^\alpha$.

CHAPTER 14

14.1)
The two critical micelle concentrations are defined as:

$$c_{m+crit} = \frac{k_{off}^+}{k_{on}^+} \qquad c_{m-crit} = \frac{k_{off}^-}{k_{on}^-}$$

(Check the table using these equations.)

The treadmilling concentration is given by:

$$c_{mtm} = \frac{k_{off}^+ + k_{off}^-}{k_{on}^+ + k_{on}^-}$$

ATP actin $c_{mtm} = 0.13\ \mu M$
ADP actin $c_{mtm} = 1.88\ \mu M$

CHAPTER 15

15.1)
Using equation (15.31) gives $E_{protein} = 1$ MPa.

15.2)
The characteristic time for stress relaxation dependent on the motion of interstitial water is:

$$\tau_1 = \frac{\delta \delta_{eq}}{\pi^2 E k} = \frac{10^{-3} \times 0.95 \times 10^{-3}}{(3.142)^2 \times 0.78 \times 10^6 \times 6 \times 10^{-13}} = 2 \text{ seconds}$$

Index

abductin 10
acid/base equilibria 4–5, 7–8, 205, 207
and polyelectrolytes 224
acidic side groups, amino acid 1, 2, 3
acids 4–5, 7–8, 207, 219
actin 145–6, 167, 207, 351
and biological motors 340, 341, 346–9
and Donnan equilibria 221
and polymer dynamics 193
self-assembly 129, 142–3, 341–6
viscoelasticity 286
adenine (DNA base) 13
adhesion 47, 154–6, 237, 280
adhesive proteins 372–3
ADP (adenosine diphosphate) 23, 339–41, 347, 349
adsorbers, diffusion to 114–16
adsorption 45, 115–16
aggrecan 20–1, 358, 366
aggregating self-assembly 129–48, 341
alanine 1, 2
alpha helices 4, 5, 6, 206, 207
Alzheimer's disease 129, 139
amino acids 1–4, 8
Amonton's law 165–6, 167, 362

amorphous materials 103
amphiphilic self-assembly 11, 130, 133
amyloids 139, 142, 147
amylopectins 16, 17, 18, 102
amylose 16, 17
anisotropic materials 257–8, 335, 343–6
cartilage, in tension 367
collagen in joints 357
see also composites
antifreeze proteins 19
aqueous ions, solvation energy of 212
aqueous solutions 32–5, 42–4
see also solutions
arginine 1, 3, 8, 205
aromatic amino acids 1, 2
Arrhenius equation 67, 124–5
arteries, elastin in 369
arthritic conditions 356–7, 358
asparagine 3, 4
aspartate/aspartic acid 3, 4, 8, 205
asperities in joints 364–5
association 224, 225, 226, 231
atomic force microscopy 306, 310–13
for frictional forces 334
microrheology with 330, 333
for myosin motion 347

ATP (adenosine triphosphate) 23, 206, 339–41
  and linear stepper motors 346, 347, 348–9
autocorrelation function 118–19, 190

B-DNA 13–15, 175, 230–1
bacteria 23, 199–200, 380
  movement 117–18, 119–21, 340, 350, 352
bacteriophages 23, 139, 141
base pairs, DNA 13–15, 378
bases 4–5, 7–8, 207, 219
beads on a string model 379–80
bending membranes 240–3, 247
bending rigidity/modulus 175–7, 241
beta sheets 4, 6, 59
bilayers 155, 156, 206
  self-assembly 130, 133, 237
biorheology 267–91
blob models 171, 180–3
blood 11, 22
  in blood vessels 267, 268, 291
  shear thinning 288, 289
  sickle cells 129, 130
blood clots 285–6, 365
bone 371–2, 373
  in joints 355–6, 357
  mechanical properties 261, 263, 264
bottle brush morphology 20–1
boundary lubrication 167, 355, 358
bovine spongiform encephalopathy 129, 139
branched morphology 249
brush morphology 20–1, 249
bulk moduli 256–7
bulk rheology 325, 326

capillarity 160–3, 373
capillary condensation 168
capillary electrophoresis 320
capillary rheometers 328–9
carbohydrates 15–18, 20, 171
  helix-coil transition 53
  self-assembly 130
  see also polysaccharides

carboxylic acid groups 205
carboxylic acids 11
carrageenan 85–6, 207, 288
cartilage 21, 167, 267, 355–68
cell division (cytokinesis) 131, 345–6
cell membranes see membranes
cell walls, plant 16, 79
cells 21–2, 237–8, 339
  DNA packing in 377–80
  morphogenesis 74, 155
cellular solids/foams 253, 261–3
cellulose 16, 78–9, 253, 288
charge
  and DNA packaging in cells 379
  point charges, forces between 207–8
  and polyelectrolyte conformation 219
  and polymer mixtures 74, 234
charge condensation 228–32, 381
charge fraction, polymer chain 228, 230–1
charged gels 285, 364
charged links, protein 64
charged polymers
  cross-linked elastomeric 358
  and helix-coil transition 58
  persistence length 233, 381
  see also polyelectrolytes
charged spheres, forces between 36–8
charged surfaces
  forces between 34–5, 359–61
  gels, friction force 364
  in liquids 33
chemical gels 283
chemical reactions, rate theories of 125–7
chirality 100, 101
  and chromosomal morphology 381
  and elastic/spring constants 91
  and liquid crystallinity 79, 81, 89–91
cholesterics 79, 82, 89–91, 101
cholesterol 11
chondrocytes 356
Chord theorem 41
chromatin 377–80
chromosomes 15, 377, 378, 379–80, 381

INDEX 409

clathrate structures 29–30
cohesive forces 25–8
collagen 9, 253, 259
  in cartilage 358, 365–6
  denatured, as gelled biopolymer 283
  helix-coil transition 100–1
  liquid crystal phases in 78, 85–6, 100–1, 103
  self-assembly 147, 148
colloids 39, 287–8
  coagulation 37–8, 46, 47
  counterion clouds around 231–2
  electrophoresis for 315
  forces in 36–8, 42–4, 209
  and Ostwald ripening 168
  and Peclet number 269
columnar phases 80
compliance 194–5, 258, 275
composites 154–5, 253, 258–61
  bone 371, 373–4
  cartilage 358, 365–6
  spider silk 368
  see also fibrous composites
compression
  of foams 262–3
  of membranes 238–9
  and phase transitions 52–3
concentrated regimes 276, 277, 278
concentration
  and diffusion 110, 112–13
  and self-assembly 133–7, 141, 147
    actin and tubulin 341–6
    lipids 237
condis crystals 79, 80
conformation 8, 79, 80, 219–20
  and solvent quality 177–83
constitutive equation 270
contact angles 156, 157, 158
  measurements 46, 47
contact value theorem 34–5, 40, 42
  application to cartilage 361
continuous phase transitions 52–3
continuum mechanics 253–65
contour length 15, 196
cork, mechanical properties 261, 263
cornea, Donnan equilibria and 221, 222

Coulombic interactions 219–20, 233
  see also electrostatics
Coulomb's law 31, 207–8
counterions 33, 34
  condensation 221, 228–33, 383
  osmotic pressure 220, 221–3, 361
  and polyelectrolytes 218–19, 220, 221, 285
covalent bonds, polarisation around 205
covalent interactions, amino acid 9
cracks, termination of 264–5
cross-bridge model 346–9
cross-linked polymer networks, collapse of 233–4
cross-links 184
  and elastomers 103, 186–7, 358
    elastin and resilin 369–70
  in gels 283–5, 286
cruciform structure (DNA) 15
crystalline phases 80
crystallisation 64–8, 218
  see also liquid crystallinity
crystals 64–8, 168
  in nacre and bone 373
  self-assembly produces 132–3
cylindrical charge distributions 227–8, 231
cylindrical membranes, stress on 245–6
cysteine 1, 2, 8, 9
cytosine (DNA base) 13
cytoskeletal filament polymerisation 142–8

damped motion 187–91
D'Arcy's law 366, 367
dashpot and spring models 273–4
Deborah number 270, 326
Debye screening radius 214, 220
Debye-Huckel theory 213–14, 227
defect textures 81, 100, 101
defects, liquid crystal 91–2, 95–100
denaturation 10
density
  and foams/cellular solids 261
  and phase transitions 52–3
dentine as a composite 259

depletion forces 42, 43, 44
depletion potential 38, 42, 43
dielectric constant of water 210–11
differential scanning calorimetry 81–2
diffusing wave spectroscopy (DWS) 330, 331–2, 333
diffusion 108–16
  facilitated, protein 383–6
  and rate of reaction 125–7
  surface tension, Marangoni effect 168
  and viscoelasticity 269
diffusion coefficients 109–10, 112, 113, 191
  colloidal solutions 269
  flexible polymer chains 192, 197
  and friction 111, 112
  Poisson motility process 121
  small globular enzyme 383–5
dilute regimes 276–7
dipole moments 19, 211–12
  from alpha helices 206, 207
dipoles 25, 26, 31, 208, 248
direct force measurements 46, 47
disclinations 96–8
disk-like adsorbers, diffusion to 115, 116
dislocations 96
disorder 79, 80
dissipation 267, 271, 273, 358
dissipative/loss moduli 270–2, 278
dissociation 7–8, 224–5
dissociation constants 225, 226
disulfide linkages 9, 10, 283
DLVO treatment 36–7, 38, 232
DNA 12–15, 377–86
  of bacteria 199, 200
  charge on 207, 219
  counterion clouds around 231, 232, 233
  electrophoresis 198, 316–19
  and electrostatic interaction 32
  globule-coil transition 59–60, 63–4
  helix-coil transition 59, 100–1
  liquid crystalline phases 85–6, 100–1, 288
  rheology 280–1
  sequencing DNA chains 317, 319
  topology of 199–201
  torus structure 63
DNA technology, recombinant 12, 370
Donnan equilibria 221–3
double helix-coil transition 59
drag flow rheometry 326–7, 334
dragline silk 255, 368, 369
durability 254
dynamic frictional effects 166–7
dynamic mechanical testing 335
dynamic scattering techniques 297–303
dynamic viscosity 271–2
dynamics
  low Reynold's number 116–19
  polyelectrolytes 220
  polymer chains 191–8, 276–80

Einstein relationship 111
elastic energy
  of disclinations 98
  storage of 271, 369, 381
elastic moduli
  of membranes 243
  semi-flexible polymer gels 285–6
  see also Young's moduli
elastic shear moduli 272, 280
elastic/spring constants 87, 270
  liquid crystals 87–9, 90–1
elasticity 183–7
  of elastin and resilin 369–70
  and liquid crystalline phases 87–92
  of membranes 243–8
elastin 9, 255, 259, 272, 369–70
elastohydrodynamic fluid films 364
elastomers 103, 183–7, 255, 283
  natural 358, 369–70
electric double layer 32–3
electric fields 208–12, 228
electric flux, Gauss' theorem and 208–9
electron microscopy of DNA 379
electron scattering 295
electrophoresis 314–21
  of DNA 198, 316–19
  of globular proteins 212
electrostatic forces 207–8, 214

INDEX 411

electrostatic interactions
 between DNA and histones 379, 381–2
 and protein globule structure 64
electrostatic persistence length 233
electrostatic/charged blobs 181, 182–3
electrostatics 30–8, 207–12
 see also Coulombic interactions
ellipsoidal adsorbers, diffusion to 115, 116
end-to-end distance 178–9
endocytosis 248
energy
 ATP hydrolysis gives 339–41
 and bending bilayer membranes 247
 cracks absorb 264
 and crystallisation/freezing 65–8
 of dipoles in electric fields 208
 of disclinations 96, 98
 dissipation 267, 271, 273, 358
 free *see* free energy
 interfacial fracture energy 154
 of ion pairs in solution 215–16
 and myosin/actin interactions 348–9
 and partition function 55
 and polymerically stabilised systems 39
 and protein folding 10, 11
 of solvation, aqueous ions 212
 and superhelical DNA 199
 torsional, DNA stores 13
 and viscoelastic materials 267, 271
 of wetting 159
energy barriers, diffusion over 123
energy storage 11, 16, 17, 23, 78
 in DNA complexation 381
 elastin and resilin for 369
 and viscoelastic materials 271
entangled regime 194–8, 277
 flexible polyelectrolytes 281, 282
enthalpy
 of hydration 216, 217
 and phase transitions 50, 51
entropy
 and globule-coil transitions 60–2
 and hydration 217
 of mixing 69–70

 of polymer chains 180, 184–6
 and self-assembly 132, 142
enzymes 8, 317, 383–6
epithelial cells 22
etched microarrays 321
excluded volume 177
exoskeletons, chitin forms 17
experimental techniques 293–335
 for diffusion 111
 intermolecular forces 44–7
 for liquid crystalline phase transitions 79–82
 motility 107
 for myosin steps 347–8
extensibility 254, 255
extrusion nozzles 339, 340, 352
eyes 71–2, 221, 222
Eyring rate theory 125

facilitated diffusion, protein 383–6
fatty acids 11, 12, 205
fibrin elasticity 285–6
fibroblast cells 22
fibrous composites 4, 253, 258
 defect textures 100, 101
 elastin/collagen, in heart walls 259
 nacre and bone as 373–4
 shear modulus 244
Fick's laws 109, 112–13
filaments, polymerisation of 142–8
filled foams, mechanical behaviour of 263
film balances 47
first order phase transitions 50, 51
 DNA histone interaction 382
first passage problem 121–5
flagellated bacteria, motility 119–21, 350
flagellin, self-assembly 129
flexible polyelectrolytes 233, 281–2
flexible polymer chains 171, 172
 conformations 179
 dynamics of 192–8, 276–80
 elasticity of 183–7
 expansion of 178
 persistence length 173
 Rouse and Zimm models 278
 thermal blob model 182

Flory approach 64, 179, 180, 184
fluctuation-dissipation theory 111, 112, 197
 and microrheology 329
fluorescence depolarisation 298–9
fluorescence intensity correlation spectroscopy 297–8
fluorescence techniques 297–9, 348
foams/cellular solids 253, 261–3
focusing techniques 296–7
folding 10, 11, 64, 131, 142, 232–3
force measurement 306–14, 347
forced resonance devices 328
forces
 and charge 34–8, 207–8, 359
 direct force measurements 46, 47
 electrostatic 207–8, 214
 and friction 165–7, 362, 363, 364
 intermembrane 248–50
 intermolecular 44–7
 mesoscopic 25–47
 steric 38–42, 248
 and surface tension 151, 153–4
 and Young's modulus 254–5
fracture 263–5
fracture stress 261
free energy 71
 in aqueous environments 30
 and association/dissociation 226
 and crystallisation/freezing 65–8
 and globule-coil transitions 60–2
 of hydration 216
 and isotropic-nematic transitions 85, 92–4
 of liquid crystals 87–92
 and mixing of liquids 69–71
 and partition function 55
 and polyelectrolyte solutions 221
 of polymer chain expansion 180
 of polymer chains, and elasticity 186
 and protein conformation 124–5
 and protein crystallisation 66–8
 and self-assembly 131–2, 136–7, 141–2
 of the superhelical state 199
 surface 66, 67, 68, 132, 168
 and surface tension 151, 153–4
 and transition between states 124–5
freezing 65–8
frequency dependence
 shear modulus 271, 278, 280
 viscoelasticity 271–2, 277, 280
 gels 284
 polyelectrolytes 282
friction 165–8, 280, 333–4
 and diffusion 111, 112
 in joints 167, 355, 357, 362–5

Gauss' theorem 208–9
Gaussian curvature 247
gel point/percolation threshold 283–4
gelation, haemoglobin 129
gels 283–7
 cartilage as 359–61
 friction and joints 362–5
 polyelectrolyte 285, 305–6
gene therapy 380–1
genetics 12, 319, 380
Gibbs phase rule 50–1
glass fibres, measurements with 306
glassy materials 290
globular enzymes, self-assembly 129
globular proteins 4, 8, 10, 65
 as colloidal systems 287
 electrophoresis of 212
 folding 64, 131, 142, 232–3
 overdamped motions of 188–90
 sedimentation 323
globule-coil transition 59–64, 129, 142
glucose, polymers of 16–17
glutamate/glutamic acid 3, 4, 8, 205
glutamine 3, 4
glycine 1, 2
glycogen, energy stored in 17
glycolipids 11
glycoproteins 20–1, 167, 249
glycosoaminoglycans 20
Gouy region, cylindrical charge and 227–8
guanine (DNA base) 13

haemoglobin 129, 130, 139, 147
hairs, adhesion of 151, 152

INDEX

heart disease 346
heat capacity 50, 51, 52
helices 4, 5, 6, 9, 95
  dipole moments 206, 207
  DNA 13–15, 53–9, 199
helicity 55, 57, 59
helix-coil transition 13, 53–9, 100–1
Henderson-Hasselbalch equation 7–8, 226
hepatitis B virus 137, 138
hexatic phases 81
histidine 1, 3, 8, 205
histones 377–80, 381
Hooke's law 87, 254
hopping (diffusion) 383, 384, 386
hyaluronic acid 18, 20, 280, 282, 357
hydration 216, 217–18
hydration shells 205, 215, 216, 217
hydrodynamic beam model 193–4
hydrodynamic interactions 44, 47, 192, 193
hydrodynamic lubrication 167, 362, 363
hydrodynamics, screened 192–3
hydrogen bonds 8, 19, 28–30
  in DNA 13, 15
  and secondary structures 5–6, 9–10
  and self-assembly 132
hydrogen ions 4–8, 207
  and polyelectrolytes 224, 225–6
hydrolysis, ATP 339–41, 349
hydrophobicity 8, 64, 132
  surfaces on plants 151, 152
  and water clathrates 29–30
hydroxyl groups, amino acid 1, 2
hydroxyl ions, water gives 4–5

ice, structure of 19, 28
inch worm motion, myosin 347, 348
induced dipoles 25, 248
information storage 17–18
intermediate scattering function 299–302
intermembrane forces 248–50
intermolecular forces, measurements of 44–7
intersegmental transfer 383, 384

ion pairs 215–16, 230
ion pumps 205, 221
ion-dipole interactions 31
ion-ion interactions 31
  see also electrostatics
ionic bonds 32, 248
ionic radius 214–18
ions 205–34
  see also polyelectrolytes
irreversible thermodynamics 267
isoelectric focusing 316, 317
isoelectric point 316
isoleucine 1, 2
isotropic-nematic transitions 81–2, 84–5
  free energy of 92–4
isotropic-nematic-smectic transitions 93

joints 167, 355–68
  hyaluronic acid in 280, 357
  and shear thinning 281, 282, 356

Kelvin model 273, 274
keratins 6, 9, 147, 193
Kramers rate theory 125
Kratky-Porod model 175, 177

lac repressor 123, 127, 383–4
lamellar ordering, defects in 95
lamin, self-assembly of 148
laminates 258, 259–60
Landau theory 58, 92, 93, 100
Landau-Pierels theorem 95, 99
Langevin equation 118, 190–1
laser deflection microrheology 330, 331–2
Legendre polynomials 80, 82–3
length, polymer chain 173–4
  see also persistence length
Lennard-Jones potential 28
leucine 1, 2
Levinthal's paradox 10, 129, 142
light scattering 295
lignins in fibrous composites 253
line imperfections/defects 95–6
line tension, measuring 248–50
linear stepper motors 339, 340, 346–9

linkages
  between carbohydrates 16, 17, 18
  disulfide 9, 10, 283
linking number 199, 200, 201
lipids 11–12, 130, 131, 133, 237
  surface charge densities 206
liquid crystallinity 77–103
  comparison with composites 259
  and defects 95–100
  DNA torus internal structure 63
  and elasticity 87–92
  and phases of condensed matter 79, 80
  polymer rheology 288–90
  rod-like polymer chains contain 171
  and surface effects 168
liquid-liquid demixing 68–74
liquid-nematic-smectic transitions 92–4
liquid-solid transition 65–8
  see also crystallisation
lock washer morphology, TMV uses 138, 139
loss modulus see dissipative/loss moduli
low Reynold's number dynamics 116–19
lubrication 167, 362, 363
  in joints 355, 357–8, 364
Lymn-Taylor scheme 346
lyotropic liquid crystals 77
lysine 1, 3, 8, 205

macromolecules 171–201, 207
  see also polymers
magnetic microrheology 330
magnetic tweezers 306, 307, 308–10
Manning charge condensation 228–32
Marangoni effect 168
Maxwell model 273–4, 278
mechanical spectroscopy 274–5
mechanics 253–65
  see also dynamics
membrane osmometers 303
membrane proteins 237–8
membranes 11, 39–42, 237–51
  and liquid crystallinity 78, 79
  self-assembly 130, 131, 133, 237
meniscus 162

mesoscopic forces 25–47
methionine 1, 2
micelles, self-assembly produces 132–7
microfibrils 6, 9, 16, 78–9
microfluidics 291
micromanipulators 155, 156
microrheology 325–6, 329–33, 348
microscopy 97–8, 379
  see also atomic force microscopy
microtubules 193, 339, 341
  self-assembly 145, 345, 346
mixing of liquids 69–71
mobility, electrophoresis for 315, 316, 318, 319
moduli 255, 256–7, 275
  bending, polymer chains 175–7, 241
  dissipative/loss 270–2, 278
  of membranes 243
  microrheology for 330, 331
  of semi-flexible polymer gels 285–6
  storage moduli 270–4, 278, 280
  see also shear moduli; Young's moduli
molecular dynamics simulations 28, 30–1
molecular weights 44, 319–20
molluscs
  adhesion by 372–3
  slugs 78, 79, 151
monomers 1, 13, 15, 171
morphogenesis 74, 131, 155
morphology 20–1, 249
  foams 262
  and mechanics 265
motility 107–27
  and actin 142, 145, 341–6
  bacteria 117–18, 119–21, 350
  energy for 339, 341
  Poisson processes 120, 121, 350
  and Reynold's number 117, 350
motor proteins 341, 346, 348
  and ratchet models 351–2
  self-assembly 131, 345
motors 339–53
moving boundary electrophoresis 315–16
mucins 21, 77–8

INDEX 415

multi-stranded filaments 145, 146–7
muscle 221, 341, 346–9
  energy dissipation in 267, 268
  low frictional losses 167
  and viscoelasticity 280
muscle cells 21
mushroom morphology 249
myosin 167, 341, 345–9
  and Donnan equilibria 221
  ratchet models 351

nacre 259, 264, 373–5
nanotechnology 339
neighbouring groups 226–7
nematic liquid crystals 80, 81–3
  free energy 87–9, 90, 91–2
  spider silk 368
  transitions 93
nematic-smectic transition 94
nerve cells 21–2
networks
  collagen 103
  fibrous, shear modulus 244
  gels as 283
  polymer, collapse of 233–4
  rubbery, elasticity 183–4, 186–7
neutron scattering 295, 297, 303
  for DNA structure 378, 379
  for persistence length 172
neutron spin echo measurements 303
NMR nuclear magnetic resonance 19, 47
non-aggregating self-assembly 130–1, 142
nuclear magnetic resonance (NMR) 19, 47
nucleation 73, 168
nucleic acids 12–15, 171, 319
  and electrostatic interaction 32
  helix-coil transition 53, 58, 59
  phosphate groups in 205
  in viruses 139
  viscoelasticity 280
  see also DNA
nucleus, cell, DNA in 377, 380

Oosawa model 343
optical microscopy, defects seen in 97–8
optical tweezers 46, 47, 306–8, 347–8
order parameters 51–2
  for globule-coil transitions 60
  for liquid crystalline phases 80, 82–4
    isotropic-nematic 92, 93, 94
    nematic-smectic 94
    side-chain polymers 101, 103
  for liquid-solid transitions 65
organs/organ systems 21
OSF theory 233, 381
osmometers 303, 306
osmotic pressure 220–1, 231, 361
  measurements with 303–6
Ostwald ripening 68, 168
overcharging 383
overdamped motion 188, 189
overlap concentration 277

packing forces, solvent molecules and 38–9
parallelised linear stepper motors 346–9
particle diffusion 108–11, 112–13
  first passage problem 121–2
particle tracking microrheology 329, 330, 331
partition function 55
patch clamp method 250
Peclet number 269–70, 326
pectins 18, 283
peptide linkages 1, 4
peptides 130, 206, 237–8
percolation threshold/gel point 283–4
peristaltic forces between membranes 42
permanent dipoles 26
persistence length 171–7
  DNA 15, 381
  and liquid crystalline phases 86, 100–1
  membranes 240–1
  of polyelectrolytes 233, 281
  and viscoelasticity 279

pH 6–8, 219, 225–6
  and helix-coil transition 58
  and isoelectric point 316
phase behaviour of DNA 377–86
phase diagrams 45, 47, 49
  DNA complexation 382
  globule-coil transitions 61, 62
  helical chain 57
  isotropic-nematic transition 86, 92
  protein molecules 64
phase lag, viscoelastic 272
phase separation, liquid-liquid 68–74
  charge reduces 234
phase transitions 49–74
  of colloids 288
  crystallisation 64–8
  globule-coil transition 59–64, 129, 142
  helix-coil transition 13, 53–9, 100–1
  liquid-liquid demixing 68–74
  and solvent quality 178
  and surface tension 53, 168
  and wetting 157, 159
phases of condensed matter 79, 80
phenylalanine 1, 2
phosphate groups 13, 205
  in ADP/ATP 23, 339–41, 349
phospholipids 11, 12, 131, 205
photon correlation spectroscopy 300
physical gels 283, 284–5
pitch, liquid crystal 89, 90, 91
plastic crystals 79, 80
plastic deformation, composites and 154–5
plectonomic structure 200, 380
point imperfections 95
Poisson equation 33–4, 210, 213
Poisson processes 120, 121, 350, 384–5
Poisson ratio 243–4, 255–6, 257
  cellular solids 261, 263
Poisson-Boltzmann equation 34, 35, 213–14
  charged plates 359–60
  cylindrical charge distributions 227–8, 231
polarisability, water 19

polarisation, induced 210
polarising optical microscopy 97–8
polyacids 219, 224, 225
polybases 219, 224
polydisperse aggregates 132
polyelectrolytes 218–21, 228–34
  complexation, DNA compaction 380–3
  gels 285, 305–6, 364
  viscoelasticity of 280–2
  see also charged polymers; ions
polymer chains 171–87
  dynamics of 191–8, 276–80
  elasticity 183–7
  flexibility 171–7
  globule-coil transition 59–64
  polyelectrolytes 228–32
  solvent effects 177–83
  topology, and super coiling 199–201
polymer networks, collapse of 233–4
polymerase chain reaction 319
polymerisation 139–48, 341–6
  and viscosity 279
polymers 1, 171
  and charged ions 205–34
  of glucose 16
  liquid crystalline phases in 85–6, 101–3
  in membranes 237–8
  separation of, phase transition 72–3
  in solution 42–4, 191–8, 276–80
  spider silk 368
  at surfaces 39, 40
  see also macromolecules; polyelectrolytes; polymer chains; proteins
polypeptides see proteins
polysaccharides 17, 18, 249
  see also carbohydrates
porous material, fluid motion through 365–7
potential, van der Waals forces and 26
potential wells, diffusion over 123–4
power spectra 190, 191

powerstroke model   347
pressure
  between membranes   42
  between surfaces   34–5, 38
  and bulk modulus   256–7
  in protein self-assembly   140–1
  and surface curvature   160–2
pressure driven rheometers   328–9
pressure flow rheometry   326
prion diseases   129, 130, 141
proline   1, 3
proteins   1–11, 12, 20, 171
  adhesive proteins   372–3
  antifreeze proteins   19
  in cell adhesion   155
  conformation, rate of change   124–5
  crystallisation   42, 64–8
  electrostatic interactions   32
  facilitated diffusion of   383–6
  and helix-coil transition   53, 56, 59
  ionic bonds, intermembrane   248
  membrane proteins   237–8, 250
  molecular weights   319–20
  in nacre and bone   373
  phase diagrams of   64
  as polyampholytes   219
  precipitation, by salts   217–18
  ratchet models for   350–1
  self-assembly of   139–42
  spider silk   368
  stress-strain properties   253–4
  structure   4, 5–7, 8, 9–10, 64
  surface active   159
proteoglycans   20–1, 32, 280
  in cartilage   365–6
  in fibrous composites   253
  and friction   167
  in joints   357
  and liquid crystallinity   78, 79, 101
  viscoelasticity   280
protocollagen, nematic phases and   78
protofilaments   147, 148
pulsed electrophoresis   319

quasi-elastic scattering   296, 299–303
quenched polyelectrolytes   223–4

radius of gyration   60, 174–5
rafts, self-assembly produces   132
ratchet models   350–2
rate theories of reactions   125–7
recombinant DNA technology   12, 370
red blood cells   129, 130
reflectivity   46, 47, 243
relaxation moduli   271, 274, 278–9
  cartilage   365–7
  gels   284
relaxation times   191, 192
  cross-linked gel   366–7
  diffusion of colloid particles   269
  entangled solutions   196, 197
  semi-flexible chains   194
replication, DNA   13
reptation   194–8, 277, 279
  and DNA electrophoresis   317–19
  sticky, and physical gels   284–5
resilience   255, 370
resilin   9–10, 255, 290, 369–70
resonance rheometers   328
restrained DNA   380
retardation effect   26
Reynold's number   116–19
rheological functions   274–6
rheology   325–33
  see also biorheology; viscoelasticity
rheometers   270, 326–9
ribosomes   12
rigid polyelectrolytes   281
rigid polymer chains   171, 172, 173
  conformations   179
  expansion of   178
RNA   12
RNA polymerase   15
rod-like polymer chains see rigid polymer chains
rotating cross-bridge model   346–9
rotational motion, diffusion and   111–12
rotatory motors   339, 340, 350, 351, 352
rouleaux structures   288, 289

Rouse model   192–3, 277, 278–9
rubbery materials *see* elastomers

salts, protein precipitation by   217–18
sap, rise of   162
scaling approach   92, 179, 180, 362
  flexible polyelectrolytes   281
scanning X-ray microdiffraction   296–7
  for disclinations   97
scattering techniques   294–303
screened electrostatic interactions   32–5
screened hydrodynamics   192–3
seaweed extracts   280
second order phase transitions   50
secondary structure   4, 5–6, 9–10, 64
sedimentation   321–5
self-assembly   129–48, 341–6
  of histones   378, 379
  of lipids   11, 130, 237
  of protein   64, 139–42
  of viruses   22, 137–9, 140
self-diffusion   111
self-organisation   131
semi-dilute regimes   276, 277, 278–9
semi-flexible polyelectrolytes   233, 281
semi-flexible polymer chains   171–2
  bending   177
  conformations   179
  dynamics of   193–5, 279–80
  persistence length   173
  topology of   199
semi-flexible polymer gels   285–7
separation distance, intermembrane   248
sequencing DNA chains   317, 319
serine   1, 2
shear flow, colloidal solutions   269
shear moduli   256, 257, 328, 329
  cartilage   357
  elastic fibrous networks   244
  gels   284
  viscoelastic materials   187, 270–2, 278, 280

shear rate   271, 273
  of colloidal solutions   288
  and frictional forces   167
  of gels   285
shear rheometers   327
shear stiffness, cartilage   365–6
shear thickening, gel   285
shear thinning   288–90
  of polyelectrolytes   281, 282
  of synovial fluid   356
shear wave propagation   328
shock absorption   280, 355–68, 369
sickle cells   129, 130, 139, 147
side-chain liquid crystalline polymers   101–3
signalling, charged molecules in   205
silks   7, 9, 255, 368–9
  and liquid crystallinity   78, 103, 368
single stranded fibres, self-assembly of   142
single stranded filaments, self-assembly of   143–5, 147
skin   267, 268
  shark, surface phenomena   151, 152
  snake, frictional behaviour   167
sliding (diffusion)   383, 384–6
slugs   78, 79, 151
small angle neutron scattering   295
  for DNA structure   379
  for persistence length   172
small angle X-ray scattering   295
  for DNA structure   379
smectic liquid crystals   79, 80, 81, 82
  defects in   98–100
  order parameters   84
  transitions   93
  in virus molecules   86, 87
snake skin, friction and   167
soft molecules, damped motion of   187–91
solid liquid crystals   103
solutions   32–5, 47, 213–18
  and diffusion   111–12
  polymers in   42–4, 191–8, 276–80
  see also colloids; polyelectrolytes

INDEX

solvation energy, aqueous ions   212
solvents   38–9, 40
  and diffusion   111–12
  and globule-coil transitions   60, 62
  and membrane interactions   239–40
  and polymer size   177–83
spherical adsorbers, diffusion to   114–16
spider silks   78, 103, 255, 368, 368–9
spinoidals   71, 72, 73
split photodiode detector   313–14
spreading coefficient   158–9
spring constants *see* elastic/spring constants
springs   87, 88, 254
  dashpot and spring models   273–4
  and motility   339, 340, 352–3
standard linear solid   273, 274
starches   16, 17, 18, 130
  as gelled biopolymer   283
  as glassy material   290
  side-chain liquid crystalline polymer   101
  smectic structures in   78
static scattering techniques   294–7
steric forces   38–42, 248
Stern region   227, 228
steroids   11, 12
sticky reptation, physical gels and   284–5
stiffness   254
  and globule-coil transitions   62–3
  of nacre biocomposites   375
  shear, of cartilage   365–6
Stoke's law   111, 214
Stoke's radius   214–15
Stokes-Einstein equation   111, 191, 269, 329
stomach, mucins in   21
storage modulus   270–4, 278, 280
strain   254–5, 270–4
  and collagen   367–8
  and fibrous composites   374
  and foam compression   262–3
  and laminates   260
  and Poisson ratio   255–6

  and velocity dependence of friction   362
  see also stress-strain relationships
strain hardening   187, 285–7
strain tensors   257–8
strength   254, 255, 262
stress   255–6, 270
  collagen   367–8
  fibrous composites   373–4
  foam compression   262–3
  membranes   244–6
  rubbery networks   186
  viscoelasticity   269–72, 274
  and viscous materials   273
  Young's modulus   254–5, 258, 259, 260
  see also stress-strain relationships
stress tensors   257
stress-strain relationships
  cancellous bone   371–2
  elastin and resilin   370
  proteins   253–4
  spider silk   369
striated muscle   341, 346–9
  and Donnan equilibria   221
  and energy dissipation   267, 268
Stribeck curves   167
strongly charged polyelectrolytes   219–20
  and counterion condensation   233
structural biomaterials   355–75
structural mechanics   254–8
structure   207
  cellulose   16
  clathrate   29–30
  DNA   13–15, 63, 377–80
  glycoproteins   20–1
  ice   19, 28
  plectonomic   200, 380
  proteins   4, 5–7, 8, 9–10, 64
  proteoglycans   20–1
  rouleaux   288, 289
  starch   17, 18, 78
  viruses   23, 138–9
sugars   13, 15–18
super coiling   13, 199–201, 316
superhelical state, DNA   199
surface charge densities   206

surface force apparatus 306, 313
surface free energy
  and crystallisation 66, 67, 68, 168
  and self-assembly 132
surface tension 151–4
  and capillarity 162–3
  measuring 162–4
  and phase transitions 53, 168
  and wetting 157
surfaces 151–68
  charged 33, 34–5, 206, 359–61
  curvature, and pressure 160–2
  experimental methods 46, 47
  polymers at, and steric forces 39
surfactants 133–7
swelling coefficient 60, 63
swinging lever arm hypothesis 346
synovial fluid 280, 282, 356
synovial joints 21, 167, 364
  see also joints
synthetic rubber, properties of 255

temperature
  and dielectric constant of water 211
  and globule-coil transitions 61, 62
  and membrane persistence length 241
  and nematic phase order parameters 83
  and protein self-assembly 139–40
  and viscosity of glasses 290
temporary dipoles 26
tendons, hierarchical structure in 9
tensile properties
  of cartilage 367–8
  of fibrous composites 373–4
tension blobs 181, 183
tertiary structure 4, 6–7, 64
thermal blobs 181–2
thermal energy 55, 215–16
thermotropic liquid crystals 77
theta solvents 178
thixotropy 289
threonine 1, 2
thymine (DNA base) 13

tilted smectics 79, 81
time constants for damped motion 188, 189
time scales
  and motility measurements 107
  and properties of water 19, 30
  dielectric constant 211
  and stress 271
  and viscoelasticity 267, 272–4
tissues 21, 74
titration curves 223–7
tobacco mosaic virus 129, 138–9
topoisomerases 200
topology of polymer chains 199–201
torsional energy, DNA stores 13
toughness 254, 255, 264
  of dentine 259
  and foams/cellular solids 261
  of nacre biocomposites 374–5
  of spider silk 368, 369
transcription 12, 15
translation 12
translational diffusion 112–13
translational self-diffusion 108–11
transport 220, 247–8, 339
treadmilling in self-assembly 344–5
tribology 333–4
tryptophan 1, 2
tubulin, self-assembly 129, 142, 341–6
twisting, DNA chain 199, 200–1
tyrosine 1, 2, 8

umbilical cord 282
undercooling, freezing from 65
underdamped motion 188, 189
undulation forces 40, 41
undulations, membrane 238–40
unscreened electrostatic interactions 30–2

valine 1, 2
van der Waals forces 25–8, 64, 248, 373
vapour pressure osmometers 306
velocity dependence of friction 362

INDEX

vesicles, transport with   247–8
vimentin, self-assembly with   148
viruses   22–3, 129, 287
  in gene therapy   381
  liquid crystalline phases in   86, 87
  self-assembly   22, 137–9, 140
  topology changes in   199–200
viscoelastic materials   267–74
  networks, shear modulus   187
viscoelasticity   267, 278, 325–35
  of cartilage in tension   367
  of gels   285–6
  of glassy material   290
  of polyelectrolytes   280–2
  of proteins in nacre   374–5
  and semi-flexible models   279–80
  see also rheology
viscosity   269, 271–2, 273
  of blood cells   288, 289
  of colloids   287–8, 289
  and compliance   195
  and diffusion   110–11
  and entangled polymer chains   197
  of glassy material   290
  and ions in solution   217
  of polymeric solutions   279, 285
viscous behaviour   273, 277

water   4–5, 18–20, 28
  Bjerrum length   216
  clathrate structures   29–30
  dielectric constant   210–11
  interaction with membranes   248
  see also solutions

weakly charged polyelectrolytes   219, 220
  and counterion condensation   232–3
wetting   156–9
wide angle neutron scattering   295
wide angle X-ray scattering   295, 378
Wilhemy plates   164
wood, mechanical properties   261
work   151, 153, 154
wrapping transitions, DNA   382
writhing, DNA chain   199, 200, 201

X-ray diffraction   97, 296–7
X-ray scattering   295, 296, 297, 303
  for counterion clouds   231, 232
  for defects   99
  for DNA structure   378, 379
  for membrane undulations   241–3

Young-Laplace equation   160
Young's equation   156, 157
Young's moduli   254–5, 257
  entangled solutions   196
  foams   261, 262–3
  layered composites   258, 259–61
  nacre biocomposites   374
  rubbery networks   186–7
  see also elastic moduli

Zimm model   171, 193, 277, 278
Zimm-Bragg (Ising) model   56–7, 58, 59
Zipper model   53–6